Singularity Theory I

Springer
Berlin
Heidelberg
New York
Barcelona
Budapest
Hong Kong
London
Milan
Paris
Santa Clara
Singapore
Tokyo

V. I. Arnold V. V. Goryunov
O. V. Lyashko V. A. Vasil'ev

Singularity Theory I

Springer

Title of the Russian edition:
Itogi nauki i tekhniki, Sovremennye problemy matematiki,
Fundamental'nye napravleniya, Vol. 6, Dinamicheskie sistemy 6
Publisher VINITI, Moscow 1988

Second Printing 1998 of the First Edition 1993, which was originally
published as Dynamical Systems VI,
Volume 6 of the Encyclopaedia of Mathematical Sciences.

Die Deutsche Bibliothek - CIP-Einheitsaufnahme

Singularity theory / by V. I. Arnol'd ... - Berlin ; Heidelberg ; New
York ; Barcelona ; Budapest ; Hongkong ; London ; Mailand ; Paris ;
Santa Clara ; Singapur ; Tokio : Springer
1. - 1. ed., 2. printing. - 1998
(Encyclopaedia of mathematical sciences ; Vol. 6)
ISBN 3-540-63711-7

Mathematics Subject Classification (1991):
Primary 58C27, Secondary 14B05, 14E15, 32S05, 32Sxx, 58C28

ISBN 3-540-63711-7 Springer-Verlag Berlin Heidelberg New York

Printed in Italy

SPIN: 10648999
46/3143-5 4 3 2 1 0 – Printed on acid-free paper.

List of Editors, Authors and Translators

Editor-in-Chief

R.V. Gamkrelidze, Russian Academy of Sciences, Steklov Mathematical Institute, ul. Gubkina 8, 117966 Moscow, Institute for Scientific Information (VINITI), ul. Usievicha 20 a, 125219 Moscow, Russia; e-mail: gam@ipsun.ras.ru

Consulting Editor

V. I. Arnold, Steklov Mathematical Institute, ul. Gubkina 8, 117966 Moscow, Russia, e-mail: arnold@mi.ras.ru; CEREMADE, Université Paris 9 - Dauphine, Place du Maréchal de Lattre de Tassigny, 75775 Paris Cedex 16-e, France, e-mail: arnold@ceremade.dauphine.fr

Authors

V. I. Arnold, Steklov Mathematical Institute, ul. Gubkina 8, 117966 Moscow, Russia, e-mail: arnold@mi.ras.ru; CEREMADE, Université Paris 9 - Dauphine, Place du Maréchal de Lattre de Tassigny, 75775 Paris Cedex 16-e, France, e-mail: arnold@ceremade.dauphine.fr

V. V. Goryunov, Moscow Aviation Institute, Volokolamskoe sh. 4, 125871 Moscow, Russia; Division of Pure Mathematics, The University of Liverpool, Liverpool L69 3BX,UK, e-mail: goryunov@liv.ac.uk

O. V. Lyashko, Airforce Engineering Academy, Leningradskij pr. 40, 125167 Moscow, Russia; e-mail: ovl@aha.ru

V. A. Vasil'ev, Steklov Mathematical Institute, ul. Gubkina 8, 117966 Moscow, Russia; e-mail: vassil@vassil.mccme.rssi.ru

Translator

A. Iacob, Mathematical Reviews, 416 Fourth Street, P.O. Box 8604, Ann Arbor, MI 48107, USA

Singularities
Local and Global Theory

V.I. Arnol'd, V.A. Vasil'ev,
V.V. Goryunov, O.V. Lyashko

Translated from the Russian
by A. Iacob

Contents

Foreword

Napoleon condemned Laplace for the attempt to "introduce the spirit of infinitesimals in government" and removed him from the post of minister.

In this two-volume survey,[1] an exposition is given of the foundations of the part of the analysis of infinitesimals that is necessary for the deliberate control of dynamical systems, for their optimization, and for understanding the behavior of complex systems that depend on several parameters.

The theory of singularities of smooth maps is an apparatus for the study of abrupt, jump-like phenomena – bifurcations, perestroikas (restructurings), catastrophes, metamorphoses – which occur in systems depending on parameters when the parameters vary in a smooth manner.

Although the applications to the theory of dynamical systems do not exhaust by far all the potential capabilities of the theory of singularities (it also has applications in geometric and physical optics, hydrodynamics, quantum mechanics, crystallography, chemistry, acoustics, sinergetics, the theory of radiowave propagation, cosmology, algebraic geometry, differential topology, and so forth), the fundamental role that the theory of singularities plays in the investigation of bifurcations of stationary and periodic regimes justifies the inclusion of this two-volume book in the series "Dynamical Systems".

In writing the survey the authors had in mind a student-reader, mathematician or physicist, who wishes to learn the modern mathematical apparatus of local mathematical analysis as an instrument for applied studies, or a specialist in the respective applied domain seeking for the needed mathematical tools and reference information. Accordingly, we replaced proofs by references to the sources where they can be found, focussing on the methods, ideas, and results, rather than the technical details of the proofs. In doing so we counted on a reader prepared to accept many details on trust, or prefering to reconstruct the proofs by himself. A thorough exposition in any of the formalized mathematical languages (be it "$\forall\varepsilon\ \exists\delta$", "Ext-Tor", or "GO TO") would have required a many times greater volume.

The first two chapters of this volume are devoted to one of the most advanced parts of the theory of singularities – the study of degeneracies of critical points of functions.

In Chapter 1 we acquaint the reader with the basic notions of the theory of singularities of smooth maps, give the initial segment of the classification of smooth functions and present the technique of reducing functions to normal forms.

In Chapter 2 we consider topological and algebro-geometric aspects of the theory of critical points of functions. Here we discuss the basic concepts of the local Picard-Lefschetz theory, that is, the discipline of branching of cycles and

[1] The second part of the survey, "Singularity Theory II. Applications", will appear as Vol. 39 of the present series (Dynamical Systems VIII).

integrals depending on parameters. A detailed study is made of the main object of this theory – the bundle of vanishing cohomology (i.e., of branching integration contours) connected with a critical point and, in particular, the base over which this bundle is defined, the complement of the discriminant of the singularity. We also consider the connection between simple singularities of functions and the classification of simple Lie groups, reflection groups, and braid groups.

Among the original results of this chapter let us mention the calculation of the cohomology groups with nonconstant coefficients of the complements of discriminants of one-dimensional singularities and their application to the theory of algorithms, the description of the stable cohomology of the complements of discriminants of arbitrary singularities, theorems on stable irreducibility of the strata of a discriminant, the noncoincidence of the dimensions of the complex and real $\mu = \text{const}$ strata of real singularities.

In Chapter 3 we give an exposition of the general theory of equivalence of maps.

Mathematical and physical problems arising in real situations lead to the investigation of the properties of maps with respect to a variety of equivalence relations. In analyzing a concrete equivalence relation one has to deal with a number of standard questions: Is the given map stable? Can one regard the map, even locally, as a polynomial, which would considerably simplify calculations? Does the map admit a versal deformation, i.e., can it be included in a family with a finite number of parameters, which contains all small deformations of the map? How much simpler does the classification become when one passes from the rigid differentiable equivalence to the less demanding topological one? For many equivalence relations the answers to these questions look the same. The statements of the corresponding theorems and of sufficient conditions for their applicability constitute the main content of the third chapter.

In the last, fourth chapter, we describe topological characteristics of singular sets of smooth maps: the cohomology classes dual to the sets of critical points and nonregular values; invariants of maps defined by these classes; the structure of the spaces of maps not having singularities of one kind or another. Apparently for the first time in the literature, we carry out the construction of characteristic classes of foliations with the help of universal complexes of singularities and multisingularities, and also the computation of the fundamental group of the space of functions with singularities no more complicated than x^3 and of the topology of complements of open swallowtails.

For the first time in monograph form, we discuss in this survey the results of S.V. Chmutov on the monodromy group of an isolated singularity in the skew-symmetric case, the theorems of O.V. Lyashko and P. Jaworski on the decomposition of simple and parabolic singularities, the estimates of the index of a polynomial vector field obtained by A.G. Khovanskiĭ, and the results of E.I. Shustin and V.I. Arnol'd on the number of the flattening points that vanish under various degeneracies of algebraic hypersurfaces.

The references within the volume are organized as follows. If the reference lies within the same chapter, we indicate the number of the corresponding section or subsection, as in the table of contents. If the reference is to a different chapter, then we precede the number of the section or subsection by the number of the chapter.

Chapters 1 and 2, with the exception of § 2.6 and subsections 2.1.11, 2.5.11, were written by O.V. Lyashko, Chapter 3 and subsections 2.1.11, 2.5.11 by V.V. Goryunov, Chapter 4 (except for § 4.4) and § 2.6 by V.A. Vasil'ev, and § 4.4 by V.I. Arnol'd.

For the second volume of the survey V.I. Arnol'd and V.A. Vasil'ev each contributed two chapters, and one chapter was written by V.V. Goryunov. The second volume gives a representation of a wide circle of problems that are currently being solved, and contains many new results.

Chapter 1
Critical Points of Functions

One of the most thoroughly studied branches of the theory of singularities is the investigation and classification of degeneracies of critical points of functions. Generic functions have only nondegenerate critical points. More complex singularities vanish under small perturbations, decomposing into nondegenerate ones.

However, in families of functions that depend on several parameters, degenerate critical points may occur in an irremovable manner. For example, the family of functions $x^3 + \lambda x$ has for the value $\lambda = 0$ of the parameter a degenerate critical point; any close one-parameter family has, for a close value of the parameter, a degeneracy of the same kind. As the number of parameters increases, in families of functions there arise degeneracies of ever increasing complexity.

In this chapter we describe the initial segment of the classification of critical points of functions. This classification of the simplest degeneracies of critical points turned out to be closely related to the classification of simple Lie groups, of reflection groups, and of braid groups.

To simplify the exposition, we restrict ourselves mainly to the case of holomorphic functions, diffeomorphisms, and so on. The theory carries over, with practically no modifications, to the case of real smooth functions; some differences arising in the real case will be pointed out. The classification of real critical points is given in subsection 2.8.

§ 1. Invariants of Critical Points

Here we give the basic definitions concerning critical points of functions.

1.1. Degenerate and Nondegenerate Critical Points

Definition. A point is said to be a *critical point* of a smooth function f if at that point the derivative of f is equal to zero.

The value that the function takes at a critical point is called a *critical value.*

Example. The function $f(x) = x^3 - \lambda x$ of the variable $x \in \mathbb{C}$ has for each $\lambda \neq 0$ the two critical points $\pm\sqrt{\lambda/3}$ with respective critical values $\mp\sqrt{4\lambda^3/27}$. For $\lambda = 0$ these two critical points "merge" into a single critical point 0.

The critical points of functions are divided into generic (general-position, or nondegenerate) critical points and degenerate critical points.

Definition. A critical point is said to be *nondegnerate* (or a *Morse critical point*) if the second differential of the function at that point is a nondegenerate quadratic form.

Example. The function $f(x) = x^3 - \lambda x$ has for $\lambda \neq 0$ the pair of nondegenerate critical points $\pm\sqrt{\lambda/3}$ and for $\lambda = 0$ the degenerate critical point 0 (Fig. 1).

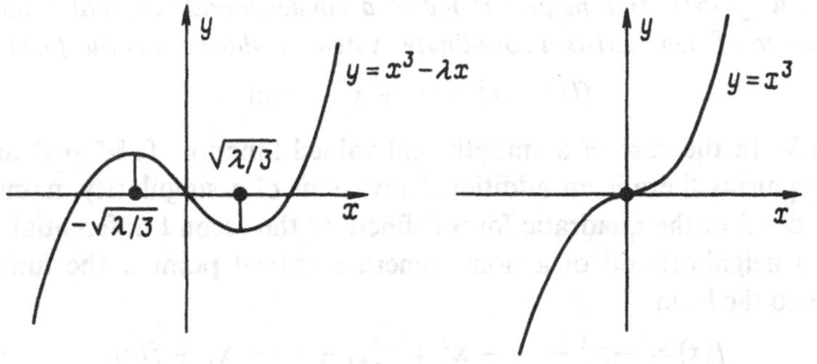

Fig. 1

The degree of degeneracy of the second differential is the simplest indicator of how degenerate a critical point is.

Definition. The *corank* of a critical point of a function is the dimension of the kernel of its second differential at the critical point.

Examples. The corank of any Morse critical point is equal to zero. The corank of the critical point 0 of the function $f = x_1^3 + x_2^2 + \cdots + x_n^2$ is equal to one.

1.2. Equivalence of Critical Points. Let us consider the set $\mathcal{O}_n$ of function-germs at the point $0 \in \mathbb{C}^n$.

Definition. Two function-germs at zero are said to be *equivalent* if one is taken into the other by a biholomorphic change of coordinates that keeps the point zero fixed.

This notion of equivalence can be alternatively described as follows. Let $\mathcal{D}_n$ denote the group of germs of biholomorphic maps $g: (\mathbb{C}^n, 0) \to (\mathbb{C}^n, 0)$. This group acts on the space $\mathcal{O}_n$ of function-germs by the rule $g(f) = f \circ g^{-1}$, where $f \in \mathcal{O}_n$, $g \in \mathcal{D}_n$. The orbits of this action are exactly the equivalence classes of function-germs.

Definition. Two critical points are said to be *equivalent* if the function-germs that define them are equivalent. The equivalence class of a function-germ at a critical point is called a *singularity*.

Example. The functions $f_1 = x^2$ and $f_2 = cx^2$, with $c \neq 0$ a constant, have the same singularity at the point $x = 0$, since $f_1 = f_2 \circ g^{-1}$, where g is the diffeomorphism given by $g(x) = \sqrt{c}x$.

Clearly, equivalent critical points have equal coranks; hence the corank is the simplest invariant of a singularity.

The behavior of a function in the neighborhood of a nondegenerate critical point is described by the Morse lemma.

Theorem ([246]). *In a neighborhood of a nondegenerate critical point $a \in \mathbb{C}^n$ of the function f there exists a coordinate system in which f has the form*

$$f(x) = x_1^2 + \cdots + x_n^2 + f(a).$$

Remark. In the case of a smooth real-valued function $f\colon \mathbb{R}^n \to \mathbb{R}$ and real diffeomorphisms there is an additional invariant of a singularity, namely, the inertia index λ of the quadratic form defined by the second differential. In this case, in a neighborhood of a nondegenerate critical point a the function is reducible to the form

$$f(x) = -x_1^2 - \cdots - x_\lambda^2 + x_{\lambda+1}^2 + \cdots + x_n^2 + f(a).$$

1.3. Stable Equivalence. For degenerate critical points, a generalization of the preceding result, the parametric Morse lemma, holds true.

Theorem ([12]). *In a neighborhood of the critical point 0 of corank k a holomorphic function $f\colon (\mathbb{C}^n, 0) \to (\mathbb{C}, 0)$ is equivalent to a function of the form*

$$\varphi(x_1, \ldots, x_k) + x_{k+1}^2 + \cdots + x_n^2,$$

where the second differential of φ at zero is equal to zero: $\varphi \in \mathfrak{m}^3 \subset \mathcal{O}_n$.

[Here and in what follows $\mathfrak{m}$ denotes the ideal of the function-germs vanishing at the origin.]

This permits one to define an equivalence relation for critical points of (functions of) different numbers of variables.

Definition. Two function-germs $f\colon (\mathbb{C}^n, 0) \to (\mathbb{C}, 0)$ and $g\colon (\mathbb{C}^m, 0) \to (\mathbb{C}, 0)$ are said to be *stably equivalent* if they become equivalent after the addition of nondegenerate quadratic forms in supplementary variables:

$$f(x_1, \ldots, x_n) + x_{n+1}^2 + \cdots + x_k^2 \sim g(y_1, \ldots, y_m) + y_{m+1}^2 + \cdots + y_k^2.$$

Theorem ([374]). *Two functions of the same number of variables are stably equivalent if and only if they are equivalent.*

Thus, the passage to stable equivalence does not affect the classification of critical points of functions of a fixed number of variables and allows one to compare degeneracies of critical points of functions of different numbers of variables.

Example. The function-germs $f(x) = x^3$ and $g(x, y, z) = x^3 + yz$ are stably equivalent at zero.

1.4. The Local Algebra and the Multiplicity of a Singularity. Let $f: (\mathbb{C}^n, 0) \to (\mathbb{C}, 0)$ be a germ of a holomorphic function with a critical point at zero. Let us consider the *gradient ideal* $I_{\nabla f} = \mathcal{O}_n\langle f_1, \ldots, f_n\rangle$ generated by the partial derivatives $f_i = \partial f/\partial x_i$ of the function f.

Definition. The *local algebra* Q_f *of the singularity of* f is defined to be the quotient of the algebra of function-germs by the gradient ideal of f:

$$Q_f = \mathcal{O}_n/I_{\nabla f}.$$

Example. Let $f(x) = x^3$ $(n = 1)$. Then $I_{\nabla f} = \mathfrak{m}^2$ and Q_f is the two-dimensional algebra of truncated polynomials of first degree.

Remark. Q_f is the local algebra of the gradient map $\partial f/\partial x$: $\mathbb{C}^n \to \mathbb{C}^n$; concerning the properties of the local algebra of a holomorphic map see subsection 3.1.10.

The algebra Q_f does not depend on the choice of local coordinate system. More precisely, the transition to another coordinate system induces a transition of the exact sequence

$$0 \to I_{\nabla f} \to \mathcal{O}_n \to Q_f \to 0$$

of $\mathbb{C}$-algebras to an isomorphic sequence. Hence, the local algebra is an invariant associated with the singularity. As a matter of fact, it is a very subtle invariant: for a quasihomogeneous singularity it completely determines the singularity, see subsection 3.3.

Definition. The *multiplicity* $\mu(f)$ *of the critical point* of the germ $f \in \mathcal{O}_n$ is the dimension of its local algebra regarded as a $\mathbb{C}$-module:

$$\mu(f) = \dim_{\mathbb{C}} Q_f.$$

A critical point is said to be *isolated* (or *of finite multiplicity*) if $\mu(f) < \infty$.

Example. The function $f(x) = x^3$ has at zero an isolated critical point of multiplicity two. The function $g(x, y) = xy^2$ has a nonisolated critical point at every point of the x-axis.

In the holomorphic case the finite-dimensionality of the local algebra is equivalent to the critical point being literally isolated, i.e., to the critical point having a neighborhood that contains no other critical points.

Theorem. *The multiplicity of an isolated critical point is equal to the number of Morse critical points into which it decomposes under a small deformation of the function.*

This result is a corollary of the theorem on the local multiplicity of a holomorphic map (subsection 3.1.10).

Remark. In the real case the local algebra and the multiplicity of a critical point are defined in analogous manner: $Q_f = \mathcal{E}_n/I_{\nabla f}$ and $\mu(f) = \dim_{\mathbb{R}} Q_f$, where $\mathcal{E}_n$ denotes the algebra of germs of C^∞-functions at zero. Notice, however, that for the C^∞-function $f(x) = \exp(-1/x^2)$, which has a unique critical point at zero, $\mu(f) = \infty$. Also, the critical point of the real function $f(x) = x^3$ has multiplicity two, but its deformation $x^3 + \lambda x$ has no real critical points at all for positive values of the parameter λ.

As we remarked at the beginning of this chapter, a generic function has only nondegenerate critical points. In a family of functions depending on parameters there may occur critical points of greater multiplicities that cannot be removed through small deformations of the family. These assertions follow from Thom's transversality theorem, given in Chapter 3. The simplest example of such a family for a critical point of multiplicity μ is its truncated miniversal deformation (subsection 1.11).

1.5. Finite Determinacy of an Isolated Singularity. The classification problem for critical points consists in describing the orbits of the action of the Lie group of germs of diffeomorphisms on the infinite-dimensional space of function-germs. It is convenient to reduce this problem to that of describing the action of a finite-dimensional Lie group on a finite-dimensional jet space.

Definition. Two function-germs $f, g \in \mathcal{O}_{n,a}$ at the point a are said to be *k-equivalent* if their difference $f(x) - g(x)$ is an infinitesimal of order higher than $\|x - a\|^k$, where $\|\cdot\|$ is some norm on $\mathbb{C}^n$. The classes of this equivalence relation on $\mathcal{O}_{n,a}$ are called *k-jets* at the point a.

For a function $f\colon \mathbb{C}^n \to \mathbb{C}$ denote by $j_a^k f$ its k-jet at the point a. If one fixes a coordinate system on $\mathbb{C}^n$, then the k-jet can be thought of as the Taylor polynomial of degree k.

Example. The 1-jet of a function of one variable is specified by the triplet of numbers x, y, dy/dx.

Definition. A k-jet is said to be *sufficient* if any two functions with that k-jet are equivalent.

Example. As follows from the Morse lemma, at a nondegenerate critical point a function is equivalent to its Taylor polynomial of degree two. Hence, the 2-jet of a function at a nondegenerate critical point is sufficient.

The next theorem, due to Tougeron, asserts that at a critical point of finite multiplicity any function is equivalent to a polynomial. It holds true in both the complex and the real case.

Theorem ([332], [8]). *The $(\mu + 1)$-jet of a function at a critical point of multiplicity μ is sufficient.*

In contrast, at a nonisolated critical point no finite jet is sufficient.

Example (Whitney [379]). The germ of the holomorphic function $f(x, y, z) = xy(x + y)(x - zy)(x - e^z y)$ at zero is not equivalent to a polynomial germ.

In fact, the plane $z = \text{const}$ intersects the level set $f = 0$ along five curves with a common point. The cross-ratios constructed from the tangents to any four of these curves depend on z. If f were equivalent to a polynomial, then each of these cross-ratios would depend algebraically on any other one. For our function f this is not so due to the presence of the multiplier e^z.

1.6. Lie Group Actions on Manifolds. Here we define the notions of the versal deformation, the transversal to an orbit, and the modality in the finite-dimensional setting of the action of a finite-dimensional Lie group on a smooth manifold. In the subsequent subsections we carry over these notions to the case of a critical point of finite multiplicity. The general definitions for singularities of maps will be given in Chapter 3.

Let G be a Lie group acting on a manifold M, and let f be a point of M.

Definition. A *deformation* of the point f with base $\Lambda = \mathbb{C}^l$ is defined to be a germ of a smooth map

$$F: (\Lambda, 0) \to (M, f)$$

of the base at zero that takes 0 into f.

We say that two deformations F_1 and F_2 of the point f with the same base Λ, are *equivalent* if one is taken into the other under the action of elements of G that depend smoothly on $\lambda \in \Lambda$, i.e., there exists a smooth map-germ $g: (\Lambda, 0) \to (G, e)$, where e is the identity element of G, such that

$$F_1(\lambda) = g(\lambda)F_2(\lambda).$$

Now let $F: \Lambda \to M$ be a deformation and let $\theta: (\Lambda', 0) \to (\Lambda, 0)$ be a smooth map. The deformation $\theta^*F = F \circ \theta: \Lambda' \to M$ is said to be *induced* from F by the map θ.

Definition. A deformation $F: (\Lambda, 0) \to (M, f)$ is said to be *versal* if every deformation of f is equivalent to one induced from F.

A versal deformation for which the base has the smallest possible dimension is called *miniversal.*

A *transversal* to the orbit of f is any deformation $F: (\Lambda, 0) \to (M, f)$ that is transverse to the orbit Gf at the point f. It is readily seen that transversality to the orbit is a necessary infinitesimal condition for the versality of a deformation. In fact, to every deformation $F: (\Lambda, 0) \to (M, f)$ there corresponds a subspace of the tangent space $T_f M$, namely, the sum $F_* T_0 \Lambda + T_f Gf$ of the image of the tangent space $T_0 \Lambda$ to the base at 0 (under the differential F_* of F) and the tangent space at f to the orbit Gf. Obviously, this subspace does not change on passing to an equivalent deformation, and the subspace corresponding to any induced deformation θ^*F is contained in the one corresponding to the original deformation F. It follows that for a versal deformation F it is necessary that this subspace coincide with the tangent space at f:

$$F_* T_0 \Lambda + T_f Gf = T_f M,$$

i.e., that F be transverse to the orbit of f (see Fig. 2).

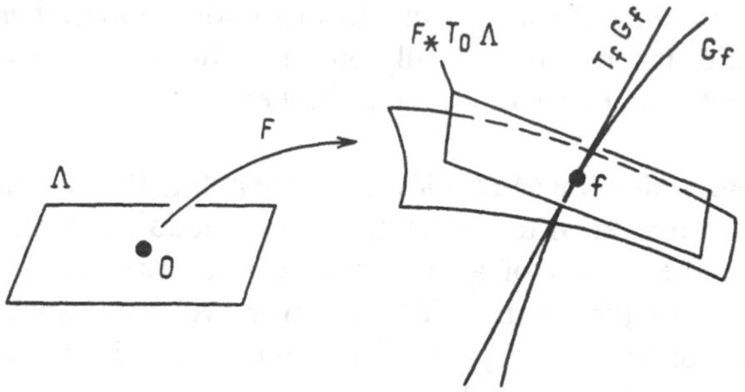

Fig. 2

It turns out that this infinitesimal condition for versality is also sufficient.

Theorem ([8]). *Any transversal [of minimal dimension] to the orbit of f is a versal [resp. miniversal] deformation.*

Definition. The *modality* of the point $f \in M$ under the action of a Lie group G is defined to be the least number m such that a sufficiently small neighborhood of f is covered by a finite number of m-parameter families of orbits.

Example. Let M be the manifold of quadruples of lines passing through the origin in $\mathbb{C}^3$. The Lie group $GL(3, \mathbb{C})$ acts on $\mathbb{C}^3$, and hence also on M. This action is transitive on the set of quadruples of lines that are not contained in a common plane. It follows that any point of M corresponding to such a generic quadruple has modality zero. Any quadruple of distinct lines lying in a single plane has modality one, because for such a quadruple the cross-ratios of the four lines is an invariant of the group action.

1.7. Versal Deformations of a Critical Point. Let us adapt the notion of a versal deformation to the case where M is the functional space of function-germs, on which the infinite-dimensional group of changes of variables acts.

A *deformation* with base $\Lambda = \mathbb{C}^l$ of the germ $f: (\mathbb{C}^n, 0) \to \mathbb{C}$ is the germ at zero of a smooth map $F: (\mathbb{C}^n \times \mathbb{C}^l, 0) \to \mathbb{C}$ such that $F(x, 0) \equiv f(x)$.

A deformation F' is *equivalent* to F if $F'(x, \lambda) = F(g(x, \lambda), \lambda)$, where $g: (\mathbb{C}^n \times \mathbb{C}^l, 0) \to (\mathbb{C}^n, 0)$, with $g(x, 0) \equiv x$, is a smooth germ (a family of diffeomorphisms depending on the parameter $\lambda \in \mathbb{C}^l$).

The deformation F' is *induced* from F if

$$F'(x, \lambda') = F(x, \theta(\lambda)),$$

where $\theta: (\mathbb{C}^{l'}, 0) \to (\mathbb{C}^l, 0)$ is a smooth germ of a mapping of the bases.

Definition. A deformation $F(x, \lambda)$ of the germ $f(x)$ is said to be *versal* if every deformation F' of $f(x)$ can be represented in the form

$$F'(x, \lambda') = F(g(x, \lambda'), \theta(\lambda')), \qquad g(x, 0) \equiv x, \qquad \theta(0) = 0,$$

i.e., if every deformation of $f(x)$ is equivalent to a deformation induced from F.

Example. The deformation $F(x, \lambda) = x^2 + \lambda$ of the germ $f(x) = x^2$ is versal. In fact, every deformation of $f(x)$ has the form

$$G(x, \mu) = \alpha(x, \mu)x^2 + \beta(\mu)x + \gamma(\mu), \qquad \alpha(x, 0) \equiv 1, \qquad \beta(0) = \gamma(0) = 0.$$

The coordinate-change diffeomorphism $g(x, \mu) = x - \beta/2\alpha$ takes G into an equivalent deformation which has a nondegenerate critical point at the origin for every value of the parameter μ. By the parametric Morse lemma, this last deformation is in turn equivalent to one of the form $x^2 + \lambda(\mu)$, induced from the deformation $F(x, \lambda) = x^2 + \lambda$.

1.8. Infinitesimal Versality. The theorem of subsection 1.6. asserting that any transversal to an orbit is a versal deformation admits an extension to the infinite-dimensional setting.

Definition. The *tangent space to the orbit of a germ f* is the linear space of rates of changes of f under the action of one-parameter families of diffeomorphisms $g(x, \tau)$, $\tau \in \mathbb{C}$, with $g(x, 0) \equiv x$.

The rate of deformation of the germ $f(x)$ is defined as

$$\frac{\partial}{\partial \tau} f(g(x, \tau))\bigg|_{\tau=0} = \sum_{i=1}^{n} f_i \frac{dg_i}{d\tau}\bigg|_{\tau=0}.$$

Thus, the tangent space to the orbit of the germ $f: (\mathbb{C}^n, 0) \to \mathbb{C}$ is the module of function-germs generated by the partial derivatives f_i of f, i.e., it coincides with the gradient ideal $I_{\nabla f}$.

The condition that the deformation $F(x, \lambda)$ of the germ f be transverse to the orbit of f is called infinitesimal versality.

Definition. A deformation $F(x, \lambda)$ of the germ $f(x)$ is said to be *infinitesimally versal* if every function-germ $\alpha(x)$ can be represented in the form

$$\alpha(x) = \sum_{i=1}^{n} h_i f_i + \sum_{j=1}^{l} c_j \frac{\partial F}{\partial \lambda_j}\bigg|_{\lambda=0}.$$

In other words, $F(x, \lambda)$ is infinitesimally versal if the images of the functions $\partial F/\partial \lambda_j|_{\lambda=0}$ under the canonical projection $\mathcal{O}_n \to Q_f = \mathcal{O}_n/I_{\nabla f}$ generate the local algebra Q_f as a $\mathbb{C}$-module.

The next versality theorem is valid also in a considerably more general context (see subsection 3.2). In a different version it was first established by Tyurina [337].

Theorem ([12], [236]). *Any infinitesimally versal deformation of a function-germ is versal.*

Example. The local algebra of the germ $f(x) = x^2$ is generated by one element: $Q_f = \mathcal{O}_n/\mathfrak{m}$. Consequently, the deformation $F(x, \lambda) = x^2 + \lambda$ of $f(x)$ is versal: $\partial F/\partial \lambda|_{\lambda=0} = 1$.

As a corollary to the last theorem we remark that the dimension of the base of a miniversal deformation of a critical point of a germ f coincides with its multiplicity $\mu(f)$.

Example. As a miniversal deformation of a critical point of a germ $f(x)$ one can take

$$F(x, \lambda) = f(x) + \sum_{j=1}^{\mu} \lambda_j \varphi_j(x),$$

where $\varphi_j(x), j = 1, \ldots, \mu$ form a basis of Q_f.

A versal deformation is unique in the following sense.

Theorem ([37]). *Every l-parameter versal deformation of a germ f is equivalent to a deformation induced from any other l-parameter versal deformation by a suitable diffeomorphism of their bases.*

1.9. The Modality of a Critical Point. The group of germs of diffeomorphisms $(\mathbb{C}^n, 0) \to (\mathbb{C}^n, 0)$ acts on the function space $\mathfrak{m} \subset \mathcal{O}_n$ of germs $f: (\mathbb{C}^n, 0) \to (\mathbb{C}, 0)$, and consequently also on the k-jet space $j^k\mathfrak{m}$. To avoid the difficulties inherent to the infinite-dimensional situation, let us define the modality of a germ $f \in \mathfrak{m}$ as the modality of its k-jet $j^k f$ for sufficiently large k. It follows from Tougeron's theorem that this number does not depend on k once k is so large that $j^k f$ is sufficient.

Definition. The *modality m* of a function-germ $f(x)$ is the modality of any of its jets $j^k f$, $k \geqslant \mu(f) + 1$.

Example. The modality of the germ $f(x) = x^2$ is zero; this is a consequence of the Morse lemma.

Remark. In defining the modality we considered the smaller function space $\mathfrak{m}$. Otherwise, every critical point would have modality larger than zero (for instance, the family of germs $F(x, \lambda) = x^2 + \lambda$ contains nonequivalent germs for distinct values of λ).

The functions of modality $m = 0, 1, 2$ are called respectively *simple*, *unimodal*, and *bimodal functions*.

Definition. The $\mu =$ const *stratum* for a function-germ f is the (connected) component containing f of the set of function-germs with the same multiplicity of the critical point as f in the germ space $\mathcal{O}_n$.

The $\mu =$ const stratum in the base of an arbitrary deformation of the germ f is defined similarly.

Theorem ([133]). *The dimension of the* $\mu =$ const *stratum in the base of a miniversal deformation of the critical point of f is one greater than the modality of f.*

From this result it follows that the codimension c of the $\mu =$ const stratum of a germ f in the space $\mathfrak{m}^2$ of germs with critical point 0 and critical value 0, the multiplicity μ of the critical point of f, and the modality m are related by

$$\mu = c + m + 1.$$

1.10. The Level Bifurcation Set. The level bifurcation set is the germ of the surface formed in the base of a versal deformation $F(x, \lambda)$ by those values of the parameter λ for which 0 is a critical value of the function $x \mapsto F(x, \lambda)$.

To give this definition an exact meaning, let us examine the picture locally. Suppose $f: (\mathbb{C}^n, 0) \to (\mathbb{C}, 0)$ is a germ with an isolated critical point at 0. Fix a versal deformation $F(x, \lambda)$ of f and choose sufficiently small neighborhoods of zero, $U = \{x: \|x\| \leqslant \rho\} \subset \mathbb{C}^n$ and $\Lambda = \{\lambda: \|\lambda\| \leqslant \delta\} \subset \mathbb{C}^\mu$.

Theorem ([247]). *For sufficiently small ρ and $\delta = \delta(\rho)$ the level set $\{x: F(x, \lambda) = 0\}$ is nonsingular on the boundary ∂U of the ball U and is transverse to ∂U for every $\lambda \in \Lambda$.*

We term a value of the parameter λ *nonsingular* if the corresponding local level set $V_\lambda = \{x: F(x, \lambda) = 0\} \cap U$ is a nonsingular manifold.

Definition. The *level bifurcation set* (or the *discriminant*) of the singularity f is the germ of the surface Σ formed in the base of its miniversal deformation by the singular values of the parameter λ.

Example 1. Let $f(x) = x^3$, $F(x, \lambda) = x^3 + \lambda_1 x + \lambda_0$. The level bifurcation set Σ for f is exactly the set of those pairs of values (λ_1, λ_0) for which the polynomial $x^3 + \lambda_1 x + \lambda_0$ has multiple roots. Hence, Σ is given by the equation $27\lambda_0^2 + 4\lambda_1^3 = 0$ (Fig. 3).

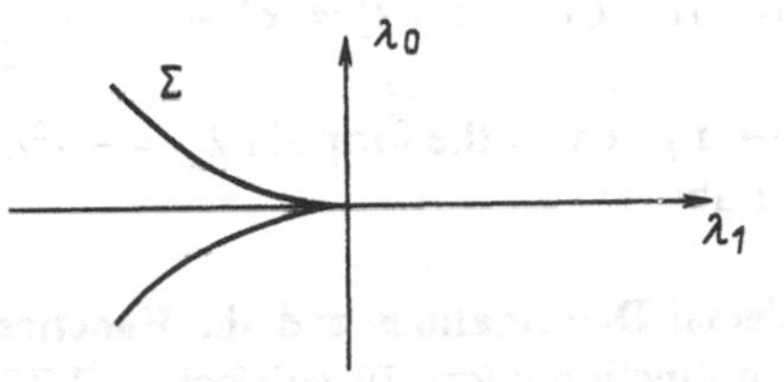

Fig. 3

Example 2. Let $f(x) = x^4$. Then the level bifurcation set Σ of f coincides with the set of all zeroes of the discriminant of the polynomial $x^4 + \lambda_2 x^2 + \lambda_1 x + \lambda_0$. The real part of this hypersurface is depicted in Fig. 4.

This hypersurface, known as a *swallowtail*, is very frequently encountered in the theory of singularities and its applications.

As will be shown later, the level bifurcation set and the function bifurcation set (defined in subsection 1.11 below) carry important information about a singularity.

Let us consider next the set $\mathfrak{m}^2 \subset \mathcal{O}_n$ of function-germs with critical point 0 and critical value 0. The group of germs of diffeomorphisms that preserve the

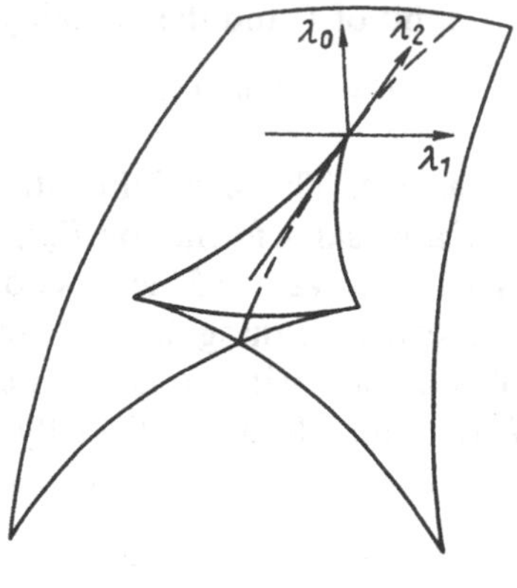

Fig. 4

origin acts on $\mathfrak{m}^2$. Consider a minimal transversal T to the orbit of f in $\mathfrak{m}^2$. T yields a deformation of the function f that is equivalent to the deformation induced from a versal one by some map of the bases $\theta: T \to \Lambda$.

Theorem ([133], [322]). *The dimension of the minimal transversal T,* dim *T, equals* $\mu - 1$. *The image of T under the map θ is exactly the level bifurcation set* $\Sigma \subset \Lambda$. *The set Σ is irreducible and admits T as a nonsingular normalization.*

Example. Let $f(x) = x^3$. As a minimal transversal in $\mathfrak{m}^2$ we can take the one-parameter deformation $G(x, t) = x^3 + tx^2$, $t \in T$, which is equivalent to the deformation

$$\tilde{G}(x, t) = G(x - t/3, t) = x^3 - \frac{t^2}{3}x + \frac{2t^3}{27}.$$

The map of bases $T \to \Lambda$ given by the formula $\lambda_1 = -t^2/3$, $\lambda_0 = 2t^3/27$ takes T into Σ (see Example 1 above).

1.11. Truncated Versal Deformations and the Function Bifurcation Set. Let $f: (\mathbb{C}^n, 0) \to (\mathbb{C}, 0)$ be a function-germ. In subsection 1.7 we defined the versal deformation of f using as the function space the space of germs of functions on $\mathbb{C}^n$. Now we confine ourselves to deformations in the class of function-germs belonging to $\mathfrak{m}$. The definitions of the notions of equivalence and versality of a deformation can be carried over without modification to the present case.

Definition. A *truncated versal deformation* of the germ f is a versal deformation of f in the class of deformations within $\mathfrak{m}$.

Theorem ([37]). *A deformation* $F'(x, \lambda')$, $\lambda' \in \Lambda'$, *of the germ* $f(x)$ *is a truncated versal deformation if and only if* $F(x, \lambda) = F'(x, \lambda') + \lambda_0$, $\lambda = (\lambda', \lambda_0) \in \Lambda' \times \mathbb{C}^1$, *is a versal deformation of* $f(x)$ *in the ordinary sense.*

The base of a truncated miniversal deformation has dimension $\mu - 1$. As an example of such a deformation one can take

$$F'(x, \lambda') = f(x) + \sum_{i=1}^{\mu-1} \lambda_i \varphi_i(x),$$

where the germs $\varphi_1(x), \dots, \varphi_{\mu-1}(x)$ form a basis of the algebra $\mathfrak{m}/I_{\nabla f}$.

We call the deformation $F(x, \lambda) = F'(x, \lambda') + \lambda_0$ with base $\Lambda = \Lambda' \times \mathbb{C}^1$ the *complete* versal deformation corresponding to the truncated deformation $F'(x, \lambda')$.

A function $h(x)$ is said to be a *Morse function* on the set U if it has only nondegenerate critical points on U and its critical values at these points are all distinct.

Definition. The *function bifurcation set* for f is the germ at zero of the hypersurface Ξ formed in the base $\Lambda' \simeq \mathbb{C}^{\mu-1}$ of a truncated miniversal deformation by those values of the parameter λ' for which the function $x \mapsto F'(x, \lambda')$ is not a Morse function in a small neighborhood of the origin.

Example. A truncated versal deformation for $f(x) = x^4$ is $F(x, \lambda') = x^4 + \lambda_2 x^2 + \lambda_1 x$. This polynomial has degenerate critical points if and only if its derivative has multiple roots, i.e., $27\lambda_1^2 + 8\lambda_2^3 = 0$. One can also verify that coincidences of critical values occur only if $F(\cdot, \lambda')$ is an even function, i.e., $\lambda_1 = 0$. The function bifurcation set for $f(x) = x^4$ is shown in Fig. 5. From the given example it is seen that the function bifurcation set is reducible and decomposes into two hypersurfaces, the caustic Ξ_1 and the Maxwell stratum Ξ_2, which correspond respectively to the functions with degenerate critical points and to those with coinciding critical values.

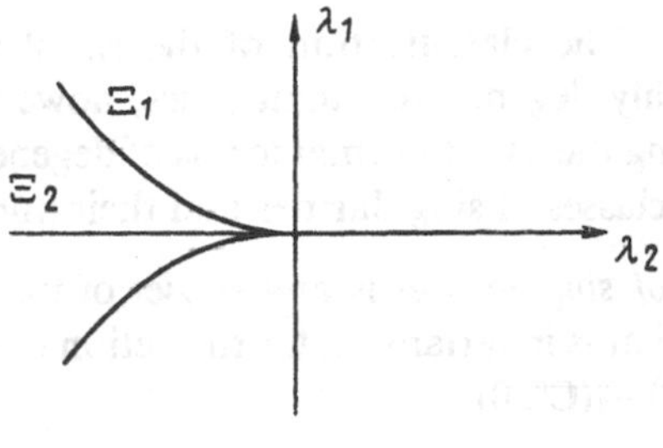

Fig. 5

Consider the projection map $\pi: \Lambda \to \Lambda'$ of the base of the complete miniversal deformation onto the base of the truncated one.

Theorem ([133], [207]). *The restriction $\pi: \Sigma \to \Lambda'$ of π to the level bifurcation set Σ is a μ-sheeted ramified covering whose set of ramification points coincides with $\Xi \subset \Lambda'$. The group of covering transformations of this covering is the symmetric group $\mathfrak{S}_\mu$.*

This theorem is illustrated by Fig. 6. Notice that the singular and the self-intersection points of the hypersurface Σ project into the caustic Ξ_1 and into the Maxwell stratum Ξ_2, respectively.

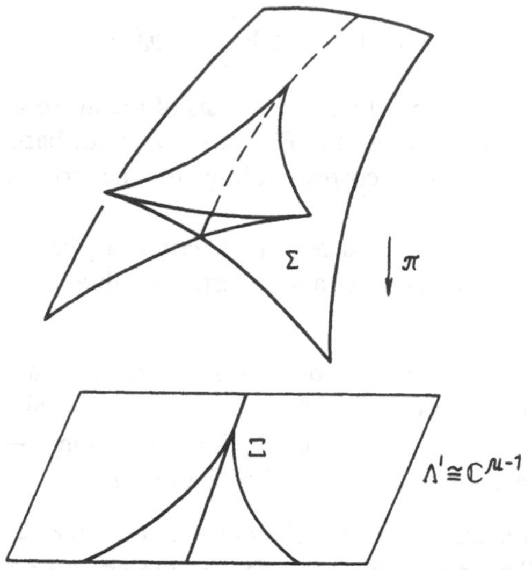

Fig. 6

§2. The Classification of Critical Points

In this section we describe the initial segment of the classification of singularities of holomorphic and smooth functions.

2.1. Normal Forms. The classification of the simplest degenerate critical points is discrete. Highly degenerate singularities, however, have moduli. For this reason, before giving the list of normal forms of degeneracies of singularities of functions we define classes of singularities and their normal forms.

Definition. A *class of singularities* is any subset of the space $\mathcal{O}_n$ of function-germs $(\mathbb{C}^n, 0) \to (\mathbb{C}, 0)$ that is invariant under the action of the group of germs of diffeomorphisms $(\mathbb{C}^n, 0) \to (\mathbb{C}^n, 0)$.

Examples of classes are the orbits of the action of the indicated group. The $\mu = \text{const}$ stratum provides another example.

Definition. Two functions are said to be *μ-equivalent* if they belong to the same $\mu = \text{const}$ stratum.

Definition. A *normal form* for a class of functions K is a smooth map $\Phi: B \to M$ of a finite-dimensional vector space B into the space of polynomials that satisfies the following conditions:

1) the image $\Phi(B)$ intersects all orbits in K;
2) the preimage of any orbit in K under Φ is a finite subset of B;
3) the preimage of the complement of K is a "meagre" set (i.e., it is contained in some proper hypersurface in B).

A normal form is said to be *polynomial* [resp. *affine*] if the map Φ is polynomial [resp. linear nonhomogeneous]. An affine normal form is said to be *simple* if Φ has the form

$$\Phi(b_1, b_2, \ldots, b_r) = \varphi(x) + \sum_{i=1}^{r} b_i x^{m_i},$$

where x^{m_i} are monomials and $\varphi(x)$ is a fixed polynomial.

The existence of a single normal form for an entire $\mu = \text{const}$ stratum is not at all obvious a priori. A surprising conclusion of the computations that lead to the classification of singularities is that such normal forms do indeed exist for all lists of singularities presently known. Most normal forms (in particular those for all 0-, 1-, and 2-modal singularities) are simple normal forms. It is not know how broad the class of functions is for which the $\mu = \text{const}$ stratum admits a simple or a polynomial normal form; one only has examples in which the $\mu = \text{const}$ stratum does not admit an affine normal form [356].

In any given list the singularities are divided into *series* labeled by capital letters with indices to indicate the $\mu = \text{const}$ strata. Although in each concrete case one can formulate rules that specify one or another of the series, no general definition of the series of singularities is known. The only clear pattern is that the series are connected with singularities of infinite multiplicity (for instance, $D \sim x^2y$ and $T \sim xyz$), so that the hierarchy of series reflects the hierarchy of nonisolated singularities.

2.2. Classes of Low Modality. From the viewpoint of applications, the most important characteristic of a class of singularities (e.g., of a $\mu = \text{const}$ stratum) is its codimension c in the space $\mathcal{O}_n$ of function-germs.

In fact, a generic function has only nondegenerate critical points of codimension $c = 0$. Degenerate singularities occur in an irremovable manner only in families of functions depending on parameters. Thus, in a family of functions depending on l parameters there may occur (in such a manner that it cannot be removed through a small perturbation of the family) only a family of singularities for which $c \leqslant l$.

Example. From the classification of singularities it follows that the singularities of codimension one are (stably) equivalent to the singularity A_2: $f(x) = x^3$. That is to say, in a generic one-parameter family of functions, for almost every value of the parameter the corresponding function has only Morse critical points, and for some isolated values of the parameter there may occur critical points of type A_2.

From the topological viewpoint, the most important characteristic of a critical point is its multiplicity μ (the number of Morse critical points into which the given "complicated" singularity disintegrates under a small perturbation).

A surprising conclusion of the computations carried out in the classification of singularities is that the algebraically most natural results are obtained not in classifying singularities by their codimension c or multiplicity μ, but rather in classifying the singularities of low modality m. We remind the reader that these numerical characteristics of a class of singularities are tied together by the relation $\mu = c + m + 1$ (see subsection 1.9).

At present a complete classification is available for all singularities for which either

(1) $c \leqslant 10$,

(2) $\mu \leqslant 16$,

or

(3) $m \leqslant 2$.

We give below the lists of 0-, 1-, and 2-modal singularities up to stable equivalence (subsection 1.3). The classification of these and other known classes of singularities is given in the papers [12], [13], [16], [17].

2.3. Singularities of Modality $\leqslant 2$

Simple singularities. There are two infinite series, A_k and D_k, and three exceptional singularities, E_6, E_7, and E_8:

$A_k, k \geqslant 1$	$D_k, k \geqslant 4$	E_6	E_7	E_8
x^{k+1}	$x^2y + y^{k-1}$	$x^3 + y^4$	$x^3 + xy^3$	$x^3 + y^5$

Unimodal singularities. There are three families of parabolic singularities, a three-index series of hyperbolic singularities, and 14 families of exceptional singularities.

Parabolic:

P_8	$x^3 + y^3 + z^3 + axyz$	$a^3 + 27 \neq 0$
X_9	$x^4 + y^4 + ax^2y^2$	$a^2 \neq 4$
J_{10}	$x^3 + y^6 + ax^2y^2$	$4a^3 + 27 \neq 0$

Hyperbolic:

$$T_{p,q,r}\colon x^p + y^q + z^r + axyz, \quad a \neq 0, \quad \frac{1}{p} + \frac{1}{q} + \frac{1}{r} < 1.$$

14 exceptional families:

E_{12}	$x^3 + y^7 + axy^5$	W_{12}	$x^4 + y^5 + ax^2y^3$
E_{13}	$x^3 + xy^5 + ay^8$	W_{13}	$x^4 + xy^4 + ay^6$
E_{14}	$x^3 + y^8 + axy^6$	Q_{10}	$x^3 + y^4 + yz^2 + axy^3$
Z_{11}	$x^3y + y^5 + axy^4$	Q_{11}	$x^3 + y^2z + xz^3 + az^5$
Z_{12}	$x^3y + xy^4 + ax^2y^3$	Q_{12}	$x^3 + y^5 + yz^2 + axy^4$
Z_{13}	$x^3y + y^6 + axy^5$	S_{11}	$x^4 + y^2z + xz^2 + ax^3z$
U_{12}	$x^3 + y^3 + z^4 + axyz^2$	S_{12}	$x^2y + y^2z + xz^3 + az^5$

Bimodal singularities. There are 8 infinite series and 14 exceptional families.

4 series of singularities of corank 2:

Name	Normal form	Restrictions	Multiplicity μ
$J_{3,0}$	$x^3 + bx^2y^3 + y^9 + cxy^7$	$4b^3 + 27 \neq 0$	16
$J_{3,p}$	$x^3 + x^2y^3 + \mathbf{a}y^{9+p}$	$p > 0, a_0 \neq 0$	$16 + p$
$Z_{1,0}$	$x^3y + dx^2y^3 + cxy^6 + y^7$	$4d^3 + 27 \neq 0$	15
$Z_{1,p}$	$x^3y + x^2y^3 + \mathbf{a}y^{7+p}$	$p > 0, a_0 \neq 0$	$15 + p$
$W_{1,0}$	$x^4 + \mathbf{a}x^2y^3 + y^6$	$a_0^2 \neq 4$	15
$W_{1,p}$	$x^4 + x^2y^3 + \mathbf{a}y^{6+p}$	$p > 0, a_0 \neq 0$	$15 + p$
$W^{\#}_{1,2q-1}$	$(x^2 + y^3)^2 + \mathbf{a}xy^{4+q}$	$q > 0, a_0 \neq 0$	$15 + 2q - 1$
$W^{\#}_{1,2q}$	$(x^2 + y^3)^2 + \mathbf{a}x^2y^{3+q}$	$q > 0, a_0 \neq 0$	$15 + 2q$

4 series of singularities of corank 3:

Name	Normal form	Restrictions	Multiplicity μ
$Q_{2,0}$	$x^3 + yz^2 + \mathbf{a}x^2y^2 + xy^4$	$a_0^2 \neq 4$	14
$Q_{2,p}$	$x^3 + yz^2 + x^2y^2 + \mathbf{a}y^{6+p}$	$p > 0, a_0 \neq 0$	$14 + p$
$S_{1,0}$	$x^2z + yz^2 + y^5 + \mathbf{a}zy^3$	$a_0^2 \neq 4$	14
$S_{1,p}$	$x^2z + yz^2 + x^2y^2 + \mathbf{a}y^{5+p}$	$p > 0, a_0 \neq 0$	$14 + p$
$S^{\#}_{1,2q-1}$	$x^2z + yz^2 + zy^3 + \mathbf{a}xy^{3+q}$	$q > 0, a_0 \neq 0$	$14 + 2q - 1$
$S^{\#}_{1,2q}$	$x^2z + yz^2 + zy^3 + \mathbf{a}x^2y^{2+q}$	$q > 0, a_0 \neq 0$	$14 + 2q$
$U_{1,0}$	$x^3 + xz^2 + xy^3 + \mathbf{a}y^3z$	$a_0(a_0^2 + 1) \neq 0$	14
$U_{1,2q-1}$	$x^3 + xz^2 + xy^3 + \mathbf{a}y^{1+q}z^2$	$q > 0, a_0 \neq 0$	$14 + 2q - 1$
$U_{1,2q}$	$x^3 + xz^2 + xy^3 + \mathbf{a}y^{3+q}z$	$q > 0, a_0 \neq 0$	$14 + 2q$

14 exceptional families:

E_{18}	$x^3 + y^{10} + \mathbf{a}xy^7$	W_{17}	$x^4 + xy^5 + \mathbf{a}y^7$
E_{19}	$x^3 + xy^7 + \mathbf{a}y^{11}$	W_{18}	$x^4 + y^7 + \mathbf{a}x^2y^4$
E_{20}	$x^3 + y^{11} + \mathbf{a}xy^8$	Q_{16}	$x^3 + yz^2 + y^7 + \mathbf{a}xy^5$
Z_{17}	$x^3y + y^8 + \mathbf{a}xy^6$	Q_{17}	$x^3 + yz^2 + xy^5 + \mathbf{a}y^8$
Z_{18}	$x^3y + xy^6 + \mathbf{a}y^9$	Q_{18}	$x^3 + yz^2 + y^8 + \mathbf{a}xy^6$
Z_{19}	$x^3y + y^9 + \mathbf{a}xy^7$	S_{16}	$x^2z + yz^2 + xy^4 + \mathbf{a}y^6$
U_{16}	$x^3 + xz^2 + y^5 + \mathbf{a}x^2y^2$	S_{17}	$x^2z + yz^2 + y^6 + \mathbf{a}zy^4$

In the last three tables $\mathbf{a} = a_0 + a_1y$.

2.4. Simple Singularities and Klein Singularities. A remarkable connection exists between the classification of simple singularities, that of regular polyhedra in three-dimensional space, and that of the Coxeter groups A_k, D_k, E_k. Although the coincidence of these classifications emerges only when classification theorems proved independently in the three indicated settings are compared, it undoubtedly goes far deeper than the mere coinciding of the lists.

A number of constructions are available that permit one to associate with one another the objects of the three different classifications. In the next chapter we shall see that the monodromy group of a simple singularity coincides with the corresponding Coxeter group, while its level bifurcation set is isomorphic to the variety of nonregular orbits of that group. Here we wish to explain how simple singularities arise from symmetry groups of regular polyhedra in $\mathbb{R}^3$ [198], [310].

As is known, the finite subgroups of $SO(3)$ are exhausted by the following list:

1) the cyclic group $\mathbb{Z}_n$;

2) the dihedral group $\mathbb{D}_{2n}$, isomorphic to the semidirect product of $\mathbb{Z}_n$ and $\mathbb{Z}_2$;

3) the groups of motions of the tetrahedron, $\mathbb{T}_{12}$, of the octahedron, $\mathbb{O}_{24}$, and of the icosahedron, $\mathbb{I}_{60}$.

Let Γ be a discrete subgroup of $SO(3)$. Consider its preimage $\Gamma^* \subset SU(2)$ under the two-sheeted covering map $SU(2) \to SO(3)$. The group Γ^* is called the *binary* group of the corresponding polyhedron and acts on $\mathbb{C}^2$ as a subgroup of $SU(2)$. Consider the algebra of polynomial invariants of this action of Γ^*. As it turns out, this algebra is generated by three invariants x, y, z, which satisfy a single relation. That relation defines a hypersurface V in the space $\mathbb{C}^3$ with coordinates x, y, z. V is naturally isomorphic to the orbit space of the action of Γ^* on $\mathbb{C}^2$ and has an isolated singular point at the origin.

For a suitable choice of generators in the algebra of invariants, the relations for the binary groups of polyhedra are as follows:

$\mathbb{Z}_n$	$x^n + yz = 0$	A_{n-1}
$\mathbb{D}^*_{2n}$	$xy^2 - x^{n+1} + z^2 = 0$	D_{n+2}
$\mathbb{T}^*$	$x^4 + y^3 + z^2 = 0$	E_6
$\mathbb{O}^*$	$x^3 + xy^3 + z^2 = 0$	E_7
$\mathbb{I}^*$	$x^5 + y^3 + z^2 = 0$	E_8

Thus, the orbit space V of the action of a binary polyhedral group on $\mathbb{C}^2$ is isomorphic to the zero level set of the corresponding singularity.

2.5. Resolution of Simple Singularities. Let V be a complex surface with a unique singular point a. A *resolution of the singularity of* V is a proper map $\pi\colon \tilde{V} \to V$ of a smooth complex surface $\tilde{V}$ onto V, which is biholomorphic outside the preimage $\pi^{-1}(a)$ of the singular point. A resolution π is said to be *minimal* if for every other resolution $\pi'\colon \tilde{V}' \to V$ there exists a holomorphic map $\beta\colon \tilde{V}' \to \tilde{V}$ such that the diagram

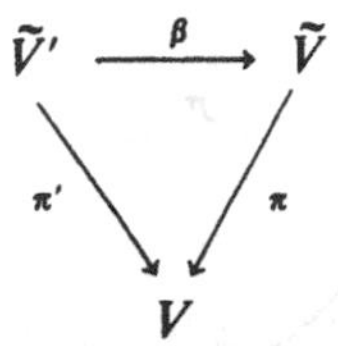

is commutative. The minimal resolution exists and is unique up to an isomorphism [57], [218].

In the case where V is the zero level set of a simple singularity of a function of three variables, the minimal resolution can be described as follows [339].

The preimage of the singular point of V is a connected union of projective lines:

$$\pi^{-1}(0) = C_1 \cup C_2 \cup \cdots \cup C_\mu, \qquad C_i \simeq \mathbb{C}P^1.$$

The self-intersection index of each component C_i is equal to -2. Pairwise intersections are described by a graph in which a vertex is assigned to each component C_i, and two vertices are or are not connected by an edge depending on whether the intersection index of the corresponding components is 1 or 0. In this manner one obtains the following graphs:

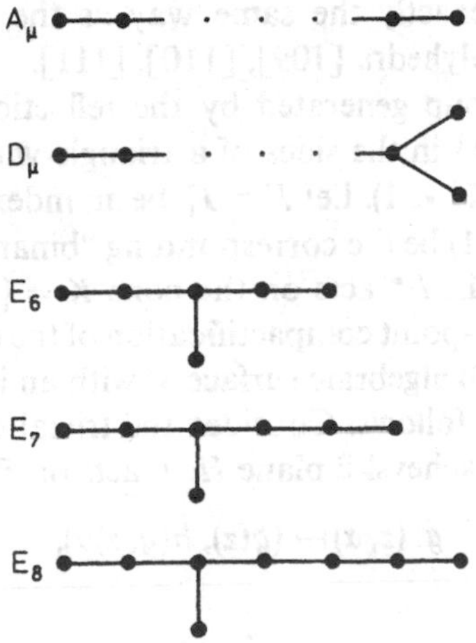

We remark that these graphs are exactly the Dynkin diagrams of the corresponding Coxeter groups (see § 2.5).

Example. In the case of the singularity D_4 the construction described above is readily followed in the real ($\mathbb{R}^3$-) picture (Fig. 7): $V = \{xy^2 - x^3 + z^2 = 0\}$.

The preimage $\pi^{-1}(0)$ of zero consists of four circles, for which the diagram looks like this:

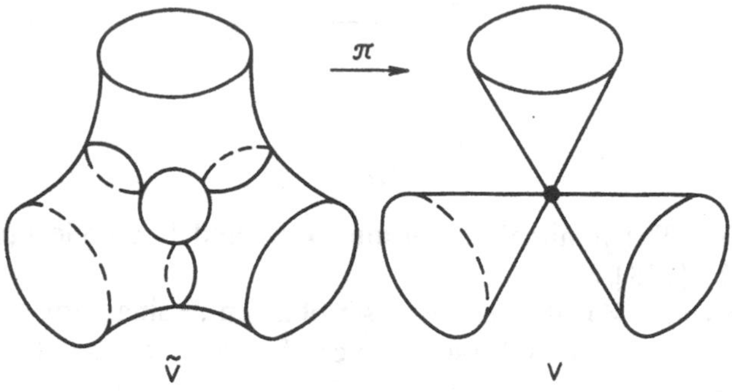

Fig. 7

The Dynkin diagram for the corresponding singularity will be derived again, by a different method, in subsection 2.5.10.

2.6. Unimodal and Bimodal Singularities. The quasihomogeneous unimodal and bimodal singularities can be obtained from automorphic forms connected with polygons in the Lobachevskiĭ plane (and with three remarkable triangles in the ordinary plane) in exactly the same way as the simple singularities are obtained from regular polyhedra [109], [110], [111].

Let Γ_0 denote the group generated by the reflections of the Lobachevskiĭ plane $H = \{z \in \mathbb{C}: |z| < 1\}$ in the sides of a triangle with the angles π/k_1, π/k_2, π/k_3 (with $k_1^{-1} + k_2^{-1} + k_3^{-1} < 1$). Let $\Gamma \subset \Gamma_0$ be an index-2 subgroup of motions of H and let $\Gamma^* \subset SU(1, 1)$ be the corresponding "binary group of the triangle". As a subgroup of $SU(1, 1)$, Γ^* acts on the cone $K = \{|z_1|^2 - |z_2|^2 > 0\} \subset \mathbb{C}^2$. There exists a natural one-point compactification of the orbit space K/Γ^*, which yields a (two-dimensional) algebraic surface V with an isolated singularity.

V can be described as follows. Consider the trivial one-dimensional bundle $E = H \times \mathbb{C}$ over the Lobachevskiĭ plane H. Γ acts on E according to the rule

$$g: (z, \alpha) \mapsto (g(z), h(g, z)\alpha),$$

where

$$g(z) = \frac{az + b}{cz + d}, \quad g \in SU(1, 1), \quad g = \begin{pmatrix} a & b \\ c & d \end{pmatrix}, \qquad \text{and } h = J^{-1} = (cz + d)^2.$$

The sections of the bundle E are the entire holomorphic 1-forms on H, and the sections invariant under the action of Γ are the entire Γ-automorphic 1-forms. The quotient E/Γ is a sheaf over H. Upon contracting its null section to a point, we obtain the sought-for two-dimensional surface V with an isolated singular point.

The algebra $\Omega(k_1, k_2, k_3)$ of entire Γ-automorphic forms on H can be identified with an algebra of functions on the surface V.

Theorem. *There exist exactly* 14 *triangles in the Lobachevskiĭ plane H, for which the algebra $\Omega(k_1, k_2, k_3)$ is generated by three elements satisfying a single* *relation. The corresponding singularities of the hypersurface V in $\mathbb{C}^3$ are exactly the 14 exceptional families of unimodal singularities.*

In a completely analogous manner, there exist exactly 6 quadrangles in H for which the generators of the algebra $\Omega(k_1, k_2, k_3, k_4)$ of automorphic forms satisfy a single relation. The hypersurface V defined by that relation leads to one of the 6 quasihomogeneous singularities contained in the list of 8 infinite series of bimodal singularities.

By considering automorphic functions of weight r (i.e., setting the factor of automorphy in the construction described above equal to $h = J^{-1/r}$), one can obtain the 14 exceptional families of bimodal singularities from 14 triangles in the Lobachevskiĭ plane, and the three parabolic unimodal families from triangles in the ordinary plane. The correspondence between singularities and triangles or quadrangles on the sphere, the Euclidean plane, or the Lobachevskiĭ plane is given in the following table:

	k_i		k_i	r
A_n	$n+1, n+1$	E_8	2, 3, 5	
D_n	$2, 2, n-1$	P_8	3, 3, 3	
E_6	2, 3, 3	X_9	2, 4, 4	
E_7	2, 3, 4	J_{10}	2, 3, 6	
Q_{10}	2, 3, 9	Q_{16}	3, 3, 9	2
Q_{11}	2, 4, 7	Q_{17}	2, 4, 13	3
Q_{12}	3, 3, 6	Q_{18}	2, 3, 21	5
S_{11}	2, 5, 6	S_{16}	3, 5, 7	2
S_{12}	3, 4, 5	S_{17}	2, 7, 10	3
U_{12}	4, 4, 4	U_{16}	5, 5, 5	2
Z_{11}	2, 3, 8	Z_{17}	3, 3, 7	2
Z_{12}	2, 4, 6	Z_{18}	2, 4, 10	3
Z_{13}	3, 3, 5	Z_{19}	2, 3, 16	5
W_{12}	2, 5, 5	W_{17}	3, 5, 5	2
W_{13}	3, 4, 4	W_{18}	2, 7, 7	3
E_{12}	2, 3, 7	E_{18}	3, 3, 5	2
E_{13}	2, 4, 5	E_{19}	2, 4, 7	3
E_{14}	3, 3, 4	E_{20}	2, 3, 11	5
$W_{1,0}$	2, 2, 3, 3	$S_{1,0}$	2, 2, 3, 4	
$J_{3,0}$	2, 2, 2, 3	$U_{1,0}$	2, 3, 3, 3	
$Q_{2,0}$	2, 2, 2, 5	$Z_{1,0}$	2, 2, 2, 4	

Example. Consider a quadrangle with the angles $\pi/2$, $\pi/2$, $\pi/2$, $\pi/3$ in the Lobachevskiĭ plane H. The action of Γ on H yields a ramified covering $H \to \mathbb{CP}^1$ with ramification scheme (2, 2, 2, 3). Suppose that the ramification points of order 2 and 3 are $z = 0, \pm 1$ and $z = \infty$ respectively ($z \in \mathbb{CP}^1$).

The algebra $\Omega(2, 2, 2, 3)$ of automorphic forms is isomorphic to the algebra generated by the forms of the type

$$\varphi(z) = P_l(z)[z^{m_1}(z-1)^{m_2}(z+1)^{m_3}]^{-1}\, dz^s,$$

where $m_i = \left[\dfrac{s}{2}\right]$ and P_l is a polynomial of degree $l \leqslant 3\left[\dfrac{s}{2}\right] + \left[\dfrac{2}{3}s\right] - 2s$. The generators

$$\alpha_1 = [z(z-1)(z+1)]^{-1}\, dz^2, \qquad \alpha_2 = [z^3(z-1)^3(z+1)^3]^{-1} z\, dz^6,$$

$$\alpha_3 = [z^4(z-1)^4(z+1)^4]^{-1}\, dz^9$$

of this algebra satisfy the single relation

$$\alpha_2^3 - \alpha_2\alpha_1^6 - \alpha_3^2 = 0,$$

which defines in $\mathbb{C}^3$ a hypersurface with an isolated singular point of type $J_{3,0}$ at the origin.

All singularities with one and two moduli are classified exactly via degeneracies of elliptic curves, the classification of which has been obtained by Kodaira (see [200]). As indicated by V.S. Kulikov, in order to obtain the above singularities from degeneracies of an elliptic curve one has to blow up one, two, or three points on the minimal resolution of the degenerate fibre and subsequently contract the original fibre to a point [200].

2.7. Adjacency of Singularities. We indicate below only those adjacencies of singularities that arise naturally in the course of the classification [17].

Definition. A class of singularities L is said to be *adjacent* to a class K, and one writes $L \to K$, if every function $f \in L$ can be deformed into a function of class K by an arbitrarily small perturbation.

In other words, an adjacency $L \to K$ occurs if the class L is contained in the closure of the class K in the function space.

The adjacencies of simple singularities are as follows:

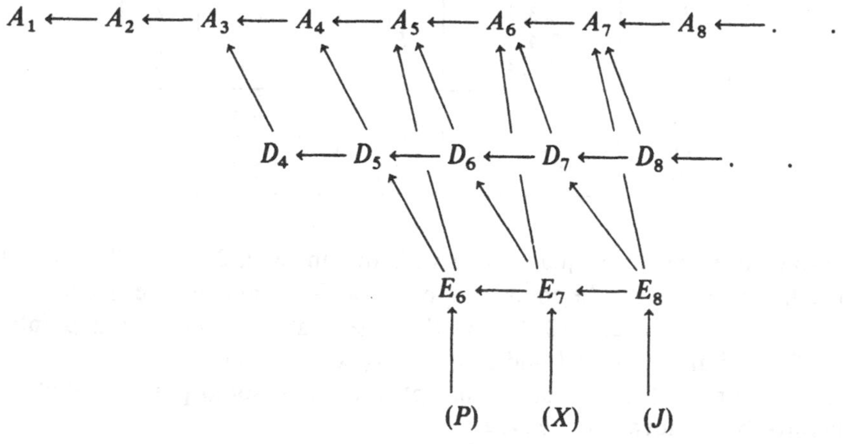

In other words, one simple singularity is adjacent to another such singularity if the Dynkin diagram of the second singularity is embedded in that of the first.

Some adjacencies of unimodal singularities are shown below:

$$J_{10} = T_{2,3,6} \longleftarrow (E_8)$$
$$\uparrow$$
$$\cdots \longrightarrow T_{2,3,8} \longrightarrow T_{2,3,7} \longleftarrow E_{12} \longleftarrow E_{13} \longleftarrow E_{14} \longleftarrow (J_3)$$

$$X_9 = T_{2,4,4} \longleftarrow (E_7)$$
$$\uparrow$$
$$\cdots \longrightarrow T_{2,4,6} \longrightarrow T_{2,4,5} \longleftarrow Z_{11} \longleftarrow Z_{12} \longleftarrow Z_{13} \longleftarrow (Z_1^1)$$
$$\uparrow \qquad\qquad \uparrow \qquad\qquad \uparrow \qquad\qquad \uparrow$$
$$\vdots\Longrightarrow T_{2,5,6} \longrightarrow T_{2,5,5} \longleftarrow W_{12} \longleftarrow W_{13} \longleftarrow (W_{1,0}; N)$$

$$P_8 = T_{3,3,3} \longleftarrow (E_6)$$
$$\uparrow$$
$$\cdots \longrightarrow T_{3,3,5} \longrightarrow T_{3,3,4} \longleftarrow Q_{10} \longleftarrow Q_{11} \longleftarrow Q_{12} \longleftarrow (Q_2)$$
$$\uparrow \qquad\qquad \uparrow \qquad\qquad \uparrow \qquad\qquad \uparrow$$
$$\vdots\Longrightarrow T_{3,4,5} \longrightarrow T_{3,4,4} \longleftarrow S_{11} \longleftarrow S_{12} \longleftarrow (S_{1,0})$$
$$\uparrow \qquad\qquad \uparrow \qquad\qquad \uparrow$$
$$\vdots\Longrightarrow T_{4,4,5} \longrightarrow T_{4,4,4} \longleftarrow U_{12} \longleftarrow (U_{1,0}; V)$$
$$\uparrow$$
$$(O)$$

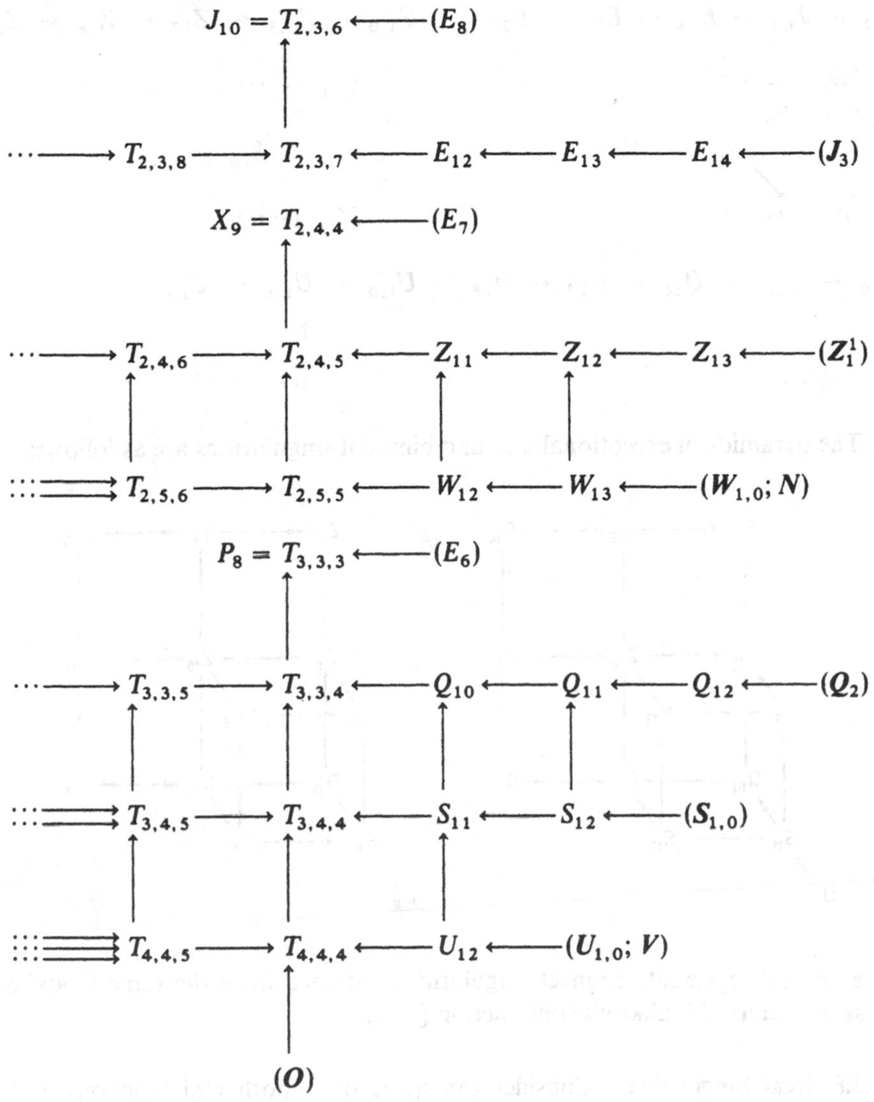

The complete list of adjacencies of all unimodal singularities was found by Brieskorn [62]. All other adjacencies (except one ($Q_{11} \to J_{10}$)) are excluded by the necessary condition adjacency of the spectra of the singularities [155]; see also subsection 2.4.8.

The adjacency $Q_{11} \to J_{10}$ is excluded due to the semicontinuity of the index of inertia of the intersection form (subsection 2.2.5).

Some adjacencies of bimodal singularities are as follows:

$$\begin{array}{ccccccccc} & & \cdots & & & & & & \\ & & \downarrow & & & & & & \\ J_{3,0} & \leftarrow & J_{3,1} & \leftarrow & E_{18} & \leftarrow & E_{19} & \leftarrow & E_{20} \end{array} \qquad \begin{array}{ccccccccc} & & \cdots & & & & & & \\ & & \downarrow & & & & & & \\ Z_{1,0} & \leftarrow & Z_{1,1} & \leftarrow & Z_{17} & \leftarrow & Z_{18} & \leftarrow & Z_{19} \end{array}$$

$$\begin{array}{ccccccc} & & W_{1,1} & \leftarrow & \cdots & & \\ & \swarrow & & \nwarrow & & & \\ W_{1,0} & & & & W_{17} & \leftarrow & W_{18} \\ & \nwarrow & & \swarrow & & & \\ & & W^{\#}_{1,1} & \leftarrow & \cdots & & \end{array} \qquad \begin{array}{ccccccc} & & S_{1,1} & \leftarrow & \cdots & & \\ & \swarrow & & \nwarrow & & & \\ S_{1,0} & & & & S_{16} & \leftarrow & S_{17} \\ & \nwarrow & & \swarrow & & & \\ & & S^{\#}_{1,1} & \leftarrow & \cdots & & \end{array}$$

$$\begin{array}{ccccccccc} Q_{2,0} & \leftarrow & Q_{2,1} & \leftarrow & Q_{16} & \leftarrow & Q_{17} & \leftarrow & Q_{18} \\ & & \uparrow & & & & & & \\ & & \cdots & & & & & & \end{array} \qquad \begin{array}{ccccc} U_{1,0} & \leftarrow & U_{1,1} & \leftarrow & U_{16} \\ & & \uparrow & & \\ & & \cdots & & \end{array}$$

The pyramids of exceptional uni- and bimodal singularities are as follows:

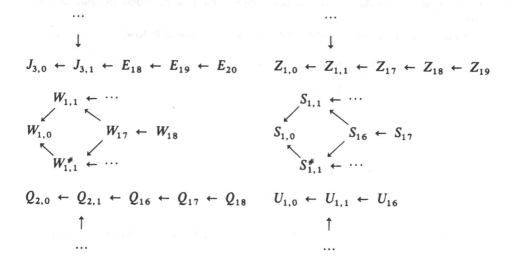

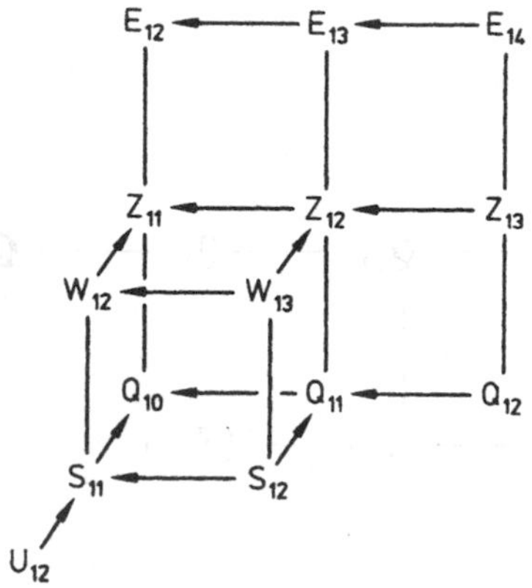

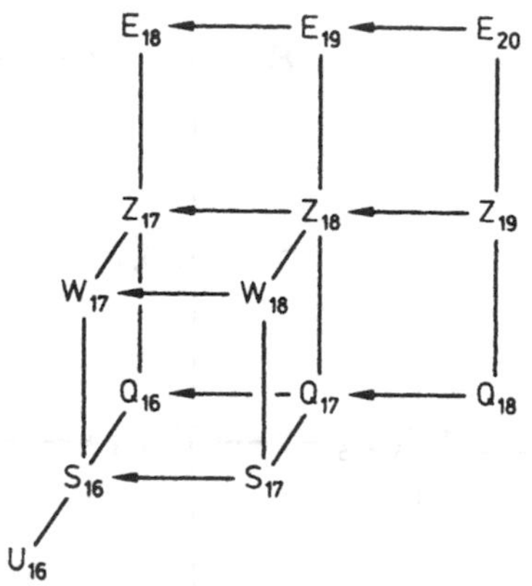

The vertical segements connect singularities obtained from the same Kodaira class by means of Kulikov's construction [200].

2.8. Real Singularities. Consider the space of smooth real functions with critical point 0 and critical value 0. By equivalence of two such functions we shall mean, as above, that they belong to the same orbit of the action of the group of germs (at zero) of real diffeomorphisms of this space; the definition of the stable equivalence is modified accordingly (see subsections 1.2, 1.3).

Example. The function-germs $f(x, y) = x^3 - y^2$ and $g(x, y, z) = x^3 + y^2 + z^2$ are stably equivalent.

We give below the classification of simple and unimodal germs up to stable equivalence.

Simple germs:

$A_k^{\pm}, k \geqslant 1$	$D_k^{\pm}, k \geqslant 4$	E_6	E_7	E_8
$\pm x^{k+1}$	$x^2y \pm y^{k-1}$	$x^3 \pm y^4$	$x^3 + xy^3$	$x^3 + y^5$

The germs A_k^+ and A_k^- are equivalent for $k = 1, 2m$; the remaining germs in the table are not equivalent. The beginning of the hierarchy of degenerate singularities of real functions is shown in the following diagram (in which we display all classes of codimension less than or equal to 5 in the space of functions with critical point 0 and critical value 0):

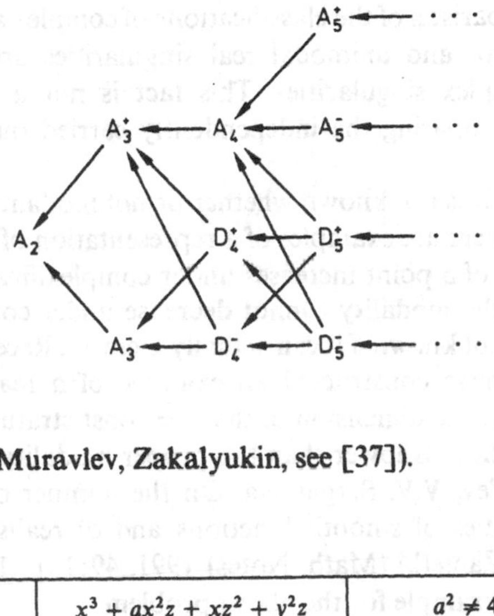

Unimodal germs (Muravlev, Zakalyukin, see [37]).

Parabolic:

$P_8 = T_{3,3,3}$	$x^3 + ax^2z \pm xz^2 + y^2z$	$a^2 \neq 4$, if $+$
$X_9 = T_{2,4,4}$	$\pm x^4 + ax^2y^2 \pm y^4$	$a^2 \neq 4$, if $++$ or $--$
$J_{10} = T_{2,3,6}$	$x^3 + ax^2y^2 \pm xy^4$	$a^2 \neq 4$, if $+$

Hyperbolic of corank 2:

$J_{10+k} = T_{2,3,6+k}$	$x^3 \pm x^2y^2 + ay^{6+k}$	$a \neq 0, k > 0$
$X_{9+k} = T_{2,4,4+k}$	$\pm x^4 \pm x^2y^2 + ay^{4+k}$	$a \neq 0, k > 0$
$Y_{r,s} = T_{2,r,s}$	$\pm x^2y^2 \pm x^r + ay^s$	$a \neq 0, r, s > 4$
$\tilde{Y}_r = \tilde{T}_{2,r,r}$	$\pm(x^2 + y^2)^2 + ax^r$	$a \neq 0, r > 4$

Hyperbolic of corank 3:

$P_{8+k} = T_{3,3,3+k}$	$x^3 \pm x^2z + y^2z + az^{k+3}$	$a \neq 0, k > 0$
$R_{l,m} = T_{3,l,m}$	$x(x^2 + yz) \pm y^l \pm az^m$	$a \neq 0, m \geqslant l > 4$
$\tilde{R}_m = \tilde{T}_{3,m,m}$	$x(\pm x^2 + y^2 + z^2) + ay^m$	$a \neq 0, m > 4$
$T_{p,q,r}$	$axyz \pm x^p \pm y^q \pm z^r$	$a \neq 0, p^{-1} + q^{-1} + r^{-1} < 1$
$\tilde{T}_{p,m} = \tilde{T}_{p,m,m}$	$x(y^2 + z^2) \pm x^p + ay^m$	$a \neq 0, p^{-1} + 2m^{-1} < 1$

Exceptional (here a is a real parameter):

E_{12}	$x^3 + y^7 \pm z^2 + axy^5$	W_{12}	$\pm x^4 + y^5 \pm z^2 + ax^2y^3$
E_{13}	$x^3 + xy^5 \pm z^2 + ay^8$	W_{13}	$\pm x^4 + xy^4 \pm z^2 + ay^6$
E_{14}	$x^3 \pm y^8 \pm z^2 + axy^6$	Q_{10}	$x^3 + y^2z \pm z^4 + axz^3$
Z_{11}	$x^3y + y^5 \pm z^2 + axy^4$	Q_{11}	$x^3 + y^2z \pm xz^3 + az^5$
Z_{12}	$x^3y + xy^4 \pm z^2 + ax^2y^3$	Q_{12}	$x^3 + y^2z \pm z^5 + az^4x$
Z_{13}	$x^3y \pm y^6 \pm z^2 + axy^5$	S_{11}	$z(x^2 + yz) \pm y^4 + ay^3z$
U_{12}	$x(x^2 \pm y^2) \pm z^4 + axyz^2$	S_{12}	$z(x^2 + yz) + xy^3 + ay^5$

Remark. A comparison of the classifications of complex and real singularities reveals that all zero- and unimodal real singularities are real forms of the corresponding complex singularities. This fact is not a priori obvious and emerges only on comparing the independently carried out real and complex classifications.

The point is that it is not known whether or not modality is preserved under complexification. There are examples of a representation of a real Lie group for which the modality of a point increases under complexification (E.B. Vinberg). For critical points the modality cannot decrease under complexification (V.V. Muravlev), but it is not known if it can actually increase. Recently V.V. Serganova and V.A. Vasil'ev have constructed an example of a real singularity whose proper modality (i.e., the dimension of the $\mu =$ const stratum in the base of the miniversal deformation) is lower than the proper modality of its complexification, see V.A. Vasil'ev, V.V. Serganova. On the number of real and complex moduli of singularities of smooth functions and of realisations of matroids, Mathematicheskie Zametki (Math. Notes) 1991, 49:1, p. 19–27. This perhaps gives also a counterexample for the above problem.

§ 3. Reduction to Normal Forms

The classification of degenerate singularities is based on computations connected with the reduction of function-germs with a given Newton diagram to normal forms.

The formal technique that allows one to handle these computations relies on the consideration of various filtrations in the algebra of functions. With each such filtration one can associate its spaces of quasijets of diffeomorphisms and its Lie algebras of quasijets of vector fields.

The simplest and most often encountered case is that of a quasihomogeneous filtration. In this case there also arises a Lie group of quasihomogeneous diffeomorphisms which in the present theory plays the same role as the general linear group in the case of usual jets.

3.1. The Newton Diagram. Consider the space $\mathbb{C}^n$ with a fixed coordinate system $x_1, \ldots, x_n$.

Definition. We shall call the function $\mathbf{x}^{\mathbf{k}} = x_1^{k_1} x_2^{k_2} \ldots x_n^{k_n}$ *monomial* with exponent $\mathbf{k} = (k_1, k_2, \ldots, k_n)$, $k_i \in \mathbb{Z}_+$.

The exponent of a monomial specifies a point of the integer lattice in the positive quadrant $\mathbb{R}_+^n$; that point will be also referred to as the exponent of the monomial.

Let $f: (\mathbb{C}^n, 0) \to (\mathbb{C}, 0)$ be a germ of holomorphic function and $\sum f_{\mathbf{k}} \mathbf{x}^{\mathbf{k}}$ its Taylor series.

Definition. The *Newton polyhedron* of a power series is the convex hull of the union of the positive quadrants $\mathbb{R}_+^n$ with vertices at the exponents of the monomials that appear in that series with nonzero coefficients. The *Newton diagram* of a series is the union of the compact faces of its Newton polyhedron.

Example. The Newton diagram and polyhedron of the function $f(x, y) = x^3y + x^2y^2 + y^5$ are shown in Fig. 8.

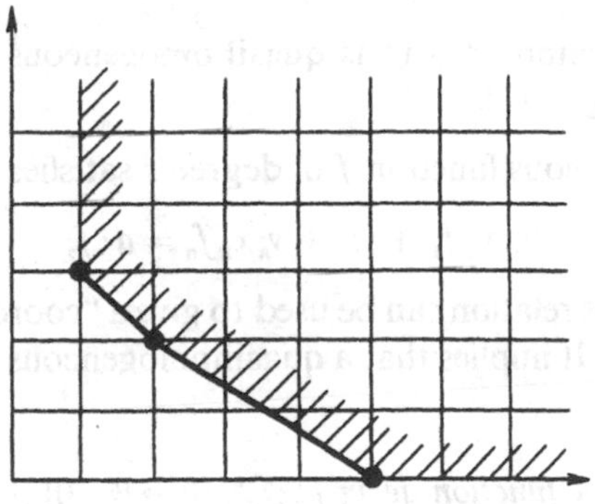

Fig. 8

Many of the characteristics attached to a singularity (multiplicity, modality, the Jordan structure and the eigenvalues of the monodromy operator, the spectrum of the singularity, etc.) are determined by its Newton diagram. More precisely, for almost all functions with a given diagram these characteristics coincide and can be expressed in terms of the geometry of the diagram.

Definition. The *support* of a power series $\sum f_{\mathbf{k}} \mathbf{x}^{\mathbf{k}}$ is the set of all exponents of monomials that appear in this series with nonzero coefficients.

Let Γ be a Newton polyhedron. Consider a holomorphic function-germ $f\colon (\mathbb{C}^n, 0) \to (\mathbb{C}, 0)$ whose support is contained in Γ. On multiplying f by a monomial $\mathbf{x}^{\mathbf{k}}$ its support gets translated by a vector with nonnegative coordinates, hence the support of $\mathbf{x}^{\mathbf{k}} f$ is also contained in Γ. It follows that the set of all functions (power series) whose supports are contained in Γ is an ideal in the ring $\mathcal{O}_n$ of function-germs [resp. in the ring of formal power series $\mathbb{C}[[x_1, x_2, \ldots, x_n]]$).

3.2. Quasihomogeneous Functions and Filtrations

Definition. A holomorphic function $f\colon (\mathbb{C}^n, 0) \to (\mathbb{C}, 0)$, defined on the numerical space $\mathbb{C}^n$, is called a *quasihomogeneous function of degree d with exponents* $v_1, \ldots, v_n$ if

$$f(\lambda^{v_1} x_1, \ldots, \lambda^{v_n} x_n) = \lambda^d f(x_1, \ldots, x_n)$$

for all $\lambda > 0$. The exponents v_i are alternatively referred to as the *weights* of the variables x_i.

In terms of the Taylor series $\sum f_{\mathbf{k}} \mathbf{x}^{\mathbf{k}}$ of f, the quasihomogeneity condition means that the exponents of the nonzero terms of the series lie in the hyperplane

$$L = \{\mathbf{k}\colon v_1 k_1 + \cdots + v_n k_n = d\}.$$

In other words, a function is quasihomogeneous if its Newton diagram is contained in L, and its support lies in the Newton diagram.

Example. The function $x^3 + y^2$ is quasihomogeneous of degree $d = 1$ with exponents $v_1 = 1/3$, $v_2 = 1/2$.

Any quasihomogeneous function f of degree d satisfies Euler's identity

$$v_1 x_1 f_1 + \cdots + v_n x_n f_n = d \cdot f,$$

where $f_i = \partial f/\partial x_i$. This relation can be used to give a "coordinate-free" definition of quasihomogeneity: It implies that a quasihomogeneous function f belongs to its gradient ideal $I_{\nabla f}$.

Theorem ([207]). *A function-germ* $f\colon (\mathbb{C}^n, 0) \to (\mathbb{C}, 0)$ *is equivalent to a quasihomogeneous function-germ if and only if* $f \in I_{\nabla f}$.

The degree d and the set of exponents v_i of a quasihomogeneous function are defined up to multiplication by a nonzero number. In what follows we shall consider quasihomogeneous functions of degree one with positive rational exponents. Such functions are automatically polynomials. The hyperplane L will be referred to as the *diagonal*. L cuts out segments of lengths $1/v_i$ on the coordinate axes.

Definition. A quasihomogeneous function f is said to be *nondegenerate* if 0 is an isolated critical point of f, i.e., $\mu(f) < \infty$.

A necessary condition for the existence of a nondegenerate quasihomogeneous function with a prescribed set of exponents is that for each axis in the space of exponents there exists an exponent $\mathbf{k} \in L$ at distance $\leqslant 1$ from that axis. That is to say, for each $i = 1, \ldots, n$ there should exist a $\mathbf{k}$ on the diagonal L such that $k_1 + \cdots + \hat{k}_i + \cdots + k_n \leqslant 1$: otherwise the x_i-axis consists entirely of critical points.

The degenerate quasihomogeneous functions form an algebraic hypersurface in the linear space of all quasihomogeneous polynomials with fixed exponents of quasihomogeneity, provided of course that in that space there is at least one nondegenerate function.

Consider again $\mathbb{C}^n$ with a fixed coordinate system $x_1, \ldots, x_n$. The algebra of formal power series[1] in the coordinates will be denoted by $A = \mathbb{C}[[x_1, \ldots, x_n]]$. We assume that a quasihomogeneity type $\boldsymbol{\nu} = (\nu_1, \ldots, \nu_n)$ is fixed. With each such $\boldsymbol{\nu}$ there is associated a filtration of the ring A, defined as follows.

Definition. The *monomial* $\mathbf{x}^{\mathbf{k}}$ is said to have *degree* (or *weight*) d if $\langle \boldsymbol{\nu}, \mathbf{k} \rangle = \nu_1 k_1 + \cdots + \nu_n k_n = d$.

The degree of any monomial is a rational number. The exponents of all monomials of degree d lie in a single hyperplane parallel to the diagonal L.

Definition. The *order d of a series* [resp. polynomial] is the smallest of the degrees of the monomials that appear in that series [resp. polynomial].

The orders of all possible series belong to a single arithmetic progression $\mathbb{Z}_+ \cdot N$, where N is the greatest common factor of the numbers ν_i (the initial segment of this progression is not necessarily filled up by the values of the orders d).

The series of order larger than or equal to d form a subspace $A_d \subset A$. The order of a product is equal to the sum of the orders of the factors. Consequently, A_d is an ideal in the ring A. The family of ideals A_d constitutes a decreasing filtration of A: $A_{d'} \subset A_d$ whenever $d' > d$. We let A_{d+} denote the ideal in A formed by the series of order higher than d.

Definition. The quotient algebra A/A_{d+} is called the *algebra of d-quasijets*, and its elements are called *d-quasijets*.

Definition. A series (polynomial) of order d with nondegenerate d-quasijet is said to be *semi-quasihomogeneous* of degree d.

In other words, a semi-quasihomogeneous function is obtained from a nondegenerate quasihomogeneous function by adding monomials whose exponents lie above the diagonal. We remark that a quasihomogeneous function is not semi-quasihomogeneous if it is degenerate.

The nondegenerate d-jet f_0 of a semi-quasihomogeneous function f is called the *quasihomogeneous part* of f.

---

[1] Most of the ensuing discussion can be carried over in straightforward manner to the case where A is the ring of convergent series over $\mathbb{R}$ or $\mathbb{C}$, or the ring of germs of smooth functions.

3.3. The Multiplicity and the Generators of the Local Algebra of a Semi-Quasihomogeneous Function. As it turns out, the local algebra Q_f has the same structure for all quasihomogeneous functions f of degree d with a given type of quasihomogeneity $\mathbf{v}$.

By abuse of language, a *basis* of the local algebra Q_f of the function f will be understood to mean any set of μ power series (polynomials) that becomes a $\mathbb{C}$-basis of Q_f after factorization by the gradient ideal $I_{\nabla f}$.

Theorem ([16]). *Suppose that the set of monomials $e_1, \ldots, e_\mu$ is a basis of the local algebra of the quasihomogeneous part f_0 of the semi-quasihomogeneous function f. Then $e_1, \ldots, e_\mu$ is also a basis of the local algebra of f.*

This theorem implies, in particular, that the multiplicity of the critical point 0 of a semi-quasihomogeneous function f is equal to the multiplicity of the critical point 0 of its quasihomogeneous part f_0: $\mu(f) = \mu(f_0)$. Moreover, as follows from the next theorem, the multiplicity μ is the same for all semi-quasihomogeneous functions with a fixed degree and a fixed type of quasi-homogeneity.

Theorem (see [37]). *The number of monomials of a given degree δ in a basis of the local algebra of a semi-quasihomogeneous function f does not depend on the choice of the monomial basis for the local algebra and is the same for all semi-quasihomogeneous functions of degree d and of a given type of quasihomogeneity $\mathbf{v}$.*

The structure of the local algebra of a quasihomogeneous singularity completely determines the type of the singularity:

Theorem ([49], [302]). *Let $f: (\mathbb{C}^n, 0) \to (\mathbb{C}, 0)$ and $g: (\mathbb{C}^n, 0) \to (\mathbb{C}, 0)$ be function-germs such that $f \in I_{\nabla f}$ and $g \in I_{\nabla g}$. Then the local algebras Q_f and Q_g are isomorphic if and only if the germs f and g are equivalent.*

3.4. Quasihomogeneous Maps. Here we give various numerical characteristics of quasihomogeneous maps, in particular the multiplicity μ and the Poincaré polynomial p.

Let $\mathbf{d} = (d_1, \ldots, d_n)$ be a vector with nonnegative components and let $\mathbf{F} = (F_1, \ldots, F_n)$ be a map $(\mathbb{C}^n, 0) \to (\mathbb{C}^n, 0)$.

Definition. $\mathbf{F}$ is said to be a *quasihomogeneous map* of degree $\mathbf{d}$ and type $\boldsymbol{\nu}$ if each component F_i is a quasihomogeneous function of degree d_i and type $\boldsymbol{\nu}$.

The *local algebra of the map* $\mathbf{F}$ is the quotient algebra $Q_\mathbf{F} = \mathcal{O}_n/I_\mathbf{F}$ of $\mathcal{O}_n$ by the ideal $I_\mathbf{F} = (F_1, \ldots, F_n)$ generated by the components of $\mathbf{F}$ (and so the local algebra Q_f of a function $f: (\mathbb{C}^n, 0) \to (\mathbb{C}, 0)$ is exactly the local algebra of the gradient map $\mathbf{x} \mapsto \nabla f(\mathbf{x})$).

Definition. A map $\mathbf{F}$ is said to be *nondegenerate* if the $\mathbb{C}$-dimension $\mu = \dim_\mathbb{C} Q_\mathbf{F}$ of its local algebra $Q_\mathbf{F}$ is finite, in which case μ is called the *multiplicity of* $\mathbf{F}$ *at* 0.

Definition. $\mathbf{F}$ is said to be a *semi-quasihomogeneous map* if $\mathbf{F} = \mathbf{F}_0 + \mathbf{F}'$, where $\mathbf{F}_0$ is a nondegenerate quasihomogeneous map and the order of each component of $\mathbf{F}'$ is higher that the order of the corresponding component of $\mathbf{F}_0$.

The map $\mathbf{F}_0$ is called the *quasihomogeneous part* of the semi-quasihomogeneous map $\mathbf{F}$.

Example. If f is a semi-quasihomogeneous function of degree d and type $\boldsymbol{\nu}$, then the map $\mathbf{x} \mapsto \nabla f(\mathbf{x})$ is semi-quasihomogeneous of degree $(d - \nu_1, \ldots, d - \nu_n)$.

The first two theorems of subsection 3.3, formulated there for gradient maps, are valid for arbitrary quasihomogeneous and semi-quasihomogeneous maps ([16], [203]).

The importance of the class of quasihomogeneous maps for the study of quasihomogeneous functions is explained by the fact that in that class one has greater freedom to construct homotopies and change variables than in the gradient case.

Definition. The *Poincaré polynomial* of a semi-quasihomogeneous map $\mathbf{F}$ of degree $\mathbf{d}$ and type $\boldsymbol{\nu}$ (where $d_i = D_i/N$ and $\nu_i = N_i/N$, with $N, N_i, D_i \in \mathbb{Z}$) is the polynomial $p_{\mathbf{F}}(t) = \sum \mu_i t^i$, where μ_i is the number of monomials of quasidegree i/N in an arbitrary monomial basis of the local algebra of $\mathbf{F}$.

Note that the polynomial $p_{\mathbf{F}}$ depends on the choice of the integer N even when the type of quasihomogeneity is fixed. Incidentally, all admissible values of N are multiples of the least common denominator of the fractions ν_i, and $p_{\mathbf{F},kN}(t) = p_{\mathbf{F},N}(t^k)$. The dimension of the local algebra of $\mathbf{F}$ is given by the formula $\mu = p_{\mathbf{F}}(1)$. The degree of $p_{\mathbf{F}}$ is equal to the largest of the quasidegrees (multiplied by N) of the monomial generators of the local algebra.

Theorem ([16], [56], [254]). *The Poincaré polynomial of a (semi-) quasihomogeneous map* $\mathbf{F}$ *of degree* $\mathbf{d}$ *and type* $\mathbf{v}$ *(where $d_i = D_i/N$, $\nu_i = N_i/N$, with D_i, N_i, $N \in \mathbb{Z}$) is given by the formula*

$$p_{\mathbf{F}}(t) = \prod_{i=1}^{n} \frac{t^{D_i} - 1}{t^{N_i} - 1}.$$

Example. Let $\mathbf{F} = \nabla f$, where f is a (semi-) quasihomogeneous function of degree 1 and type $\boldsymbol{\nu}$. Then

$$p_{\mathbf{F}}(t) = \prod_{i=1}^{n} \frac{t^{N-N_i} - 1}{t^{N_i} - 1}.$$

We next list a number of useful corollaries of the above theorem.

1° ([16], [56], [248]). *The dimension of the local algebra of a semi-quasihomogeneous map is given by the "generalized Bézout formula"*

$$\mu = \prod_{i=1}^{n} \frac{d_i}{\nu_i}.$$

2° ([16], [286]). *The local algebra of a semi-quasihomogeneous map* $\mathbf{F}$ *has exactly one basis monomial of maximal degree*

$$d_{\max} = \sum_{i=1}^{n} (d_i - v_i);$$

all monomials of higher degree belong to the ideal $I_{\mathbf{F}}$ *generated by the components of* $\mathbf{F}$.

3° ([16], [56], [248]). *The dimension of the local algebra of a semi-quasihomogeneous function* f *of type* $\mathbf{v}$ *and degree* 1 *is given by the formula*

$$\mu = \prod_{i=1}^{n} \left(\frac{1}{v_i} - 1\right).$$

4° ([16], [286]). *A monomial basis of the local algebra of a semi-quasihomogeneous function* f *of type* $\mathbf{v}$ *and degree* 1 *has exactly one generator of degree*

$$d_{\max} = \prod_{i=1}^{n} (1 - 2v_i);$$

all monomials of higher degree belong to the gradient ideal $I_{\nabla f}$.

5° ([16]). *The Poincaré polynomial of a semi-quasihomogeneous map is always a reflexive polynomial*:

$$\mu_i = \mu_{k-i}, \quad \textit{where} \quad k = \sum_{i=1}^{n} (D_i - N_i).$$

6° ([16]). *A necessary condition for the existence of a nondegenerate quasihomogeneous map of degree* $\mathbf{d}$ *and type* $\mathbf{v}$ *is that the polynomial* $\prod_{i=1}^{n} (t^{D_i} - 1)$ *be divisible by* $\prod_{i=1}^{n} (t^{N_i} - 1)$.

7° ([16]). *A necessary condition for the existence of a nondegenerate quasihomogeneous function of degree* 1 *and type* $\mathbf{v}$ *is that* $\prod_{i=1}^{n} (t^{N-N_i} - 1)/(t^{N_i} - 1)$ *be a polynomial.*

Remark. In the case of functions of two or three variables the reducibility of the fraction $\prod_{i=1}^{n} (t^{N-N_i} - 1)/(t^{N_i} - 1)$ is not only necessary, but also sufficient for the existence of a nondegnerate quasihomogeneous function with the exponents N_i/N [16]. For functions of four variables this is no longer the case, as demonstrated by the following example of B.M. Ivlev:

$$N = 265, \qquad N_1 = 1, \qquad N_2 = 24, \qquad N_3 = 33, \qquad N_4 = 58.$$

In this example the quotient in question is a polynomial with nonnegative coefficients, but all quasihomogeneous functions with exponents N_i/N are degenerate.

3.5. Quasihomogeneous Diffeomorphisms and Vector Fields. Several Lie groups and algebras are associated with the filtration defined in the ring A of power series by the type of quasihomogeneity $\mathbf{v}$. In the case of ordinary homogeneity these are the general linear group, the group of k-jets of diffeomorphisms, its subgroup of k-jets with $(k - 1)$-jet equal to the identity, and their quotient groups. Their analogues for the case of a quasihomogeneous filtration are defined as follows [16].

As above, we use the notations of subsection 3.2 and we fix a type of quasihomogeneity $\mathbf{v} = (v_1, \ldots, v_n)$.

Definition. A *formal diffeomorphism* $g: (\mathbb{C}^n, 0) \to (\mathbb{C}^n, 0)$ is a set of n power series $g_i \in A$ without constant terms for which the map $g^*: A \to A$ given by the rule $g^*f = f \circ g$ is an algebra isomorphism.

Definition. The diffeomorphism g is said to have *order d* if for every s $(g^* - 1)A_s \subset A_{s+d}$.

The set of all diffeomorphisms of order $d \geqslant 0$ is a group G_d. The family of groups G_d yields a decreasing filtration of the group G of formal diffeomorphisms; indeed, for $d' > d \geqslant 0$, $G_{d'} \subset G_d$ and is a normal subgroup in G_d.

The group G_0 plays the role in the quasihomogeneous case that the full group of formal diffeomorphism plays in the homogeneous case. We should emphasize that in the quasihomogeneous case $G_0 \neq G$, since certain diffeomorphisms have negative orders and do not belong to G_0.

Definition. The *group of d-quasijets* of type $\mathbf{v}$ is the quotient group of the group of diffeomorphisms G_0 by the subgroup G_{d+} of diffeomorphisms of order higher than d:

$$J_d = G_0/G_{d+}.$$

Remark. In the ordinary homogeneous case our numbering differs from the standard one by 1: for us J_0 is the group of 1-jets, and so on.

J_d acts as a group of linear transformations on the space A/A_{d+} of d-quasijets of functions.

A special importance is attached to the group J_0, which is the quasihomogeneous generalization of the general linear group.

Definition. A *diffeomorphism* $g \in G_0$ is said to be *quasihomogeneous of type* $\mathbf{v}$ if each of the spaces of quasihomogeneous functions of degree d (and type $\mathbf{v}$) is invariant under the action of g^*.

The set of all quasihomogeneous diffeomorphisms is a subgroup of G_0. This subgroup is canonically isomorphic to J_0, the isomorphism being provided by the restriction of the canonical projection $G_0 \to J_0$.

The infinitesimal analogues of the concepts introduced above look as follows.

Definition. A *formal vector field* $v = \sum v_i \partial_i$, where $\partial_i = \partial/\partial x_i$, is said to have *order d* if differentiation in the direction of v raises the degree of any function by at least d: $L_v A_s \subset A_{s+d}$.

We let $\mathfrak{g}_d$ denote the set of all vector fields of order d. The filtration arising in this way in the Lie algebra $\mathfrak{g}$ of vector fields (i.e., of derivations of the algebra A) is compatible with the filtrations in A and in the group of diffeomorphisms G:

1°. $f \in A_d$, $v \in \mathfrak{g}_s \Rightarrow fv \in \mathfrak{g}_{d+s}$, $L_v f \in A_{d+s}$.

2°. The module $\mathfrak{g}_d$, $d \geqslant 0$, is a Lie algebra with respect to the Poisson bracket of vector fields.

3°. The Lie algebra $\mathfrak{g}_d$ is an ideal in the Lie algebra $\mathfrak{g}_0$.

4°. The Lie algebra $\mathfrak{j}_d$ of the Lie group J_d of d-quasijets of diffeomorphisms is equal to the quotient algebra $\mathfrak{g}_0/\mathfrak{g}_{d+}$.

5°. The quasihomogeneous vector fields of degree 0 form a finite-dimensional Lie subalgebra of the Lie algebra $\mathfrak{g}_0$; this subalgebra is canonically isomorphic to the Lie algebra $\mathfrak{j}_0$ of the group of 0-jets of diffeomorphisms.

The order of a vector field $v = \sum v_i \partial_i$ is connected with the order of its components as follows: v has order d (and type $\mathbf{v}$) if and only if each v_i is a series of order $d + v_i$.

3.6. The Normal Form of a Semi-Quasihomogeneous Function. We consider the local algebra of a nondegenerate or quasihomogeneous function f of degree d, in which we fix some set of monomials that form a basis.

Definition. A monomial is called an *upper* monomial [resp. a *lower* monomial, a *diagonal monomial*] (or is said *to lie above* [resp. below, on] *the diagonal*) if its degree is larger than d [resp. smaller than d, equal to d] for the given exponents of homogeneity.

The number of upper, lower, and diagonal monomials in a basis does not depend on the choice of the basis (subsection 3.3).

Let $e_1, \ldots, e_s$ be the set of all upper monomials in the fixed basis of the local algebra of the function f_0.

Theorem ([16]). *Every semi-quasihomogeneous function f with quasihomogeneous part f_0 is equivalent to a function of the form $f_0 + \sum c_k e_k$, where the c_k are constants.*

Examples. 1. If $f_0 = x^2y + y^k$, then $f \sim f_0$.

2. If $f_0 = x^4 + y^6$, then $f \sim x^4 + y^6 + cx^2y^4$, as shown by Fig. 9. The shaded region in the plane of exponents represents the ideal $I_{\nabla f}$; the fixed basis of Q_f

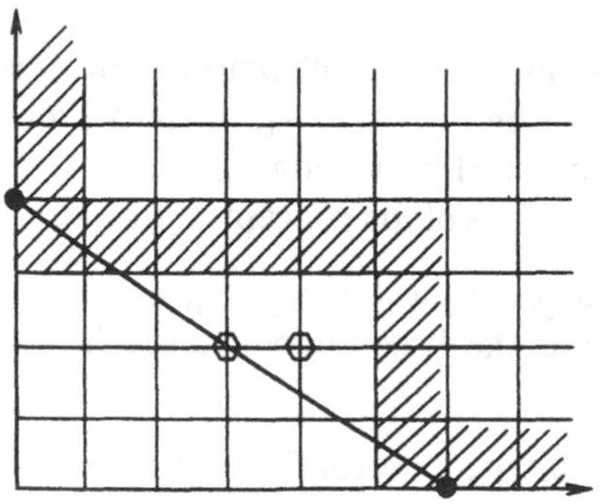

Fig. 9

contains exactly one diagonal monomial, x^2y^3, and exactly one upper monomial, x^2y^4.

The proof of the theorem relies on a procedure whereby the monomials of degree $d' > d$ that do not belong to the basis of Q_f are "killed" by means of formal diffeomorphisms of degree $d' - d$. Tougeron's theorem on the finite determinacy of isolated singularities (subsection 1.5) then guarantees that such a reduction to normal form can be achieved by means of a genuine diffeomorphism.

3.7. The Normal Form of a Quasihomogeneous Function. In order to reduce quasihomogeneous functions to normal form it is necessary to classify the orbits of the action of the group of quasihomogeneous diffeomorphisms on the spaces of quasihomogeneous functions. The computations are based on two general theorems about quasihomogeneous functions.

Definition. The *support of a quasihomogeneous function* of degree d and type $\mathbf{v}$ is the set of all points $\mathbf{k}$ with nonnegative integer coordinates on the diagonal

$$L = \{\mathbf{k}: \langle \mathbf{k}, \mathbf{v} \rangle = d\}.$$

The support is said to be *complete* if it is not contained in any affine subspace of $\mathbb{C}^n$ of dimension less than $n - 1$.

Quasihomogeneous functions can be regarded as functions given on their supports: $\sum f_{\mathbf{k}} \mathbf{x}^{\mathbf{k}}$ assumes at $\mathbf{k}$ the value $f_{\mathbf{k}}$. The set of all such functions is a linear space $\mathbb{C}^r$, where r is the number of points in the support. Both the group of quasihomogeneous diffeomorphisms (of type $\mathbf{v}$) and its Lie algebra $\mathfrak{a}$ act on this space.

The Lie algebra $\mathfrak{a}$ of quasihomogeneous vector fields of degree 0 is spanned, as a $\mathbb{C}$-linear space, by all monomial fields $\mathbf{x}^{\mathbf{p}}\partial_i$ for which $\langle \mathbf{p}, \mathbf{v} \rangle = v_i$ (subsection 3.5). For example, the n fields $x_i\partial_i$ belong to $\mathfrak{a}$ for any $\mathbf{v}$.

Definition. A *root of the quasihomogeneous algebra* $\mathfrak{a}$ is any nonzero vector $\mathbf{m}$ in the space of exponents that lies in the plane $\langle \mathbf{m}, \mathbf{v} \rangle = 0$ and has the form $\mathbf{m} = \mathbf{p} - \mathbf{1}_i$, where $\mathbf{1}_i$ is the vector with the ith component equal to 1 and the others equal to 0, and where $\mathbf{p}$ has nonnegative integer components.

In other words, $\mathbf{m}$ is a root if $\mathbf{x}^{\mathbf{p}}\partial_i$ is a monomial vector field belonging to $\mathfrak{a}$ and is different from $x_i\partial_i$.

Note that i is uniquely determined by the root $\mathbf{m}$, since the vector $\mathbf{m}$ has exactly one negative component $m_i = -1$ (in view of the equality $\langle \mathbf{m}, \mathbf{v} \rangle = 0$ the components of $\mathbf{m}$ cannot be all nonnegative).

Theorem ([17]). *Suppose that the support is complete. Then the action of the Lie algebra* $\mathfrak{a}$ *on the space* $\mathbb{C}^r$ *of functions defined on the support is uniquely determined by the class of affine equivalence of the pair (support, root system).*

Theorem ([17]). *The quasihomogeneous Lie algebra* $\mathfrak{a}$ *is uniquely specified, up to a finite number of variants, by its root system (regarded as a subset of the linear space spanned by the roots) and its dimension.*

That is to say, if one does not distinguish between algebras that are obtained from one another by adding a trivial (commutative) algebra as a direct summand, then there are but a finite number of nonisomorphic Lie algebras $\mathfrak{a}$ with linearly equivalent root systems.

Remarks. 1. In problems arising in the classification of the known list of singularities this finite number is always equal to one.

2. The affine equivalences of supports and the linear equivalences of root systems entering into the above theorems do not necessarily map into themselves either the coordinate simplex $\mathbb{R}^n_+ \cap L$ in the diagonal $\langle \mathbf{k}, \mathbf{v} \rangle = d$, or the lattice of nonnegative integer exponents $\mathbf{k}$ in $\mathbb{C}^n$. Under the hypotheses of the two theorems, the groups of quasihomogeneous diffeomorphisms and their orbits in the spaces of quasihomogeneous functions do not necessarily coincide; however, the connected components of the orbits coincide.

The orbit of the action of the Lie algebra $\mathfrak{a}$ on the tangent space at a nondegenerate quasihomogeneous function f_0 is transverse to the subspace spanned by the diagonal monomials in a basis of the local algebra of f_0:

Theorem ([16]). *Every quasihomogeneous function sufficiently close to* f_0 *is equivalent to a function of the form* $f_0 + \sum c_k e_k$, *where* $e_1, \ldots, e_n$ *is the set of all diagonal monomials in a basis of the local algebra of* f_0.

Example. Let $f_0 = x^3 + y^3 + z^3$. Then every (quasi)homogeneous function sufficiently close to f_0 is equivalent to $x^3 + y^3 + z^3 + cxyz$. In Fig. 10 on the

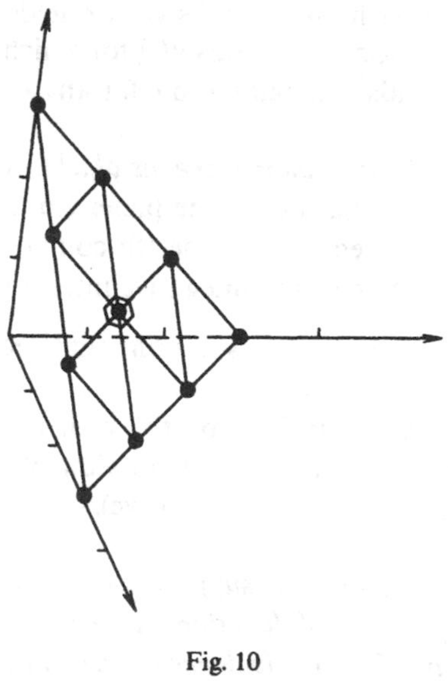

Fig. 10

plane L in the space of exponents, we have marked the unique diagonal basis monomial xyz of the local algebra of f_0.

Definition. The *intrinsic modality* m_0 of a quasihomogeneous function is the total number of diagonal and upper monomials in a monomial basis of its local algebra.

A consequence of the preceding theorem and the analysis in subsection 3.6 is that the modality m of any semi-quasihomogeneous function is not smaller than the intrinsic modality m_0 of its quasihomogeneous part.

Theorem ([347]). $m = m_0$ (see subsection 2.4.8).

3.8. The Newton Filtration. Often it is useful to work with a filtration in which the role of the diagonal is played by the Newton diagram. The formal definition is as follows.

Let Γ be a Newton diagram. Each face Γ_j of Γ specifies a quasihomogeneity type $\mathbf{v}_j$, in which the degree of the monomials with exponents lying on Γ_j is equal to one: $\langle \mathbf{k}, \mathbf{v}_j \rangle = 1$ is the equation of an (affine) hyperplane containing Γ_j.

Let us fix a Newton diagram Γ.

Definition. A monomial $x^{\mathbf{k}}$ is said to have *Newton degree d* if $d = \min_j \langle \mathbf{k}, \mathbf{v}_j \rangle$.

In other words, the Newton degree of a monomial is the smallest of its degrees in any of the quasihomogeneous filtrations defined by the faces of the diagram Γ. The monomials of Newton degree d are exactly those whose exponents lie in the diagram $d\Gamma$ obtained from Γ through a homothetic transformation of ratio d.

Definition. The *Newton order d of a power series* is the smallest of the Newton degrees of the monomials that appear in it.

The series of order at least d form an ideal A_d in the ring A. The ideals A_d yield the *Newton filtration* in the ring of power series.

The sum f_0 of the terms of Newton degree d of a series f of order d will be referred to as the *principal part* of f.

Definition. A polynomial whose monomials have Newton degree d is called a *Newton-homogeneous function* of degree d.

Analogous notions are defined for vector fields: the degree of the monomial field $x^{\mathbf{p}}\partial/\partial x_i$ is defined to be $\min_j \langle \mathbf{p} - \mathbf{1}_i, \mathbf{v}_j \rangle$.

Groups of diffeomorphisms of order d, groups of d-jets of functions and of d-jets of diffeomorphisms and the corresponding Lie algebras are defined exactly as in the case of quasihomogeneous filtrations. Only the group of quasi-homogeneous diffeomorphisms has no analogue for Newton filtrations.

Definition. A Newton-homogeneous function f_0 of degree d is said to satisfy *Condition A* if for every function g of order $d + \delta > d$ that belongs to the ideal spanned by the partial derivatives of f_0 there is a vector field v of order δ such that

$$g = L_v f_0 + g',$$

where g' is of order higher than $d + \delta$.

Every quasihomogeneous function satisfies Condition A.

Let $e_1, \ldots, e_\mu$ be a basis of the local algebra of a Newton-homogeneous function f_0 of finite multiplicity μ.

Definition. A basis $e_1, \ldots, e_\mu$ consisting of homogeneous elements is said to be *regular* if for every δ the basis elements of degree δ are independent modulo the sum of ideals $I_{\nabla f} + A_{\delta+}$ (where $A_{\delta+}$ denotes the space of functions of degree higher than δ). As it turns out, a regular basis always exists; moreover, one can always choose such a basis consisting of monomials.

Theorem ([16]). *Suppose the principal part f_0 of the function f satisfies Condition A and has finite multiplicity μ. Then f can be reduced by means of a diffeomorphism to the form $f_0 + \sum c_k e_k$, where $e_1, \ldots, e_s$ are the upper monomials in a regular basis.*

Example. Let $f_0 = x^a + \lambda x^2 y^2 + y^b$, where $a \geqslant 4$, $b \geqslant 5$, $\lambda \neq 0$. One can show that the set of monomials $1, x, \ldots, x^{a-1}, y, \ldots, y^b, xy$ is a regular basis of the local algebra of f_0 and that the filtration defined by the Newton diagram Γ of f_0 satisfies Condition A. Since in this example the basis contains no upper monomials, it follows from the preceding theorem that every function with principal part f_0 is equivalent to f_0. In Fig. 11 the shaded region in the plane of exponents indicates the ideal $I_{\nabla f_0}$. Modulo this ideal, one has the equalities $x^a \sim x^2y^2 \sim y^b$, $x^{a-1} \sim xy^2$, $x^2y \sim y^{b-1}$, where the sign $\sim$ denotes equality up to a multiplicative constant.

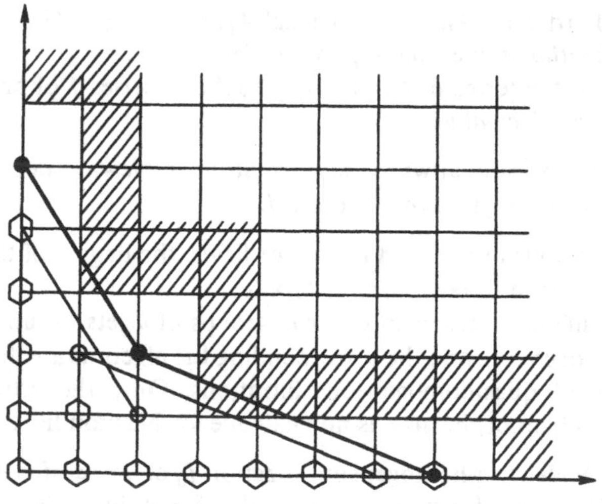

Fig. 11

3.9. The Spectral Sequence. Here we describe a method of reduction to normal form based on a spectral sequence, constructed from the filtration of the Koszul complexes which is defined by the partial derivatives of the function under study. The correspondence between our constructions and the usual algebraic constructions [227] is described in [18].

Let f be a series in A and let $\mathfrak{g}$ denote, as above, the Lie algebra of formal vector fields. The function f defines the map of A-modules

$$\partial\colon \mathfrak{g} \to A, \qquad v \mapsto L_v f$$

We introduce the following notations:

$I_f = \operatorname{Im} \partial$, the gradient ideal for f,

$S_f = \operatorname{Ker} \partial$, the isotropy algebra for f,

$Q_f = A/I_f$, the local algebra for f.

We describe below a method for computing S_f and Q_f by successive approximations. To this end, we fix a quasihomogeneity type $\mathbf{v}$. $\mathbf{v}$ defines filtrations in A and $\mathfrak{g}$ (see subsections 3.2, 3.5),

$$A = A_0 \supset A_1 \supset \cdots \supset \quad \text{and} \quad \mathfrak{g} \supset \cdots \supset \mathfrak{g}_{-1} \supset \mathfrak{g}_0 \supset \mathfrak{g}_1 \supset \cdots.$$

Suppose $f \in A_N$. Then $\partial \mathfrak{g}_0 \subset A_N$. We shall denote $\mathfrak{g}_0$ and A_N by $\mathfrak{g}^+$ and A^+, respectively. The restriction of ∂ to $\mathfrak{g}^+$ defines a map of A-modules $\partial^+\colon \mathfrak{g}^+ \to A^+$. We introduce the notations:

$I_f^+ = \operatorname{Im} \partial^+$, the upper gradient ideal for f,

$S_f^+ = \operatorname{Ker} \partial^+$, the upper isotropy algebra for f,

$Q_f^+ = A^+/I_f^+$, the upper local algebra for f.

For the purposes of the ensuing discussion it is convenient to define filtrations in the A-modules $\mathfrak{g}^+$ and A^+ as follows: $\mathfrak{a}_p = \mathfrak{g}_p$ for $p \geqslant 0$, $\mathfrak{a}_p = \mathfrak{g}_0$ for $p \leqslant 0$, $\mathscr{A}_p = A_{N+p}$ for $p \geqslant 0$, $\mathscr{A}_p = A_N$ for $p \leqslant 0$. The map ∂^+ respects these filtrations: $\partial^+ \mathfrak{a}_p \subset \mathscr{A}_p$.

Remark. Our successive approximations are given by the spectral sequence of the filtered complex whose differential is defined by the sequence

$$0 \to \mathfrak{g}^+ \xrightarrow{\partial^+} A^+ \to 0.$$

Let $r \geqslant 0$, $p \geqslant 0$. We identify the quotient spaces $S_p^0 = \mathfrak{a}_p/\mathfrak{g}_{p+1}$ and $A_p^0 = \mathscr{A}_p/\mathscr{A}_{p+1}$ with the spaces of quasihomogeneous vector fields of degree p and of polynomials of degree $N + p$, respectively. The spaces S_p^0 and A_p^0 are the 0-th approximation to the "p-components" of the A-modules S_f^+ and Q_f^+, respectively. The next approximations are defined below.

Let $f = f_0 + f_1 + \cdots$ be the decomposition of the series f into quasihomogeneous components of degrees $N, N + 1, \ldots$.

Definition 1. The *r-th approximation to the p-component of the isotropy algebra* S_p^+ is defined to be the space of quasihomogeneous vector fields s_p of degree

p that admit "prolongations" to vector-polynomials $s_p + \cdots + s_{p+r-1}$ satisfying the system of r equations

$$s_p f_0 = 0, \qquad s_p f_1 + s_{p+1} f_0 = 0, \ldots, s_p f_{r-1} + \cdots + s_{p+r-1} f_0 = 0.$$

Definition 2. The "differential" d^r acts on the quasihomogeneous vector field $s_p \in S_p^r$ according to the rule

$$d^r s_p = s_p f_r + \cdots + s_{p+r} f_0 \bmod I_{p+r}^r,$$

where the s_q satisfy the conditions of Definition 1 and I_{p+r}^r is defined below (for $r > 0$ one can take $s_{p+r} = 0$).

Definition 3. The $(r+1)$*-st aproximation to the p-component of the gradient ideal*, I_p^{r+1}, is defined to be the set of all quasihomogeneous polynomials of degree $N + p$ that are representable in the form $s_{p-r} f_2 + \cdots + s_p f_0$, where the quasihomogeneous vector fields s_q of the indicated degrees satisfy the r conditions

$$s_{p-r} f_0 = 0, \qquad s_{p-r} f_1 + s_{p-r+1} f_0 = 0, \ldots, s_{p-r} f_{r-1} + \cdots + s_{p-1} f_0 = 0$$

and belong to g^+ (i.e., $s_q = 0$ for all $q < 0$).

Definition 4. The *r-th approximation to the p-component of the local algebra* is defined to be

$$A_p^r = A_p^0 / I_p^r \qquad (r > 0).$$

Proposition. *The following equalities hold:*

$$S_p^{r+1} = \mathrm{Ker}(d^r\colon S_p^r \to A_{p+r}^r), \qquad A_{p+r}^{r+1} = A_{p+r}^r / d^r S_p^r.$$

Example 1. $d^0\colon S_p^0 \to A_p^0$ is defined by the equality $d^0 s_p = s_p f_0$. Consequently, I_p^1 is the homogeneous $(N + p)$-component of the gradient ideal of the quasihomogeneous part f_0 of degree N of the function $f = f_0 + \cdots$. A_p^1 can therefore be identified with the $(N + p)$-component of the local algebra Q_{f_0} of the quasihomogeneous function f_0. Furthermore, S_p^1 is the p-component of the isotropy algebra of f_0.

Example 2. $d^1\colon S_p^1 \to A_p^1$ is defined by the equality $d^1 s_p = s_p f_1 \bmod \{s_{p+1} f_0\}$, where $s_p f_0 = 0$. Consequently

$$S_p^2 = \{s_p\colon s_p f_0 = 0, \exists s_{p+1} \text{ such that } s_p f_1 = s_{p+1} f_0\}$$

and

$$I_p^2 = \{s_p f_0\} + \{s_{p-1} f_1\colon s_{p-1} f_0 = 0\}.$$

Thus, both for $r = 0$ and $r = 1$, the value $d^r s_p$ of the differential d^r is defined as a coset of the polynomial $s_p f_r$. This simple formula for the differential does not hold for larger values of r.

Example 3. $d^2\colon S_p^2 \to A_{p+2}^2$ is defined by the equality

$$d^2 s_p = s_p f_2 + s_{p+1} f_1 \bmod \{s_{p+2} f_0\} + \{s_{p+1} f_1\colon s_{p+1} f_0 = 0\},$$

where $s_p f_0 = 0$, $s_p f_1 + s_{p+1} f_0 = 0$.

3.10. Theorems on Normal Forms for the Spectral Sequence.

1°. Convergence of the successive approximations.

Theorem. *For each $p \geqslant 0$ the sequences S_p^r and A_p^r stabilize for large enough r, i.e., $S_p^r = S_p^\infty$ and $A_p^r = A_p^\infty$. The limit spaces S_p^∞ and A_p^∞ coincide with the spaces of the initial p-forms of the elements of the upper isotropy algebra and the upper local algebra for f, respectively:*

$$S_p^\infty \cong (S_f^+ \cap \mathfrak{a}_p)/(S_f^+ \cap \mathfrak{a}_{p+1})$$

and

$$A_p^\infty \cong \mathscr{A}_p/[(\mathscr{A}_p \cap I_f^+) + \mathscr{A}_{p+1}].$$

The isomorphisms indicated above are defined by the natural maps $\mathscr{A}_p \to \mathscr{A}_p^r$ and $S_f^+ \cap \mathfrak{a}_p \to S_p^r$.

2°. The normal form for the terms of degree p. Let us fix the numbers $p > r \geqslant 0$.

Theorem T$_{r,p}$. *Let $e_1, \ldots, e_s$ be quasihomogeneous polynomials of degree $N + p$ whose images under the natural map $\mathscr{A}_p \to A_p^{r+1}$ generate A_p^{r+1} over $\mathbb{C}$. Then there exists a formal diffeomorphism*

$$y = x + v(x), \qquad v\colon \mathbb{C}^n \to \mathbb{C}^n, \qquad \sum v_i \partial_i \in \mathfrak{a}_{p-r},$$

such that, following the substitution of y, the series $f = f_0 + f_1 + \cdots$ takes the form

$$f(y) = f_0(x) + \cdots + f_{p-1}(x) + \sum c_i e_i(x) + R,$$

where $R \in \mathscr{A}_{p+1}$ and the c_i are numbers.

Here, as usual, $f = f_0 + f_1 + \cdots$ denotes the decomposition of f into quasihomogeneous components of degrees $N, N+1, \ldots$.

3°. The normal form for the r-th approximation. Let us fix the number $r \geqslant 0$.

Theorem T$_r$. *Let $e_1, e_2, \ldots$ be quasihomogeneous polynomials of the various degrees $N + p$, where $p > r$, whose images under the natural maps $\mathscr{A}_p \to A_p^{r+1}$ $\mathbb{C}$-generate all the spaces A_p^{r+1}. Then there exists a formal diffeomorphism*

$$y = x + v(x), \qquad v\colon \mathbb{C}^n \to \mathbb{C}^n, \qquad \sum v_i \partial_i \in \mathfrak{g}^+,$$

such that, following the substitution of y, the series $f = f_0 + f_1 + \cdots$ takes the form

$$f(y) = f_0(x) + \cdots + f_r(x) + \sum c_i e_i,$$

where $\deg e_i > N + r$ and the c_i are numbers.

4°. Conditions B and C. Consider the principal quasihomogeneous part f_0 of the series $f = f_0 + f_1 + \cdots$.

Definition. The series f is said to *satisfy Condition B* if the isotropy Lie algebra of the point f_0 under the action of the Lie algebra of quasihomogeneous vector fields on the space of quasihomogeneous polynomials of degree $N = \deg f_0$ is trivial (i.e., equal to $\{0\}$).

In other words, f satisfies Condition B if $S_0^1 = 0$. Thus, Condition B is actually a condition only on f_0.

Theorem BT. *If f satisfies Condition B, then Theorem $T_{r,p}$ holds for $r = p \geqslant 1$.*

Definition. The *negative Lie algebra* $\mathfrak{g}^-$ of type $\mathbf{v}$ is defined to be the algebra of vector fields of the form $\sum v_i \partial_i$ such that all monomials of each of the polynomials v_i are of degree strictly smaller than the degree of the monomial x_i (i.e., than v_i).

Note that $\mathfrak{g}^-$ is a finite-dimensional Lie algebra.

Definition. The series $f = f_0 + f_1 + \cdots$ is said to *satisfy Condition C* if the isotropy algebra of the point f_0 under the action of the negative Lie algebra $\mathfrak{g}^-$ on the space of polynomials of degree no higher than $N = \deg f_0$ is trivial (i.e., equal to $\{0\}$).

Note that Condition C, too, is imposed only on f_0.

Theorem CT. *If f satisfies Condition C, then $I_f^+ = A \cap I_f$.*

Corollary. *Let f satisfy Condition C and let $e_1, e_2, \ldots$ be quasihomogeneous polynomials of all possible degrees $N + p$, $p \geqslant 0$, whose images under the natural maps $\mathscr{A}_p \to A_p^\infty$ form bases in the spaces A_p^∞ of the spectral sequence. Then the images of $e_1, e_2, \ldots$ in the local algebra $Q_f = A/I_f$ are $\mathbb{C}$-linearly independent.*

In other words, the tangent space of the deformation $f + \sum \lambda_i e_i$ intersects the tangent space to the orbit of f at a single point.

Chapter 2
Monodromy Groups of Critical Points

Morse theory studies the restructurings, perestroikas, or metamorphoses that the level set $f^{-1}(x)$ of a real function $f\colon M \to \mathbb{R}$, defined on a manifold M, undergoes as x passes through the critical values of f. The Picard-Lefschetz theory is the complex analogue of Morse theory. In the complex case the set of critical values does not divide the range $\mathbb{C}$ of a complex-valued function into connected components, and no restructurings occur: all level manifolds close to a critical one are topologically identical. For this reason, in the complex case, rather than passing through a critical value, one has to go around it in the plane $\mathbb{C}$ where the function takes its values.

If we fix a small circle that goes around the critical value, then to each point of the circle there corresponds a nonsingular level manifold of the function. The set of all such levels is a fibration over the circle. Going around the circle defines a map of the homology of the fibre [over the initial point] of this fibration into itself. This map is called the monodromy corresponding to the critical value of the singularity. It is precisely the monodromy that represents the complex analogue of the restructurings in Morse theory.

§ 1. The Picard-Lefschetz Theory

Here we define the monodromy groups and the related notions of vanishing cycles and Dynkin diagrams, and then we describe the Picard-Lefschetz formulas.

1.1. Topology of the Nonsingular Level Manifold. Consider a holomorphic function $f: \mathbb{C}^n \to \mathbb{C}$, $z \mapsto t$, which has an isolated critical point $z = a$ with critical value $f(a) = \alpha$. Pick a sufficiently small ball $U \subset \mathbb{C}^n$ centered at a such that U contains no other critical points of f. Then the level set $f^{-1}(\alpha)$ is an $(n-1)$-dimensional complex manifold, nonsingular everywhere in U except for the point a.

Theorem ([247]). *The manifold $f^{-1}(\alpha)$ is transverse to the boundary ∂U of the ball U for all sufficiently small values of the radius of U.*

Fix U such that the assertion of the theorem holds true for U and for all balls centered at a and contained in U and choose a sufficiently small neighborhood $T \subset \mathbb{C}$ of the critical value α, such that for every $t \in T$, $t \neq \alpha$, the level manifold $f^{-1}(t)$ is nonsingular inside U and transverse to ∂U.

In this way there arises inside U a family $V_t = f^{-1}(t) \cap U$, $t \in T$, of complex hypersurfaces with boundary $\partial V_t = V_t \cap \partial U$.

Definition. The set V_t $(t \neq \alpha)$ is called a *nonsingular level set* of f near the critical point a.

The family of hypersurfaces V_t, $t \in T$, fills the domain $V_T = f^{-1}(T) \cap U$. Let T' and V'_T denote the punctured neighborhood $T \setminus \{\alpha\}$ of the critical value α and its preimage $f^{-1}(T') = V_T \setminus V_\alpha$, respectively. The function f induces maps $V_T \to T$ and $V'_T \to T'$ (Fig. 12)

It follows from the implicit function theorem that the maps $f: V'_T \to T'$ and $f|_{\partial U}: \partial V_T \to T$, where $\partial V_T = V_T \cap \partial U$, are locally trivial fibrations.

Remark. Since the disk T is contractible, the fibration $\partial V_T \to T$ is trivial. Moreover, the direct product structure in this fibration is unique up to homotopy.

Let μ be the multiplicity of the critical point a. The topology of the nonsingular level manifold V_t is described by the following theorem.

Theorem ([247]). *V_t is homotopy equivalent to a wedge (bouquet) of $(n-1)$-dimensional spheres.*

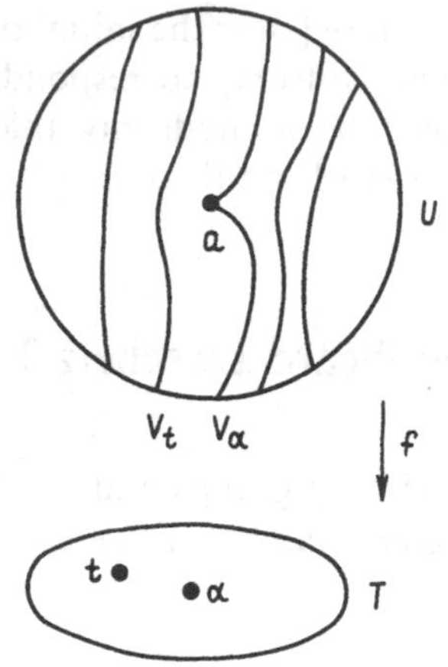

Fig. 12

Example. Consider the function $f: \mathbb{C}^2 \to \mathbb{C}$, $f(z, w) = z^2 + w^2$. It has a Morse critical point at the origin. To the critical value 0 there corresponds a singular level set $V_0 = \{(z, w): z^2 + w^2 = 0\}$, i.e., a pair of complex lines passing through the origin. All the other level sets V_t, $t \neq 0$, are topologically identical and are homeomorphic to the cylinder $S^1 \times \mathbb{R}$: indeed, the Riemann surface of the function $w = \sqrt{t - z^2}$ is constructed from two copies of the complex z-plane glued along the cut from $-\sqrt{t}$ to $+\sqrt{t}$ (Fig. 13).

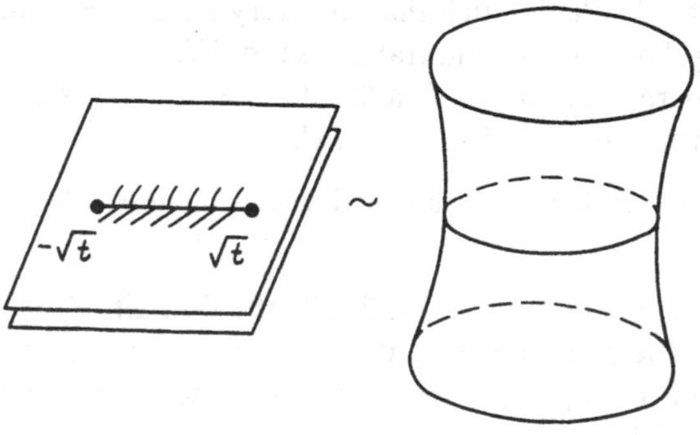

Fig. 13

From the above theorem it follows that the homology group $H_k(V_t)$ with coefficients in $\mathbb{Z}$ is equal to zero for $k \neq n - 1$, whereas $H_{n-1}(V_t) = \mathbb{Z}^\mu$ is a free abelian group with μ generators.

Remark. Here and in all that follows it is assumed that the absolute [resp. relative] homology groups are reduced modulo a point [resp. modulo a fundamental cycle]. If not otherwise stipulated, all homologies are with integer coefficients.

1.2. The Classical Monodromy and the Variation Operator. Fix a noncritical value α_* on the boundary of the disk T and the corresponding nonsingular level manifold $V_* = V_{\alpha_*}$. Consider a loop γ: $[0, 1] \to T'$, $\gamma(0) = \gamma(1) = \alpha_*$, which makes one anticlockwise circuit along the boundary of T. Then [the homotopy class of] γ is a generator of the fundamental group $\pi_1(T', \alpha_*) = \mathbb{Z}$.

By the covering homotopy property, going around γ generates a continuous family of maps Γ_τ: $V_* \to V_{\gamma(\tau)}$ such that $\Gamma_0 = \mathrm{id}$: $V_* \to V_*$. As it turns out, the family Γ_τ can be chosen to be compatible with the direct product structure $\partial V = \partial V_* \times T$, and so one can assume that the map $h_\gamma = \Gamma_1$: $V_* \to V_*$ is the identity on the boundary ∂V_* of the nonsingular level set V_*.

Definition. The action of the map h_γ on the integer homology group of the nonsingular fibre V_*,

$$h_*: H_{n-1}(V_*) \to H_{n-1}(V_*),$$

is called the *classical-monodromy operator* of the singularity f.

In a similar manner, the monodromy map h_γ defines an automorphism h_*^r: $H_{n-1}(V_*, \partial V_*) \to H_{n-1}(V_*, \partial V_*)$.

Remark. The operators h_* and h_*^r defined in the indicated way are independent of the choice of the trivialization Γ_τ, and are uniquely determined by the homotopy class of the loop γ in $\pi_1(T', \alpha_*) = \mathbb{Z}$.

Consider now a relative cycle δ of the pair $(V_*, \partial V_*)$. Since h_γ is the identity on ∂V_*, the cycles δ and $h_\gamma \delta$ have the same boundary, and therefore the difference $h_\gamma \delta - \delta$ yields an absolute cycle in V_*.

Definition. The homomorphism

$$\mathrm{Var}_f: H_{n-1}(V_*, \partial V_*) \to H_{n-1}(V_*),$$

defined by the correspondence $\delta \mapsto h_\gamma \delta - \delta$, is called the *variation operator* of the singularity f.

Example. Consider the function $f(z, w) = z^2 + w^2$ in the ball $U_R = \{z\bar{z} + w\bar{w} \leqslant R\}$. The homology group of the nonsingular fibre $V_* = \{(z, w): z^2 + w^2 = 1\}$ is of rank one, $H_1(V_*) = \mathbb{Z}$, and is generated by the homology class of the "going-around-the-throttle" circle Δ (Fig. 14).

The homology group $H_1(V_*, \partial V_*)$ is also of rank one and is generated by the relative cycle ∇, the vertical generatrix of the cylinder.

Now consider the loop $\gamma(\tau) = \exp(2\pi i \tau)$, $\tau \in [0, 1]$, and follow the evolution of the fibre $V_{\gamma(\tau)}$ as one moves along γ. As τ increases from 0 to 1, both ramification points $z = \pm\exp(\pi i \tau)$ make an anticlockwise half-circuit around the point 0.

One can readily construct a family of homomorphisms Γ_τ: $V_* \to V_{\gamma(\tau)}$, depending continuously on τ and equal to the identity map outside some disk, and then follow the "fate" of the cycles Δ and ∇ (see Fig. 15). Under the action of the transformation $h = h_\gamma$, the cycle Δ goes into itself, whereas the relative cycle ∇ winds once around the cylinder. Thus,

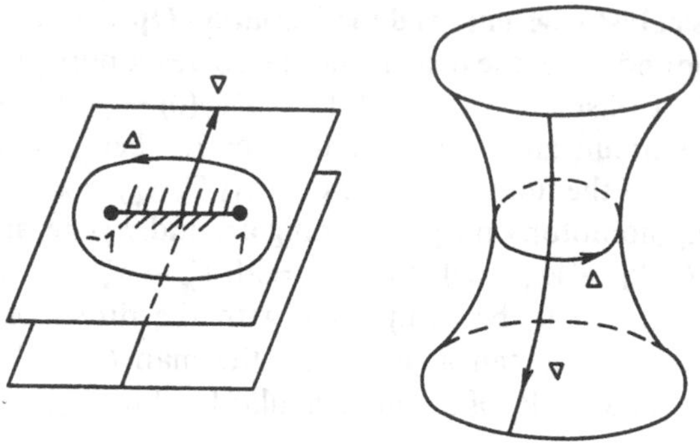

Fig. 14

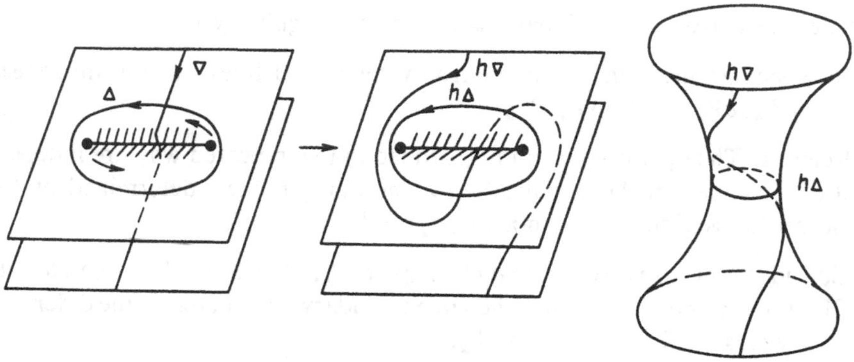

Fig. 15

$$h\Delta = \Delta, \qquad h\nabla = \nabla - \Delta,$$

and so in our case $h_* = \mathrm{id}$ and Var: $\delta \mapsto (\Delta \circ \delta)\Delta$, where $\delta \in H_1(V_*, \partial V_*)$ and $(\cdot \circ \cdot)$ denotes the intersection index.

1.3. The Monodromy of a Morse Singularity. By analogy with the case examined above, one can describe the monodromy and the variation operator in the general case of a critical point of multiplicity $\mu = 1$.

By the Morse lemma, in a neighborhood U of the nondegenerate critical point a of the function f there exists a coordinate system $z_1, \ldots, z_n$ in which f has the form

$$f(z_1, \ldots, z_n) = \alpha + z_1^2 + \cdots + z_n^2$$

(one may assume that the fixed ball U is sufficiently small).

Consider a path $\varphi(\tau)$, $\tau \in [0, 1]$ in T that connects the marked point α_* with the critical value α: $\varphi(0) = \alpha_*$, $\varphi(1) = \alpha$ (Fig. 16). Next, in the manifold $V_{\varphi(\tau)}$ consider the oriented sphere

$$S_\tau = \sqrt{\varphi(\tau) - \alpha}\, S^{n-1},$$

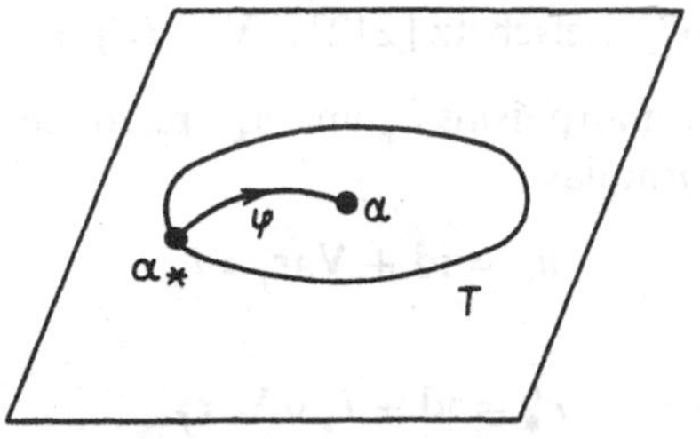

Fig. 16

where $S^{n-1} = \{z: z_1^2 + \cdots + z_n^2 = 1, \operatorname{Im} z_i = 0, i = 1, \ldots, n\}$ is the standard unit sphere. The constructed family of spheres $S_\tau \subset V_{\varphi(\tau)}$ degenerates to a point for $\tau = 1$ (Fig. 17).

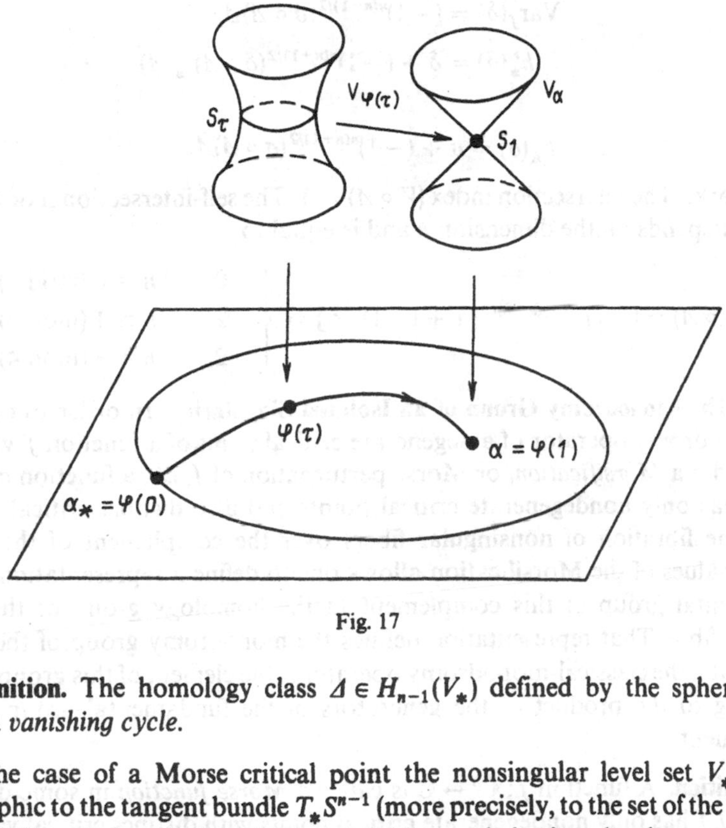

Fig. 17

Definition. The homology class $\Delta \in H_{n-1}(V_*)$ defined by the sphere S_0 is called a *vanishing cycle*.

In the case of a Morse critical point the nonsingular level set V_* is diffeomorphic to the tangent bundle $T_* S^{n-1}$ (more precisely, to the set of the tangent vectors of length less than one). The vanishing cycle is given by the zero section of this bundle and is a generator of the group $H_{n-1}(V_*) = \mathbb{Z}$. The generator ∇ of the relative homology group $H_{n-1}(V_*, \partial V_*)$ is defined by an $(n-1)$-dimensional ball, namely, the fibre of the bundle. Such a description enables us to carry out all calculations to the end and to describe the monodromy and variation operators.

Theorem (Picard [267]; Lefschetz [212]). $\mathrm{Var}_f(\nabla) = (-1)^{n(n+1)/2}\Delta$.

The monodromy automorphisms h_* and h_*^r are connected with the variation operator through the formulas

$$h_* = \mathrm{id} + \mathrm{Var}_f \circ i_*$$

and

$$h_*^r = \mathrm{id} + i_* \circ \mathrm{Var}_f,$$

respectively; here $i_*: H_{n-1}(V_*) \to H_{n-1}(V_*, \partial V_*)$ is the homomorphism induced by the embedding $V_* \to (V_*, \partial V_*)$. These relations, in conjunction with the preceding theorem, allow us to obtain the following Picard-Lefschetz formulas [267], [212].

Theorem. *Let* $\delta \in H_{n-1}(V_*, \partial V_*)$ *and* $\sigma \in H_{n-1}(V_*)$. *Then*

$$\mathrm{Var}_f(\delta) = (-1)^{n(n+1)/2}(\delta \circ \Delta)\Delta,$$

$$h_*^r(\delta) = \delta + (-1)^{n(n+1)/2}(\delta \circ \Delta) i_*(\Delta),$$

and

$$h_*(\sigma) = \sigma + (-1)^{n(n+1)/2}(\sigma \circ \Delta)\Delta.$$

Remark. The intersection index $(\nabla \circ \Delta) = 1$. The self-intersection index of the cycle Δ depends on the dimension n and is equal to

$$(\Delta \circ \Delta) = (-1)^{(n-1)(n-2)/2}[1 + (-1)^{n-1}] = \begin{cases} 0, & n \equiv 0 \pmod 2, \\ 2, & n \equiv 1 \pmod 4, \\ -2, & n \equiv 3 \pmod 4. \end{cases}$$

1.4. The Monodromy Group of an Isolated Singularity. In order to describe the monodromy operator of a degenerate critical point of a function f we need to consider a *Morsification*, or Morse perturbation of f, i.e., a function close to f that has only nondegenerate critical points and also distinct critical values. Then, the fibration of nonsingular fibers over the complement of the set of critical values of the Morsification allows one to define a representation of the fundamental group of this complement in the homology group of the nonsingular fibre. That representation defines the monodromy group of the given singularity; the classical-monodromy operator is the element of this group corresponding to the product of the generators of the fundamental group of the complement.

Definition. A function $f: \mathbb{C}^n \to \mathbb{C}$ is called a *Morse function* in some domain U if in U f has only nondegenerate critical points with distinct critical values.

Suppose that $f: \mathbb{C}^n \to \mathbb{C}$ has a degenerate critical point a of multiplicity μ. Perturb f slightly to $f_\varepsilon = f + \varepsilon g$. For a suitable choice of g (for instance, one can take g to be a linear function in general position), the function f_ε of variable z is a Morse function in a neighborhood U of the critical point a for all sufficiently small values of the parameter ε.

If ε is small enough, then inside the ball U f_ε has exactly μ critical points a_1, …, a_μ with pairwise distinct critical values $\alpha_1 = f(a_1), \ldots, \alpha_\mu = f(a_\mu)$; moreover, all α_i lie inside a disk $T \subset \mathbb{C}$ and the level sets $V_t = f_\varepsilon^{-1}(t) \cap U$ are transverse to the boundary ∂U of U. Note also that the level manifold V_t is nonsingular for t in the complement $T' = T \backslash \{\alpha, \ldots, \alpha_n\}$ of the critical values (Fig. 18).

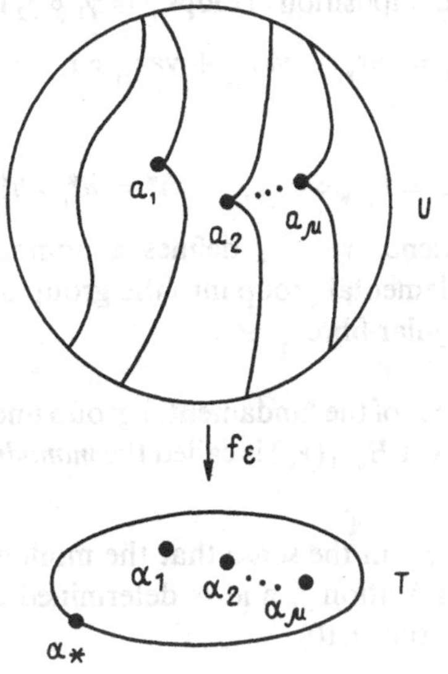

Fig. 18

The nonsingular fibres V_t of the function f_ε form a locally trivial fibration with base T' and fibre homeomorphic to the nonsingular fibre of the unperturbed function f. Similarly, $f_\varepsilon^{-1}(T) \cap \partial U$ is a trivial fibration over T with fibre ∂V_t.

The construction of the monodromy and variation operators described in subsection 1.2 is applicable without modifications in the present situation. In fact, fix a noncritical value α_* on the boundary of the disk T and assign to each loop γ in [a given homotopy class of] the fundamental group $\pi_1(T', \alpha_*)$ a family of maps $\Gamma_\tau \colon V_* \to V_{\gamma(\tau)}$, defined on the nonsingular fibre over α_*, that is compatible with the direct product structure $\partial V = \partial V_* \times T$. Such a family defines a monodromy transformation $h_\gamma = \Gamma_1 \colon V_* \to V_*$, and hence the corresponding *variation* operator and *monodromy* operator *along the loop* γ,

$$\mathrm{var}_\gamma \colon H_{n-1}(V_*, \partial V_*) \to H_{n-1}(V_*)$$

and respectively

$$h_{\gamma*} \colon H_{n-1}(V_*) \to H_{n-1}(V_*),$$
$$h_\gamma^* \colon H_{n-1}(V_*, \partial V_*) \to H_{n-1}(V_*, \partial V_*).$$

Remark. The homotopy class of the monodromy transformation $h_\gamma \colon V_* \to V_*$, and consequently the variation and monodromy operators along the loop γ, are independent of the choice of the loop within a given homotopy class in the fundamental group $\pi_1(T', \alpha_*)$, i.e., are uniquely determined by that class.

The variation and monodromy operators along γ are again related as in subsection 1.3:

$$h_{\gamma *} = \mathrm{id} + \mathrm{var}_\gamma \circ i_*, \qquad h_\gamma^* = \mathrm{id} + i_* \circ \mathrm{var}_\gamma.$$

Furthermore, under the composition of loops $\gamma = \gamma_1 \circ \gamma_2$ they behave as follows:

$$\mathrm{var}_\gamma = \mathrm{var}_{\gamma_1} + \mathrm{var}_{\gamma_2} + \mathrm{var}_{\gamma_1} \circ i_* \circ \mathrm{var}_{\gamma_2}$$

and

$$h_{\gamma *} = h_{\gamma_1 *} \circ h_{\gamma_2 *}, \qquad h_\gamma^* = h_{\gamma_1}^* \circ h_{\gamma_2}^*.$$

Thus, the correspondence $\gamma \mapsto h_{\gamma *}$ defines a homomorphism $\pi_1(T', \alpha_*) \to \mathrm{Aut}\, H_{n-1}(V_*)$ of the fundamental group into the group of automorphisms of the homology of the nonsingular fibre.

Definition. The image Γ of the fundamental group under the indicated homomorphism $\pi_1(T', \alpha_*) \to \mathrm{Aut}\, H_{n-1}(V_*)$ is called the *monodromy group* of the singularity f.

This definition is correct in the sense that the monodromy does not depend on the choice of a Morsification f_ε and is determined only by the type of the singular point (see subsection 1.10).

1.5. Vanishing Cycles and Distinguished Bases. Any path connecting a noncritical value of some fixed Morsification f_ε of the singularity f with one of its critical values allows us to define a vanishing cycle in the homology of the nonsingular fibre; a choice of such paths for all critical values provides a basis of cycles in $H_{n-1}(V_*)$.

In the disk T consider a path $\varphi(\tau)$, $\tau \in [0, 1]$, connecting the marked point $\alpha_* = \varphi(0) \in \partial T$ with one of the critical values $\alpha_i = \varphi(1)$; assume that for $\tau < 1$ the path φ avoids all critical values of the Morsification f_ε. By analogy to subsection 1.3, in a neighborhood of the critical point a_i there is a coordinate system in which

$$f_\varepsilon(z) = \alpha_i + z_1^2 + \cdots + z_n^2.$$

These coordinates allow us to single out in the nonsingular fibre $V_{\varphi(\tau)}$, for τ close to 1, the sphere

$$S_\tau = \sqrt{\varphi(\tau) - \alpha_i}\, S^{n-1}.$$

As $\tau \to 1$, S_τ degenerates to a point. Using the covering homotopy property, the family of spheres $S_\tau \subset V_{\varphi(\tau)}$ can be continued along the whole path $\varphi(\tau)$.

Definition. The homology class $\Delta_\varphi \in H_{n-1}(V_*)$, defined by the sphere S_0 in the nonsingular fibre V_*, is called a *vanishing cycle* (along the path φ; see Fig. 19).

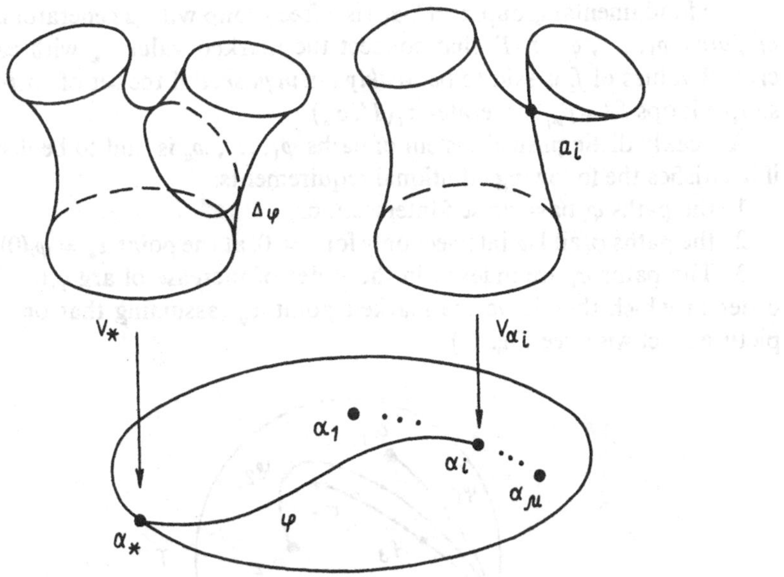

Fig. 19

A vanishing cycle does not depend on the choice of the coordinate system in a neighborhood of the critical point and the way in which the family S_τ is continued along the path φ. Up to orientation, Δ_φ is uniquely determined by the homotopy class of φ in the set of all paths connecting α_* with the critical value α_i in the base T' of the fibration.

Now consider a loop in T' obtained by going along φ from α_* to the critical value α_i, then going anticlockwise around α_i, and finally returning to α_* along φ (Fig. 20).

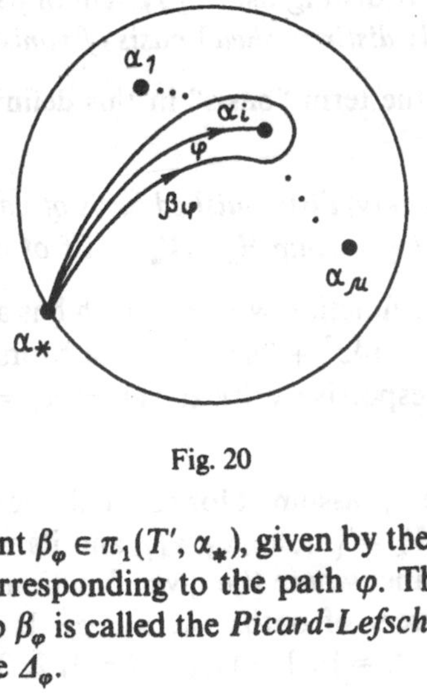

Fig. 20

Definition. The element $\beta_\varphi \in \pi_1(T', \alpha_*)$, given by the loop described above, is called the *simple loop* corresponding to the path φ. The monodromy operator $h_\varphi(=h_{\beta_\varphi *})$ along the loop β_φ is called the *Picard-Lefschetz operator* corresponding to the vanishing cycle Δ_φ.

The fundamental group $\pi_1(T', \alpha_*)$ is a free group with μ generators. A *system of paths* $\varphi_1, \ldots, \varphi_\mu$ in T' that connect the marked value α_* with each of the critical values of f_ε is said to be *weakly distinguished* if the set of corresponding simple loops $\beta_i(=\beta_{\varphi_i})$ generates $\pi_1(T', \alpha_*)$.

A weakly distinguished system of paths $\varphi_1, \ldots, \varphi_\mu$ is said to be *distinguished* if it satisfies the following additional requirements:

1. the paths φ_i have no self-intersections;
2. the paths φ_i and φ_j intersect only for $\tau = 0$, at the point $\alpha_* = \varphi_i(0) = \varphi_j(0)$;
3. The paths φ_i are indexed in the order of increase of $\arg \varphi_i'(0)$, i.e., in the order in which they leave the marked point α_* (assuming that one reads the picture clockwise; see Fig. 21).

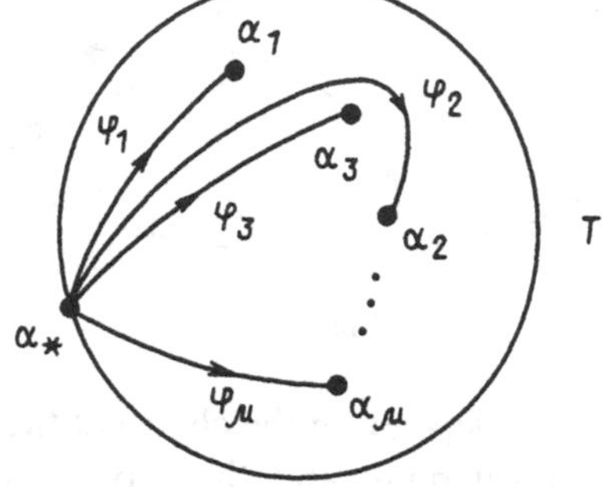

Fig. 21

Remark. The set of critical values is not a priori ordered, so one can index it as one finds suitable, for example, in accordance with the indexation of the paths.

Definition. The set of vanishing cycles $\Delta_1, \ldots, \Delta_\mu$ ($\Delta_i = \Delta_{\varphi_i}$) defined by a distinguished [resp. weakly distinguished] system of paths $\varphi_1, \ldots, \varphi_\mu$ is called a *distinguished* [resp. *weakly distinguished*] *basis of vanishing cycles.*

The reason for using the term "basis" in this definition is explained by the following result:

Theorem ([58]). *A (weakly) distinguished basis of vanishing cycles $\Delta_1, \ldots, \Delta_\mu$ forms a basis of the homology group $H_{n-1}(V_*) \cong \mathbb{Z}^\mu$ of the nonsingular fibre.*

Example. Consider the function $w = z^4$, which has an A_3-singularity at zero. Its Morsification $w = z^4 - 14z^2 + 24z$ has three Morse critical points: $a_1 = 2$, $a_2 = 1$, $a_3 = -3$, with respective critical values $\alpha_1 = 8$, $\alpha_2 = 11$, $\alpha_3 = -135$ (Fig. 22).

Fix a noncritical value α_*, assumed for convenience to be real: $\alpha_1 < \alpha_* < \alpha_2$. The nonsingular fibre is $V_* = \{z_1, z_2, z_3, z_4\}$, and its reduced homology group $H_0(V_*) = \mathbb{Z}^3$. In the w-plane (where the given function assumes its values) consider a distinguished system of paths $\varphi_1, \varphi_2, \varphi_3$. It yields the distinguished basis of vanishing cycles $\Delta_i = [z_i] - [z_{i+1}]$, $i = 1, 2, 3$, of $H_0(V_*)$. The mono-

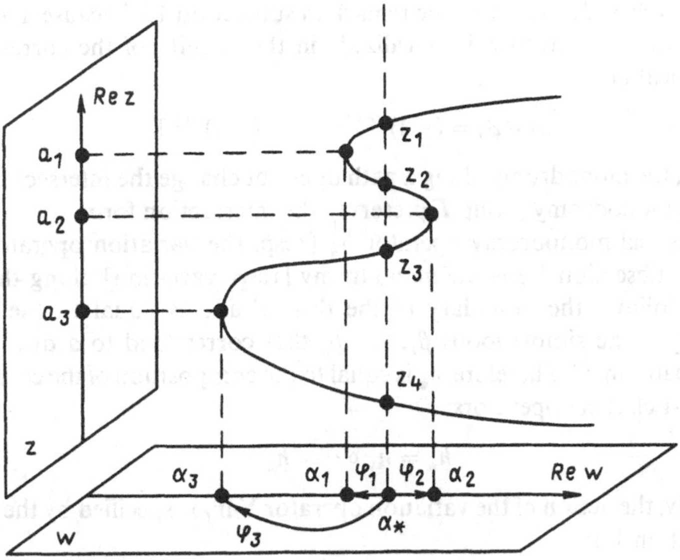

Fig. 22

dromy along the simple loop corresponding to φ_i permutes the points z_i and z_{i+1} in the fibre V_*.

A basis of vanishing cycles is an ordered set. Note that while a weakly distinguished basis remains weakly distinguished for any relabeling of its elements, distinguished bases do not enjoy such a property.

1.6. The Intersection Matrix of a Singularity. The monodromy along a simple loop is "localized" in the vicinity of one of the critical values. As a consequence, the monodromy and variation operators along such a loop are described by the Picard-Lefschetz formulas (the theorem of subsection 1.3), with no modifications.

The set of simple loops corresponding to a (weakly) distinguished system of paths $\varphi_1, \ldots, \varphi_\mu$ generates $\pi_1(T', \alpha_*)$. Consequently, the Picard-Lefschetz operators

$$h_i\colon \delta \mapsto \delta + (-1)^{n(n+1)/2}(\delta \circ \Delta_i)\Delta_i$$

corresponding to the vanishing cycles Δ_i of a (weakly) distinguished basis $\Delta_1, \ldots, \Delta_\mu$ generate the monodromy group Γ.

Thus, in order to describe the monodromy group it suffices to know the structure of the $\mathbb{Z}$-bilinear form on $H_{n-1}(V_*)$ defined by the intersection index.

Definition. The matrix $[(\Delta_i \circ \Delta_j)]_{ij}$ of the intersection form in a (weakly) distinguished basis $\Delta_1, \ldots, \Delta_\mu$ is called the *intersection matrix of the singularity f*.

The bilinear intersection form (matrix) associated with a singularity is symmetric for n odd and skew-symmetric for n even. The self-intersection index of a vanishing cycle is described by the remark in subsection 1.3 because a vanishing cycle, like the monodromy, is "localized" in the vicinity of the corresponding Morse critical point:

$$(\Delta \circ \Delta) = (-1)^{n(n+1)/2}[1 + (-1)^{n+1}].$$

Clearly, the monodromy along a path does not change the intersection index, and so the monodromy group Γ preserves the intersection form.

The classical-monodromy operator h_* [resp. the variation operator Var_f] defined in subsection 1.2 is the monodromy [resp. variation] along the closed path that follows the boundary of the disk T and is equal to the product $\beta_1 \circ \cdots \circ \beta_\mu$ of the simple loops $\beta_1, \ldots, \beta_\mu$ that correspond to a distinguished system of paths in T'. Therefore, h_* is equal to the composition of the corresponding Picard-Lefschetz operators:

$$h_* = h_1 \circ \cdots \circ h_\mu.$$

Similarly, the action of the variation operator Var_f is specified by the formula (see subsection 1.4)

$$\mathrm{Var}_f = \mathrm{var}_{\beta_1 \circ \cdots \circ \beta_\mu} = \sum_{r=1}^{\mu} \sum_{i_1 < i_2 \cdots < i_r} \prod_{j=1}^{r} \mathrm{var}_{i_j},$$

where

$$\mathrm{var}_k = \mathrm{var}_{\beta_k}\colon \delta \mapsto (-1)^{n(n+1)/2}(\delta \circ \Delta_k)\Delta_k$$

and the product is to be understood as the composition of the variation operators with the natural embedding $H_{n-1}(V_*) \to H_{n-1}(V_*, \partial V_*)$ taken into account.

The formulas given above allow us to express the classical-monodromy and variation operators in terms of the intersection matrix of the singularity. The converse is also true: the classical-monodromy [resp. variation] operator determines the variation [resp. classical-monodromy] operator and the intersection form. In fact, we have the following:

Theorem.

(i) ([172], [205]). *The variation operator of the singularity,*

$$\mathrm{Var}_f\colon H_{n-1}(V_*, \partial V_*) \to H_{n-1}(V_*),$$

is an isomorphism.

(ii) ([205]). *The intersection form on $H_{n-1}(V_*)$ is expressible in terms of the variation operator as*

$$(\cdot \circ \cdot) = -(\mathrm{Var}_f^{-1} \cdot \circ \cdot) - (\cdot \circ \mathrm{Var}_f^{-1} \cdot).$$

(iii) ([205]). *The classical-monodromy operator is expressible in terms of the variation operator as*

$$h_* = (-1)^n\, \mathrm{Var}_f (\mathrm{Var}_f^{-1})^{\mathrm{T}}.$$

(iv) ([208]). *The matrix of the classical-monodromy operator in a distinguished basis uniquely determines the matrix of its variation operator via relation (iii).*

Example. The intersection form of the A_3-singularity given in the example of subsection 1.5 is symmetric ($n = 1$). The intersection matrix in the distinguished basis $\Delta_1, \Delta_2, \Delta_3$ is

$$\begin{bmatrix} 2 & -1 & 0 \\ -1 & 2 & -1 \\ 0 & -1 & 2 \end{bmatrix}.$$

The Picard-Lefschetz and classical-monodromy operators corresponding to the cycles Δ_i are given by the matrices

$$h_1\colon \begin{bmatrix} -1 & -1 & 0 \\ 0 & 1 & 0 \\ 0 & 0 & 1 \end{bmatrix}, \quad h_2\colon \begin{bmatrix} 1 & 0 & 0 \\ -1 & -1 & -1 \\ 0 & 0 & 1 \end{bmatrix}, \quad h_3\colon \begin{bmatrix} 1 & 0 & 0 \\ 0 & 1 & 0 \\ 0 & -1 & -1 \end{bmatrix}$$

and

$$h_*\colon \begin{bmatrix} 0 & 0 & -1 \\ -1 & 0 & 1 \\ 0 & -1 & -1 \end{bmatrix}.$$

1.7. Stabilization of Singularities. Let $f\colon (\mathbb{C}^n, 0) \to (\mathbb{C}, 0)$ be a function of n variables $z_1, \ldots, z_n$.

Definition. The function $\tilde{f}\colon (\mathbb{C}^{n+k}, 0) \to (\mathbb{C}, 0)$,

$$\tilde{f} = f + z_{n+1}^2 + \cdots + z_{n+k}^2,$$

of the variables $z_1, \ldots, z_n, z_{n+1}, \ldots, z_{n+k}$, is called a *stabilization* of f.

Two function-germs of different numbers of variables are said to be *stably equivalent* if they admit equivalent stabilizations (see subsection 1.1.3).

Consider a germ $f\colon (\mathbb{C}^n, 0) \to (\mathbb{C}, 0)$, $f = f(z)$, with singularity at zero, and a stabilization $\tilde{f}(z, w) = f(z) + w_1^2 + \cdots + w_k^2$ of f. To any given Morsification f_ε of the singularity f there corresponds the Morsification $\tilde{f}_\varepsilon = f_\varepsilon + w_1^2 + \cdots + w_k^2$ of its stabilization $\tilde{f}$, with the same set of critical values $\alpha_1, \ldots, \alpha_\mu$.

Let $\Delta_1, \ldots, \Delta_\mu$ be a distinguished basis of vanishing cycles of the singularity of f, constructed from a distinguished system of paths. The same system of paths defines (up to orientation) a distinguished basis $\tilde{\Delta}_1, \ldots, \tilde{\Delta}_\mu$ of the singularity of the stabilization $\tilde{f}$. The connection between the intersection matrices of the singularities of f and of $\tilde{f}$ is described by the following result.

Theorem ([172]). *For a suitable orientation of the distinguished basis $\tilde{\Delta}_1, \ldots, \tilde{\Delta}_\mu$ the intersection matrix is given by the formulas*

$$(\tilde{\Delta}_i \circ \tilde{\Delta}_j) = [\operatorname{sign}(j - i)]^k (-1)^{kn + k(k-1)/2} (\Delta_i \circ \Delta_j), \qquad i \neq j.$$

Thus, the intersection matrices of stably equivalent singularities determine one another. Another corollary of the theorem is that in a class of stably equivalent singularities there are exactly four distinct intersection forms: for

$k \equiv 0 \pmod 4$ [resp. $k \equiv 2 \pmod 4$] the intersection indices coincide [resp. differ by their sign]. Consequently, with each singularity are associated two symmetric and two skew-symmetric bilinear forms, which differ by sign, and two monodromy groups, corresponding to the symmetric and skew-symmetric cases.

1.8. Dynkin Diagrams. The theorem of subsection 1.7 permits us, in investigating intersection matrices of singularities, to confine ourselves to the case of a dimension that has a fixed residue mod 4. The choice most commonly used in singularity theory is $n \equiv 3 \pmod 4$, and will be ours from now on, unless otherwise stipulated.

Under the above assumption on the number of variables, the bilinear intersection form is symmetric, and the self-intersection index of vanishing cycles is equal to -2. The Picard-Lefschetz operator corresponding to the vanishing cycle Δ,

$$h_\Delta \colon \delta \mapsto \delta + (\delta \circ \Delta)\Delta,$$

is in the present situation the reflection in the hyperplane of the space $H_{n-1}(V_*)$ that is orthogonal to the cycle Δ with respect to the inner product $\langle \cdot, \cdot \rangle = -(\cdot \circ \cdot)$. Hence, the monodromy group Γ is a reflection group.

Definition. The quadratic form on $H_{n-1}(V_*)$ associated with the bilinear intersection form for $n \equiv 3 \pmod 4$ is called the *quadratic form of the singularity*.

It is convenient to "encode" the information describing an arbitrary reflection group Γ in a graph, called the *Dynkin diagram* of Γ. The graph is constructed in terms of a distinguished basis of vanishing cycles $\Delta_1, \ldots, \Delta_\mu$ as follows : to each Δ_i one assigns a vertex labeled i; two vertices, $\langle i \rangle$ and $\langle j \rangle$, are connected by a [dotted] edge with index $|k|$ if the intersection index of the cycles Δ_i and Δ_j equals $k > 0$ [resp. $k < 0$].

The Dynkin diagram allows one to recover the intersection form, the monodromy group, and all the rest.

Example. In the case $n \equiv 3 \pmod 4$ the intersection matrix of the A_3-singularity and its Dynkin diagram are as follows (see the example in subsection 1.6):

$$\begin{bmatrix} -2 & 1 & 0 \\ 1 & -2 & 1 \\ 0 & 1 & -2 \end{bmatrix} \Rightarrow$$

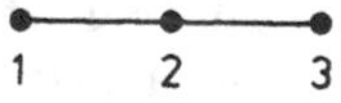

Remark. A basis obtained by permuting the elements of a distinguished basis of vanishing cycles is not necessarily distinguished, so that on relabeling the vertices of the Dynkin diagram one may obtain a graph that is not Dynkin.

1.9. Transformations of a Basis and of its Dynkin Diagram. A distinguished basis of vanishing cycles and its Dynkin diagram are not uniquely defined; rather, they depend on the choice of a distinguished system of paths and on that of an orientation of the vanishing cycles. Let us describe two types of elementary changes of basis and the associated transformations of the Dynkin diagram.

The operation denoted by s_i $(i = 1, \ldots, \mu)$ changes the orientation of the cycle Δ_i in the given basis to the opposite orientation:

$$s_i: (\Delta_1, \ldots, \Delta_\mu) \mapsto (\Delta_1, \ldots, \Delta_{i-1}, -\Delta_i, \Delta_{i+1}, \ldots, \Delta_\mu).$$

In the Dynkin diagram the operation s_i changes all simple edges emanating from the vertex $\langle i \rangle$ into dotted edges, and vice versa.

Example. s_2: 

Definition. The *braid group* $\mathrm{Br}(\mu)$ of μ strands is the fundamental group of the space of monic polynomials in one variable with no multiple roots. The elements of $\mathrm{Br}(\mu)$ are called *braids.*

A point in the indicated space of polynomials may be thought of as a copy of the complex line $\mathbb{C}$ with a marked unordered subset of μ distinct points (the roots of the corresponding polynomial); accordingly, a closed path in that space may be regarded as a family $\mathbb{C}_\tau$, $\tau \in [0, 1]$ of complex lines together with a set of μ distinct points of $\mathbb{C}_\tau$, depending continuously on the parameter τ, which coincide for $\tau = 0$ and $\tau = 1$ (Fig. 23).

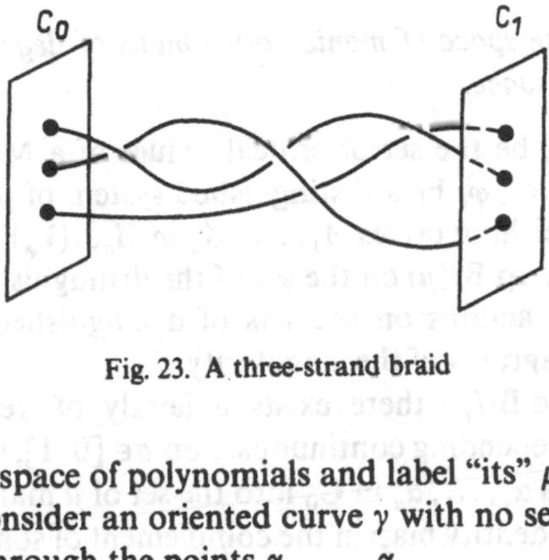

Fig. 23. A three-strand braid

Fix a point in the space of polynomials and label "its" μ roots as, say, $\alpha_1, \ldots, \alpha_\mu$. In the plane $\mathbb{C}$ consider an oriented curve γ with no self-intersections which passes successively through the points $\alpha_1, \ldots, \alpha_\mu$.

Now define the braid t_i for $i = 1, \ldots, \mu - 1$ as follows: as τ increases from 0 to 1, the points $\alpha_i(\tau)$ and $\alpha_{i+1}(\tau)$ move towards one another along γ, to the left and respectively to the right of γ, so that eventually they switch places: $\alpha_i(1) = \alpha_{i+1}(0)$ and $\alpha_{i+1}(1) = \alpha_i(0)$. The remaining marked points α_j are kept fixed during this process (Fig. 24).

Theorem ([60]). *The braids* $t_1, \ldots, t_{\mu-1}$ *constitute a system of generators of the braid group* $\mathrm{Br}(\mu)$, *subject to the defining relations*

$$t_i t_j = t_j t_i \qquad \text{for } |i - j| \geqslant 2,$$

$$t_i t_{i+1} t_i = t_{i+1} t_i t_{i+1} \qquad \text{for } i = 1, \ldots, \mu - 2.$$

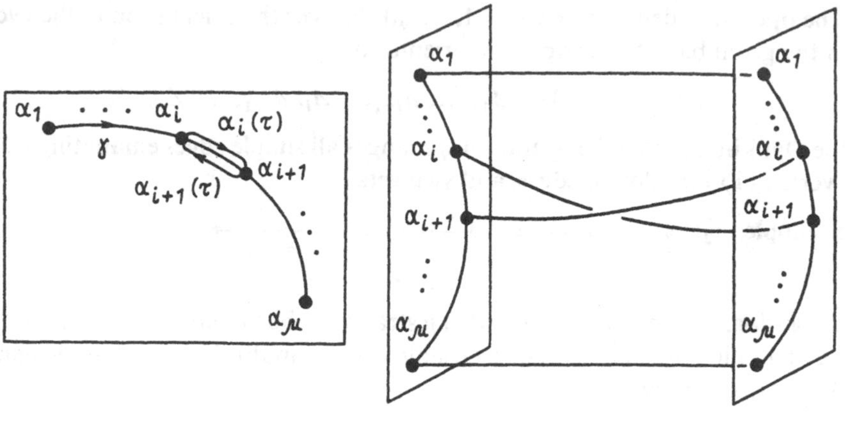

Fig. 24

Definition. Let G be a group. A topological space X is called an *Eilenberg-Mac Lane space of type* $K(G, n)$, or simply a $K(G, n)$*-space*, if the homotopy groups of X are $\pi_i(X) = 0$ for $i \neq n$, $\pi_n(X) = G$.

Remark. For $n > 1$, a necessary condition for the existence of a $K(G, n)$-space is that the group G be abelian.

Theorem ([9]). *The space of monic polynomials of degree* μ *with no multiple roots is a* $K(\mathrm{Br}(\mu), 1)$*-space.*

Now let $\alpha_1, \ldots, \alpha_\mu$ be the set of critical values of a Morsification f_ε of the singularity f. Let $\varphi_1, \ldots, \varphi_\mu$ be a distinguished system of paths, with the corresponding basis of vanishing cycles $\Delta_1, \ldots, \Delta_\mu$ in $H_{n-1}(V_*)$. We define below an action of the braid group $\mathrm{Br}(\mu)$ on the set of the distinguished systems of paths, and hence associated actions on the sets of distinguished bases of vanishing cycles and Dynkin diagrams of the singularity f.

For each braid $t \in \mathrm{Br}(\mu)$ there exists a family of (real) diffeomorphisms $\theta_\tau: \mathbb{C}_0 \to \mathbb{C}_\tau$, $\theta_0 = \mathrm{id}$, depending continuously on $\tau \in [0, 1]$, such that θ_τ maps the set of μ marked points $\alpha_1, \ldots, \alpha_\mu$ in $\mathbb{C}_0$ into the set of μ marked points $\alpha_1(\tau), \ldots, \alpha_\mu(\tau)$ in $\mathbb{C}_\tau$, and is the identity map in the complement of some disk. With no loss of generality we may assume that the marked value α_* of the Morsification f_ε is "remote", i.e., $\theta_\tau(\alpha_*) = \alpha_*$ for all $\tau \in [0, 1]$.

Definition. The image of the system of paths φ_i, $i = 1, \ldots, \mu$, under the action of the braid $t \in \mathrm{Br}(\mu)$ is the system of paths $t(\varphi_i) = \theta_1 \circ \varphi_i$, $i = 1, \ldots, \mu$, where $\theta_1: \mathbb{C}_0 \to \mathbb{C}_1$ is the terminal diffeomorphism in the family described above.

A braid t takes a distinguished system of paths again into such a system; moreover, the homotopy classes of the paths $t(\varphi_i)$ are independent of the choice of the family of diffeomorphisms θ_τ.

The correspondence

$$(\varphi_1, \ldots, \varphi_\mu) \mapsto (t(\varphi_1), \ldots, t(\varphi_\mu))$$

defines a transitive action of the braid group $\mathrm{Br}(\mu)$ on the set of homotopy classes of distinguished systems of paths.

The generator t_i of $\mathrm{Br}(\mu)$ takes the distinguished system $\varphi_1, \ldots, \varphi_\mu$ into the system $t_i(\varphi_j)$:

$$t_i\colon (\varphi_1, \ldots, \varphi_\mu) \mapsto (\varphi_1, \ldots, \varphi_{i-1}, \varphi_{i+1} \circ \beta_i^{-1}, \varphi_i, \varphi_{i+2}, \ldots, \varphi_\mu),$$

where β_i is the simple loop corresponding to the path φ_i (Fig. 25).

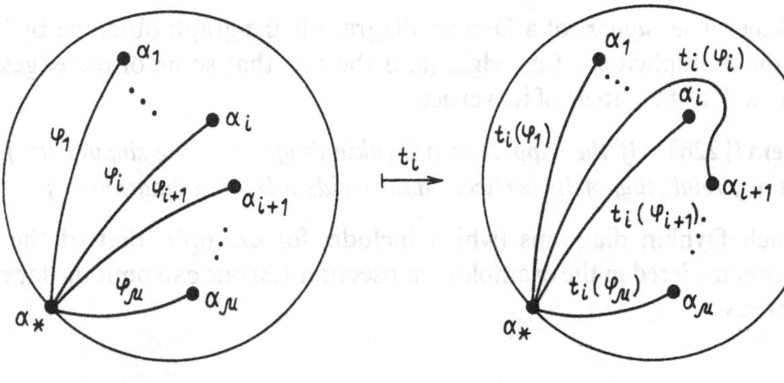

Fig. 25

The system of paths $t_i(\varphi_1), \ldots, t_i(\varphi_\mu)$ defines a basis of vanishing cycles $t_i(\Delta_1)$, $\ldots, t_i(\Delta_\mu)$, which can be expressed (for one of the possible ways of orienting the cycles) through the basis $\Delta_1, \ldots, \Delta_\mu$ via the Picard-Lefschetz formulas.

By definition, the operation t_i acts on the distinguished basis of vanishing cycles $\Delta_1, \ldots, \Delta_\mu$ in $H_{n-1}(V_*)$ according to the rule $t_i\colon (\Delta_1, \ldots, \Delta_\mu) \mapsto (\Delta_1, \ldots, \Delta_{i-1}, \Delta_{i+1} + (\Delta_{i+1} \circ \Delta_i)\Delta_i, \Delta_i, \Delta_{i+2}, \ldots, \Delta_\mu)$ (recall that the number of variables $n \equiv 3 \pmod 4$).

The operations t_i, $i = 1, \ldots, \mu - 1$, define an action of the braid group $\mathrm{Br}(\mu)$ on the set of distinguished bases of vanishing cycles, as well as an action on the set of Dynkin diagrams of the singularity f.

Example. The braid $t = t_1 \circ t_2 \circ t_1$, applied to the Dynkin diagram A_3 of the example considered in subsection 1.8, leads to the following diagram:

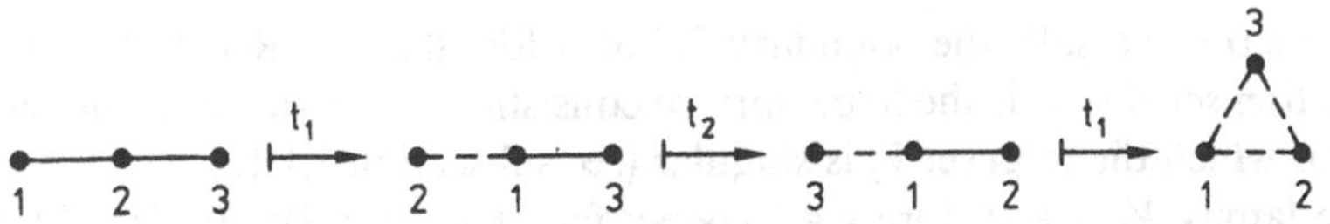

The operators s_i, $i = 1, \ldots, \mu$, and t_j, $j = 1, \ldots, \mu - 1$, together generate a group of operations acting on the set of all distinguished bases of vanishing cycles and Dynkin diagrams. This group is the "semidirect" product of the groups $\mathbb{Z}_2^\mu$ (the s_i-operations) and $\mathrm{Br}(\mu)$ (the t_j-operations):

$$s_i \circ t_j = t_j \circ s_i \qquad \text{for } i \neq j, j+1$$

and

$$s_i \circ t_i = t_i \circ s_{i+1}, \qquad s_{i+1} \circ t_i = t_i \circ s_i .$$

Its action is transitive thanks to the transitivity of the action of the braid group on the set of distinguished systems of paths.

At the end of subsection 1.8 we mentioned that a diagram obtained from a Dynkin diagram by relabeling its vertices is not necessarily a Dynkin diagram of the same singularity.

Definition. The *support* of a Dynkin diagram is the graph obtained by "forgetting" the multiplicities of its edges (and the fact that some of the edges are dotted) as well as the labels of its vertices.

Theorem ([226]). *If the support of a Dynkin diagram of the singularity f is a tree, then any relabeling of its vertices again yields a Dynkin diagram of f.*

For such Dynkin diagrams (which include, for example, that of the A_3-singularity considered in the example of subsection 1.8) one can omit the labeling of the vertices.

1.10. The Milnor Fibration over the Complement of the Level Bifurcation Set. Let us give another description of the monodromy group of a singularity, which is equivalent to the one given earlier. The new description is invariant in the sense that it does not depend on the choice of a Morsification of the original function f.

We begin by considering a miniversal deformation

$$F(z, \lambda) = f(z) + \sum_{i=0}^{\mu-1} \lambda_i \varphi_i(z), \qquad \varphi_0(z) \equiv 1,\ \lambda \in C^\mu,$$

of the function $f\colon (\mathbb{C}^n, 0) \to (\mathbb{C}, 0)$, $f = f(z)$, with a singular point of multiplicity μ at zero (see the example in subsection 1.1.8).

Fix a small ball $U \subset \mathbb{C}^n$ centered at zero and choose a sufficiently small ball $\Lambda \subset \mathbb{C}^\mu$ centered at the origin in the base $\mathbb{C}^\mu$ of the deformation F such that the level set

$$V_\lambda = \{z \in U\colon F(z, \lambda) = 0\}$$

intersects transversally the boundary ∂U of U for all $\lambda \in \Lambda$. Recall that the level bifurcation set $\Sigma \subset \Lambda$ is the hypersurface consisting of the values of the parameter λ for which the level set V_λ is singular (see subsection 1.1.10).

The family V_λ, $\lambda \in \Lambda$, forms a hypersurface $V_\Lambda = \{(z, \lambda)\colon F(z, \lambda) = 0\}$ in the direct product $U \times \Lambda \subset \mathbb{C}^{n+\mu}$.

Definition. The *Whitney map* $\pi\colon V_\Lambda \to \Lambda$ is defined to be the restriction of the canonical projection $U \times \Lambda \to \Lambda$ to V_Λ.

By the implicit function theorem, the restriction of the Whitney map to the intersection of the hypersurface V_Λ with the boundary of the ball U is a locally trivial fibration $V_\Lambda \cap \partial U \to \Lambda$, which is actually trivial thanks to the contractibility of the base Λ.

The preimage V'_Λ of the complement $\Lambda' = \Lambda \setminus \Sigma$ of the level bifurcation set under the Whitney map is the union of the nonsingular fibres V_λ. The restriction of the Whitney map to V'_Λ is also a locally trivial fibration with base Λ'.

Definition. The fibration $\pi\colon V'_\Lambda \to \Lambda'$ is called the *Milnor fibration* of the singularity f.

The fibre of the Milnor fibration is homeomorphic to the nonsingular level set of f and is homotopy equivalent to a wedge of μ $(n-1)$-dimensional spheres.

Fix a nonsingular value $\lambda_* \in \Lambda'$ and the corresponding fibre $V_* = V_{\lambda_*}$. By the covering homotopy property, the Milnor fibration defines a representation of the fundamental group of the base, $\pi_1(\Lambda', \lambda_*)$, in the group of homotopy classes of maps $V_* \to V_*$ of the singular fibre V_* into itself.

The image of this homomorphism is contained in the subgroup of the homotopy classes of maps of V_* that are the identity on the boundary ∂V_*, i.e., the lifting of homotopies can be made compatible with the direct product structure $V_\Lambda \cap \partial U \cong \partial V_* \times \Lambda$ on ∂U. In this way one obtains a representation of the fundamental group $\pi_1(\Lambda', \lambda_*)$ in the homology [resp. cohomology] group $H_{n-1}(V_*)$ [resp. $H^{n-1}(V_*)$], $\Gamma\colon \pi_1(\Lambda', \lambda_*) \to \operatorname{Aut}(H_{n-1}(V_*))$ [resp. $\Gamma\colon \pi_1(\Lambda', \lambda_*) \to \operatorname{Aut}(H^{n-1}(V_*))$].

Definition. The *monodromy group* of the singularity f is the image of the representation of the fundamental group of the Milnor fibration of f in the (co)homology group of its fibre.

As the next argument shows, this definition of the monodromy group is indeed equivalent to that given in subsection 1.4.

The versal deformation $F(z, \lambda)$ has the form $F'(z, \lambda') + \lambda_0$, where $\lambda = (\lambda', \lambda_0) \in \mathbb{C}^{\mu-1} \times \mathbb{C}$. Any Morsification f_ε of the function f is equivalent, for ε small enough, to a function of the form $F'(\cdot, \lambda'(\varepsilon)) + \lambda_0(\varepsilon)$, since the one-parameter deformation f_ε of f is equivalent to one induced from the versal deformation F.

Now consider the line $L = \{\lambda\colon \lambda' = \lambda'(\varepsilon)\}$ in the base Λ of the deformation (see Fig. 26). L is in general position with respect to the hypersurface Σ and intersects the latter in μ points, whose λ_0-coordinates are precisely the critical values of the function $F'(\cdot, \lambda'(\varepsilon))$ (up to a minus sign).

The monodromy group, as defined in subsection 1.4, results from a representation of the fundamental group $\pi_1(L \setminus L \cap \Sigma)$ in homology. By a theorem of Zariski, the generators of $\pi_1(L \setminus L \cap \Sigma)$ also generate $\pi_1(\Lambda')$; this establishes the equivalence of the two definitions of the monodromy group.

Definition. A loop in [a homotopy class of] $\pi_1(\Lambda', \lambda_*)$ is said to be *simple* if it goes from the base point λ_* to a nonsingular point of the bifurcation set, then goes once in anticlockwise direction around the hypersurface Σ inside a complex line transverse to Σ, and finally returns to λ_* along the same path.

The bifurcation set Σ is an irreducible hypersurface (see subsection 1.1.10). Consequently, in the group $\pi_1(\Lambda', \lambda_*)$ any two simple loops are conjugate. It

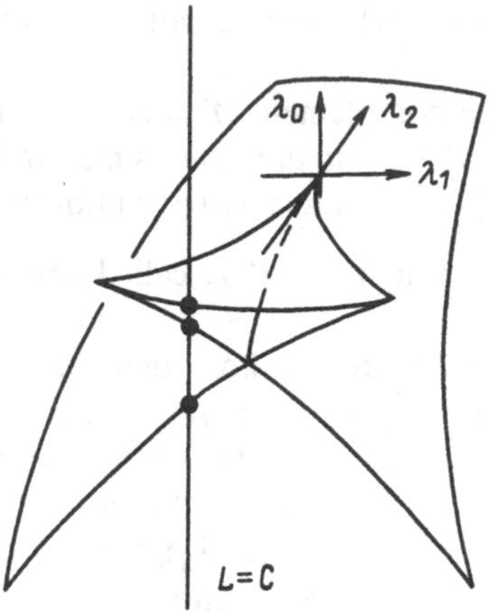

Fig. 26

follows that the monodromy group acts transitively on the set of vanishing cycles (up to orientation) and hence that one has the following:

Theorem ([133], [208]). *The Dynkin diagram of any singularity is connected.*

This result allows one to obtain several important corollaries.

Theorem ([133], [209]). *If under a one-parameter deformation of the function f its critical point decomposes into several critical points, then the corresponding critical values are not all equal.*

Theorem ([1], [172]; "irreducibility" of the classical monodromy). *Let $E \subset H_{n-1}(V_*)$ denote the linear span of some subset of a distinguished basis of vanishing cycles that is invariant under the classical-monodromy operator h_*. Then either $E = \{0\}$ or $E = H_{n-1}(V_*)$. In particular, if $h_* = \pm\mathrm{id}$, then the singularity is nondegenerate.*

1.11. The Topological Type of a Singularity Along the μ-Constant Stratum. Deformations that preserve the multiplicity of a singularity also preserve a series of invariants of the singularity.

Consider a deformation of a germ f in the class $\mu = \text{const}$, i.e., a family of holomorphic germs $F(\cdot, t)\colon (\mathbb{C}^n, 0) \to (\mathbb{C}, 0)$, depending smoothly on a real p-dimensional parameter t, which all have a singularity of the same multiplicity at zero. Let $U \subset \mathbb{C}^n$ be a sufficiently small ball centered at zero.

Theorem ([210]). *Let $n \neq 3$. Then the topological type of the pair (U, singular level set in U) and the differentiable type of the Milnor fibration, are independent of t.*

The restriction $n \neq 3$ is imposed because the proof relies on the h-cobordism theorem.

Theorem ([330]). *The topological type of a function does not change under deformations of constant multiplicity (provided $n \neq 3$).*

In other words, for any such deformation one can find neighborhoods $\tilde{U}$, U_0, and D of the origins in $\mathbb{C}^n \times \mathbb{R}^p$, $\mathbb{C}^n$, and $\mathbb{R}^p$, respectively, and a homeomorphism $\alpha\colon \tilde{U} \to U_0 \times D$, $\alpha(0, t) = (0, t)$, such that the diagram

$$\begin{array}{ccc} \tilde{U} & \xrightarrow{\alpha} & U_0 \times D \\ {\scriptstyle F}\downarrow & & \downarrow{\scriptstyle f \times \mathrm{id}} \\ \mathbb{C} & \xrightarrow{\pi} & \mathbb{C} \times D \end{array}$$

commutes, where π denotes the projection onto the first factor. Note that [259], devoted to a proof of this assertion for $n = 3$, contains errors.

One can readily see that the intersection matrix and the monodromy group of the singularity do not change along the μ = const stratum.

However, other characteristics of the singularity, of a "more analytical" nature, may change. For instance, Pham [263] showed that under a deformation of constant multiplicity the topology of the level bifurcation set of the singularity – more precisely, its partition into parts according to the singularities of the zero level set – may indeed change.

Example. Consider the μ = const deformation

$$x^3 + y^9 + t_1 xy^6 + t_2 xy^7, \qquad t_1, t_2 \in \mathbb{C},$$

of the singularity $x^3 + y^9$. Only for $t_1 = 0$ does the critical point of this family decompose in such a manner that on the zero level set the singularities E_6 and E_8 arise simultaneously.

Guseĭn-Zade and Nekhoroshev [174] showed that under μ = const deformations of a nondegenerate homogeneous polynomial of degree 22 in two variables the largest of the multiplicities of adjacent A_k-singularities changes.

In the base of the truncated miniversal deformation of a function of two variables, the μ = const stratum is a nonsingular manifold [61]. For a higher number of variables this is no longer the case [222].

Another well-known question concerning the structure of the μ = const stratum is whether the complex and the real modalities of a real singularity of finite multiplicity necessarily coincide (see [37], §17), or whether this can be asserted at least about the analogous proper modalities, i.e., the dimensions of the μ = const strata in the bases of the complex and real miniversal deformations. It is readily seen that any counterexample answering this question in the negative also provides an example of a nonsmooth μ = const stratum. Such a counterexample does indeed exist, and is based on the notion of a matroid (see [376]). Specifically, V.V. Serganova constructed a rank-3 matroid with 59 generators, realized by configurations of 59 planes in $\mathbb{R}^3$ (and in $\mathbb{C}^3$, too), such that the classes of GL-equivalent configurations in $\mathbb{C}^3$ realizing the matroid are naturally parametrized by the points of some Zariski-open subset U of the space $\{x \in \mathbf{C}^9: (x_1 - 3)^2 + (x_2 - 4)^2 = 0\}$. The set $U \cap \mathbb{R}^9$ is nonempty and corresponds to the real configurations. Moreover, if $x \in U \setminus \mathbb{R}^9$, then the union of the planes in the configuration corresponding to x is not invariant under complex conjugation, regardless of the choice of coordinates in $\mathbb{C}^3$. The desired example is provided by the function $f + f'$, where f and f' are homogeneous polynomials on $\mathbb{R}^3$ of respective degree 59 and 60, such that f vanishes on the planes realizing Serganova's matroid and f' is in general position with respect to f.

§2. Dynkin Diagrams and Monodromy Groups

In this section we describe some methods of computing intersection matrices of singularities. These methods allow us to obtain the Dynkin diagrams for the initial segment of the classification of critical points. We shall conclude the section by formulating a number of results that describe the monodromy group in terms of the integer lattice determined in the homology of the nonsingular fibre by the intersection form.

2.1. Intersection Matrices of Singularities of Functions of Two Variables. Here we present a method of Guseĭn-Zade [170], [171] and A'Campo [2], [3] that allows us to construct the Dynkin diagram of a function of two variables directly from the level-curves portrait of its real Morsification.

Definition. A holomorphic function $f: (\mathbb{C}^n, 0) \to (\mathbb{C}, 0)$ is said to be *real* if it is the complexification of the function $f|_{\mathbb{R}^n}: (\mathbb{R}^n, 0) \to (\mathbb{R}, 0)$.

Theorem. *Any singularity of a function of two variables is μ-equivalent to a singularity of a real function.*

Remark. It is not known whether this assertion is true for $n > 2$.

The intersection matrix is an invariant of the $\mu = \text{const}$ stratum (subsection 1.11). Hence, it suffices to construct the Dynkin diagram of a real function that belongs to this stratum.

Theorem. *Let f be a real function of two variables with an isolated singularity. Then there exists a real Morsification f_ε of f such that all its critical points are real and f_ε takes the value zero [resp. negative, positive values] at all saddle points [resp. points of local minima and maxima].*

We shall refer to such an f_ε as a *purely real Morsification.*

Example. The real Morsification $f_\varepsilon = (z + \varepsilon^2)(z^2 + w^2 - \varepsilon w^2)$, $\varepsilon \in (0, 4/27)$, of the real function $f = z^3 + zw^3$ with an E_7-singularity at zero is purely real (Fig. 27).

Fix a sufficiently small disk $T \subset \mathbb{R}^2$ centered at zero. Then, for sufficiently small ε, all μ critical points of the purely real Morsification f_ε lie inside T, the

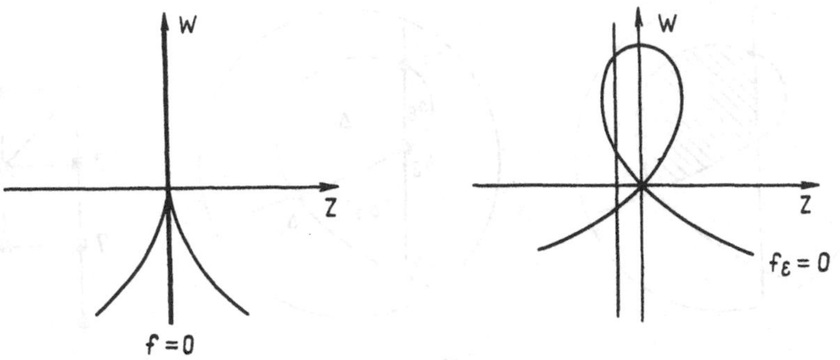

Fig. 27

curve $f_\varepsilon = 0$ is transverse to ∂T, and its only singularities are a finite number of simple double points.

The zero level set $f_\varepsilon = 0$ divides the disk T into a number of domains. There exists a one-to-one correspondence between the relatively compact (i.e., not intersecting ∂T) domains and the extremum points of f_ε. A domain containing a minimum [resp. maximum] will be termed *negative* [resp. *positive*].

Consider a set of formal generators Δ_i, $i = 1, \ldots, \mu$, corresponding to the self-intersection points of the curve $f_\varepsilon = 0$ and to the relatively compact domains. A generator corresponding to a negative [resp. positive] domain will be termed *negative* [resp. *positive*]; we call the generators corresponding to the self-intersection points *null generators*.

Label the generators according to the following rule: The index of any negative generator is less than the index of any null generator, which in turn is less than the index of any positive generator.

Now define a symmetric bilinear form on the lattice spanned by the generators Δ_i by specifying the intersection matrix as follows:

(i) for generators Δ_i, Δ_j of "identical signs", $(\Delta_i \circ \Delta_j) = -2\delta_{ij}$;

(ii) the intersection index of a non-null and a null generator is equal to the number of vertices of the respective domain that coincide with the corresponding point (it can take the values 0, 1, 2);

(iii) the intersection index of a positive and a negative generator is equal to minus the number of common edges of the respective domains.

Theorem. *The intersection matrix of any stabilization of a singularity f in $n \equiv 3 \pmod 4$ variables, calculated in some distinguished basis of vanishing cycles, coincides with the intersection matrix defined by a purely real Morsification of f.*

Example. The Dynkin diagram E_7 is constructed from the purely real Morsification given in the preceding example (see Fig. 28, in which the negative domain is shaded).

Example. The Dynkin diagram A_μ. Consider a Morsification f_ε of the function $f = z^{\mu+1}$, shown in Fig. 29, and the corresponding purely real Morsification $g_\varepsilon = f_\varepsilon + w^2$ of the function $g = z^{\mu+1} + w^2$ (Fig. 29 shows the case in which μ is even).

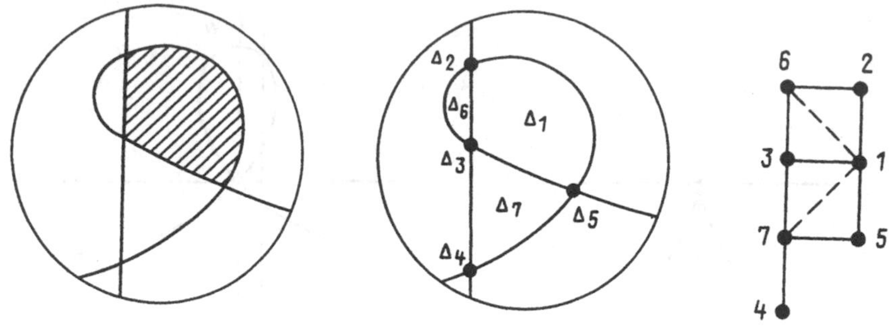

Fig. 28

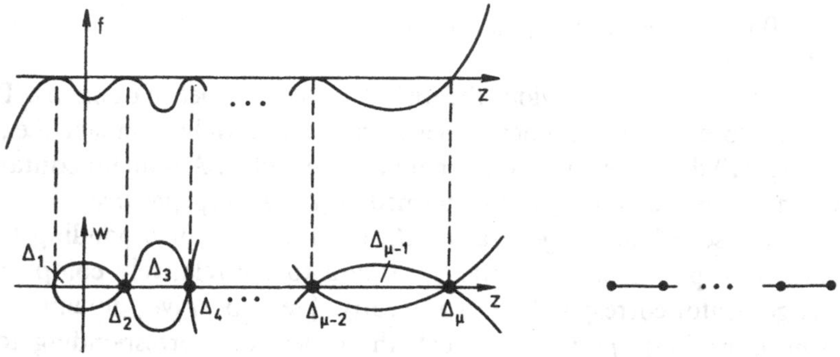

Fig. 29

The labeling of the vertices is omitted, since the diagram obtained is a tree and hence can be labeled in an arbitrary manner (see subsection 1.9).

Let γ_0 and γ_1 be curves in T that are transverse to ∂T and have as singularities only simple double self-intersection points.

Definition. A *smooth homotopy* γ_τ, $\tau \in [0, 1]$, from γ_0 to γ_1 is said to be *admissible* if for every $\tau \in (0, 1)$ the curve γ_τ is transverse to ∂T and has as singularities only double and triple self-intersection points at which the tangents to the local components of the curve are distinct.

Consider a purely real Morsification f_ε of the singularity f, and a real function g.

Theorem. *Suppose that there exists an admissible homotopy from the curve* $f_\varepsilon = 0$ *to the curve* $g = 0$. *Then* g *is a purely real Morsification of the singularity* f.

Therefore, the intersection matrix and the Dynkin diagram defined by the real function g are the intersection matrix and respectively the Dynkin diagram of the singularity f in some distinguished basis.

Example. An admissible homotopy of the zero level set of the Morsification of the singularity E_7 (Fig. 30) leads to the standard Dynkin diagram E_7 (see subsection 2.5).

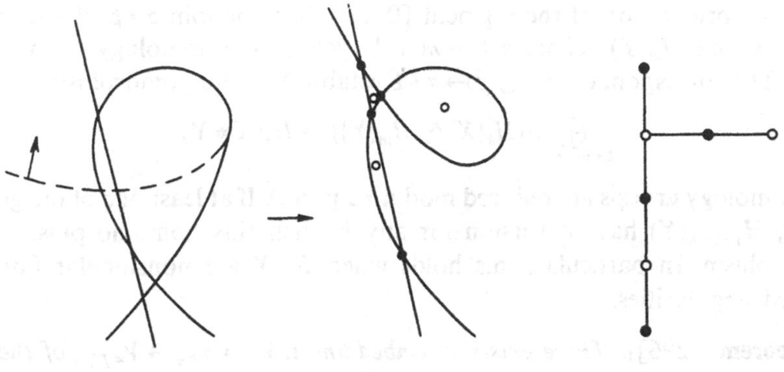

Fig. 30

2.2. The Intersection Matrix of a Direct Sum of Singularities. Let $f: (\mathbb{C}^n, 0) \to (\mathbb{C}, 0)$, $f = f(z)$, and $g: (\mathbb{C}^m, 0) \to (\mathbb{C}, 0)$, $g = g(w)$, be function-germs with isolated singularities at zero.

Definition. The *direct sum* $f \oplus g$ *of the singularities* f *and* g is defined to be the singularity of the function $f(z) + g(w): (\mathbb{C}^{n+m}, 0) \to (\mathbb{C}, 0)$.

Gabrielov [132] obtained a description of the intersection matrix of $f \oplus g$ in terms of the intersection matrices of the summands f and g.

Definition. The *join* $X * Y$ of the topological spaces X, Y is the quotient of the direct product $X \times Y \times [0, 1]$ by the equivalence relation

$$(x, y_1, 0) \sim (x, y_2, 0), \qquad (x_1, y, 1) \sim (x_2, y, 1)$$

for any $x, x_1, x_2 \in X$ and any $y, y_1, y_2 \in Y$.

Example. The join $S^1 * S^0$ of the circle and a pair of points is the two-dimensional sphere S^2 (Fig. 31).

Fig. 31

Fix an orientation of the segment [0, 1]. Then the join $\alpha * \beta$ of two cycles $\alpha \in H_k(X)$, $\beta \in H_m(Y)$ defines a $k + m + 1$ cycle in the homology of the space $X * Y$. The correspondence $\alpha \otimes \beta \mapsto \alpha * \beta$ establishes a homomorphism

$$\bigoplus_{k+m=n-1} (H_k(X) \otimes H_m(Y)) \to H_n(X * Y)$$

(the homology groups are reduced modulo a point). If at least one of the groups $H_k(X)$, $H_{n-1-k}(Y)$ has no torsion for any k, then this homomorphism is an isomorphism. In particular, this holds when X, Y are nonsingular fibres of isolated singularities.

Theorem ([296]). *There exists an embedding $i: V_{*f} * V_{*g} \to V_{*f\oplus g}$ of the join of the nonsingular fibres of f and g in the nonsingular fibre of $f \oplus g$, such that i is a homotopy equivalence.*

This embedding induces an isomorphism $i_*: H_{n-1}(V_{*f}) \otimes H_{m-1}(V_{*g}) \to H_{n+m-1}(V_{*f\oplus g})$; moreover, if $\Delta_1, \ldots, \Delta_{\mu(f)}$ and $\Delta'_1, \ldots, \Delta'_{\mu(g)}$ are distinguished bases of vanishing cycles for the singularities f and g, respectively, then the set of cycles $\Delta_{kl} = i_*(\Delta_k * \Delta'_l)$, written in lexicographical order, is a distinguished basis for the direct sum $f \oplus g$.

Corollary. $\mu(f \oplus g) = \mu(f)\mu(g)$.

Theorem ([107]).

$$\mathrm{Var}_{f\oplus g} = (-1)^{nm}\, \mathrm{Var}_f \otimes \mathrm{Var}_g.$$

This result permits us to calculate the intersection matrix of $f \oplus g$ in the basis Δ_{ij}.

Theorem ([132]). *The intersection matrix of the singularity $f \oplus g$ is expressible in terms of the intersection matrices of the summands f and g by the formulas:*

$$(\Delta_{ij_1} \circ \Delta_{ij_2}) = (\Delta'_{j_1} \circ \Delta'_{j_2}),$$

$$(\Delta_{i_1 j} \circ \Delta_{i_2 j}) = (\Delta_{i_1} \circ \Delta_{i_2}),$$

$$(\Delta_{i_1 j_1} \circ \Delta_{i_2 j_2}) = -(\Delta_{i_1} \circ \Delta_{i_2})(\Delta'_{j_1} \circ \Delta'_{j_2}) \qquad \text{for } (i_2 - i_1)(j_2 - j_1) > 0,$$

and

$$(\Delta_{i_1 j_1} \circ \Delta_{i_2 j_2}) = 0 \qquad \text{for } (i_2 - i_1)(j_2 - j_1) < 0.$$

Remark. In this theorem we considered intersection matrices of stabilizations of functions f, g, $f \oplus g$ with a number of variables $\equiv 3 \pmod 4$.

The theorem yields the following description of the Dynkin diagram of $f \oplus g$. First, its vertex set is the direct product of the vertex sets of the diagrams of f and g. Second, two vertices, $\langle i_1, j_1 \rangle$ and $\langle i_2, j_2 \rangle$, are connected by:

(i) an edge of the same multiplicity as the one connecting the vertices j_2 and j_1 in the diagram of g, if $i_1 = i_2$;

(ii) an edge of the same multiplicity as the one connecting the vertices i_1 and i_2 in the diagram of f, if $j_1 = j_2$;

(iii) an edge of multiplicity equal to minus the product of the multiplicities of the edges connecting i_1 and i_2 in the diagram of f, and j_1 and j_2 in the diagram of g, if $(i_2 - i_1)(j_2 - j_1) > 0$.

Example. The Dynkin diagram E_6 of the function $f \oplus g = z^4 + w^3$ is obtained in the indicated manner from the diagrams A_3 and A_2 of the functions $f = z^4$ and $g = w^3$, respectively (Fig. 32).

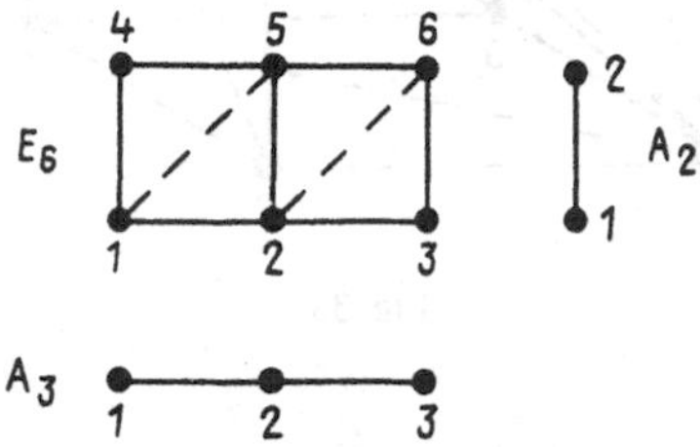

Fig. 32

Theorem ([296]). *The classical-monodromy operator of the singularity $f \oplus g$ is equal to*

$$h_{*f\oplus g} = h_{*f} \otimes h_{*g}.$$

2.3. Pham Singularities. The first calculations of intersection forms and classical-monodromy operators were carried out by Pham [262] for singularities of the form

$$f(z) = z_1^{k_1} + \cdots + z_n^{k_n}, \qquad k_i \geqslant 2.$$

Such a singularity has multiplicity $\mu(f) = \prod (k_i - 1)$. A *Pham singularity* is defined to be an arbitrary direct sum of A_{k_i-1}-singularities $f_i = z_i^{k_i}$.

It follows from the theorem of subsection 2.2 that there exists a distinguished basis of vanishing cycles Δ_i, $i = (i_1, \ldots, i_n)$, $0 \leqslant i_s \leqslant k_s - 2$, lexicographically ordered, for which

1) $(\Delta_i \circ \Delta_i) = -2$;

2) $(\Delta_i \circ \Delta_j) = (-1)^{1+|i-j|}$ for $|i - j| = \sum |i_s - j_s| > 0$, $i_s \leqslant j_s \leqslant i_s + 1$ for all s (or $j_s \leqslant i_s \leqslant j_s + 1$ for all s);

3) $(\Delta_i \circ \Delta_j) = 0$ for the remaining cases.

Example. Figure 33 shows the Dynkin diagram of the P_8-singularity $f = x^3 + y^3 + z^3$.

2.4. The Polar Curve and the Intersection Matrix. This subsection presents the results of the paper [135], which allow us to calculate the intersection matrix of a singularity $f: (\mathbb{C}^n, 0) \to (\mathbb{C}, 0)$ in terms of the intersection matrix of the restriction $f|_{w=0}$ and of the invariants of the polar curve of f relative to a linear function w on $\mathbb{C}^n$.

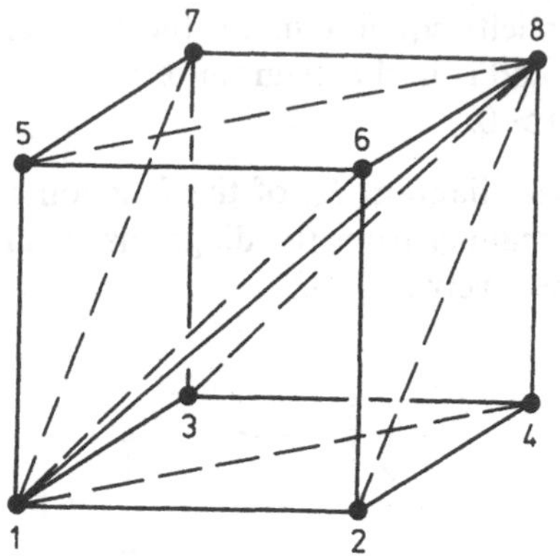

Fig. 33

Definition. The *polar curve* $\gamma_w(f)$ of a function $f: (\mathbb{C}^n, 0) \to (\mathbb{C}, 0)$ relative to a linear function $w: (\mathbb{C}^n, 0) \to (\mathbb{C}, 0)$ is the set of critical points of the map $(f, w): (\mathbb{C}^n, 0) \to (\mathbb{C}^2, 0)$.

In other words, the polar curve is the union of the critical points of the family of functions $f - \lambda w$ for different values of the parameter λ.

Example. The polar curve $\gamma_w(f)$ of the function $f = w^2 z - z^{\mu-1}$ with a D_μ-singularity at the origin is given by the equation $w^2 - (\lambda - 1)z^{\mu-2} = 0$.

Let us examine a number of properties of the polar curve; for a more detailed discussion the reader is referred to [323].

Let $\gamma_w(f) = \bigcup \gamma_i$ be the decomposition of the polar curve $\gamma_w(f)$ into irreducible components. Assume that w is a linear function in general position. Then, if f has an isolated singularity at zero, so does its restriction $f|_{w=0}: (\mathbb{C}^{n-1}, 0) \to (\mathbb{C}, 0)$. Let $\mu = \mu(f)$ and $\mu' = \mu(f|_{w=0})$ denote the corresponding multiplicities. From the fact that the singularity of $f|_{w=0}$ is isolated it follows that none of the components γ_i of the polar curve is contained in the hyperplane $\{w = 0\}$, and that $f|_{\gamma_i} \neq 0$ for all γ_i. Let $c_i w^{\beta_i}$ be the first nonzero term in the Puiseux expansion of $f|_{\gamma_i}$ in the variable w. Then $\beta_i > 1$, since the critical points of $f - \lambda w$ lying on γ_i are determined from the equation $f'_w = c_i \beta_i w^{\beta_i - 1} + \cdots = \lambda$ and tend to zero when $\lambda \to 0$.

Consider the one-parameter deformation $F_\varepsilon = f|_{w=\varepsilon}$ of the function $f|_{w=0}$. For a small generic value of ε, the function F_ε has μ' distinct critical points; these are precisely the points at which the hyperplane $\{w = \varepsilon\}$ intersects the polar curve $\gamma_w(f)$. The critical values of F_ε at points on γ_i have the asymptotics (as $\varepsilon \to 0$)

$$c_i \varepsilon^{\beta_i} + o(\varepsilon^{\beta_i}).$$

Example. The polar curve $\gamma_w(f)$, $f = w^2 z - z^{\mu-1}$, is irreducible if μ is odd and has two components if μ is even. In both cases the critical values of F_ε have the same asymptotics $c\varepsilon^\beta + o(\varepsilon^\beta)$, with $c = (\mu - 1)^{1/(2-\mu)} - (\mu - 1)^{(\mu-1)/(\mu-2)}$ and $\beta = 2(\mu - 1)/(\mu - 2)$.

Suppose β belongs to the set of exponents β_i of the polar curve $\gamma_w(f)$. In the plane $\mathbb{C}$ where f assumes its values consider an annulus $\{t \in \mathbb{C}: r_\beta < |t| < R_\beta\}$ containing all critical values of F_ε with asymptotic exponent β. For sufficiently small ε, one can choose the annuli corresponding to distinct asymptotic exponents to be disjoint.

Let $\psi(|t|)$ be a continuous nonincreasing function that is equal to $\beta - 1$ in the annulus corresponding to the exponent β. The collection of sets $T_0 = \mathbb{C} \supset T_1 \supset T_2 \supset \cdots$, where

$$T_m = \{t: \arg t + 2\pi\psi(|t|) \geqslant \pi(2m-1)\},$$

has the following property: the annulus corresponding to the exponent β is contained in T_m for $m \leqslant \beta - 1$ and does not intersect T_m for $m \geqslant \beta$.

Fix a value of ε for which the critical values of F_ε lie neither on the negative half-line $\mathbb{R} \subset \mathbb{C}$ nor on the boundaries of the sets T_m. Let φ_i, $i = 1, \ldots, \mu'$, be a distinguished system of paths in $\mathbb{C} \setminus \mathbb{R}_-$, which connect the regular value 0 with the critical values of F_ε.

Definition. The distinguished *system of paths* φ_i is said to be *admissible* if each φ_i is entirely contained in the set T_m in which its terminal point lies.

Let $\Delta_1, \ldots, \Delta_{\mu'}$ be the distinguished basis of vanishing cycles of the singularity $f|_{w=0}$ defined by an admissible system of paths $\varphi_1, \ldots, \varphi_{\mu'}$, and put $M_i = \max\{m: \varphi_i \in T_m\}$.

Theorem ([135]). *There exists a distinguished basis Δ_{mi}, $i = 1, \ldots, \mu'$, $m = 1, \ldots, M_i$, of the singularity f, with the lexicographical order, whose intersection matrix is given by the relations*

$$(\Delta_{mi} \circ \Delta_{mj}) = (\Delta_i \circ \Delta_j)$$

$$(\Delta_{mi} \circ \Delta_{ki}) = 1, \qquad \textit{if } |m-k| = 1,$$

$$(\Delta_{mi} \circ \Delta_{kj}) = -(\Delta_i \circ \Delta_j), \qquad \textit{if } |m-k| = 1 \textit{ and } (m-k)(i-j) < 0,$$

and

$$(\Delta_{mi} \circ \Delta_{kj}) = 0, \quad \textit{if } |m-k| > 1 \textit{ or } |m-k| = 1 \textit{ and } (m-k)(i-j) > 0.$$

Remark. This theorem is concerned with the intersection matrices of the stabilizations of f and $f|_{w=0}$ with a number of variables $\equiv 3 \pmod 4$.

In terms of the Dynkin diagram of the singularity f the assertion of the theorem reads as follows: The vertex set of the diagram coincides with the set of pairs (m, i), $i = 1, \ldots, \mu'$, $m = 1, \ldots, M_i$; two vertices, (m, i) and (k, j), are connected by:

(i) an edge of the same multiplicity as the one connecting the pair i, j in the diagram of $f|_{w=0}$, if $m = k$;

(ii) an edge of multiplicity 1, if $i = j$ and $|m-k| = 1$;

(iii) an edge of minus the same multiplicity as the one connecting the pair i, j in the diagram of $f|_{w=0}$, if $|m-k| = 1$ and $(m-k)(i-j) < 0$.

Example. For the D_μ-singularity $f = w^2 z - z^{\mu-1}$ there is only one value of the exponent β, and we accordingly put $\psi(|t|) \equiv \beta - 1$. We then partition the range $\mathbb{C}$ of f into the $\mu - 2$ sectors

$$S_k = \left\{t\colon -\pi + \frac{2\pi(k-1)}{\mu-2} \leqslant \arg t \leqslant -\pi + \frac{2\pi k}{\mu-2}\right\}.$$

Then $T_0 = T_1 = \mathbb{C}$, $T_2 = S_{\mu-2} \cup S_{\mu-3}$, and $T_3 = \varnothing$. For small generic values of ε there is exactly one critical value of $F_\varepsilon = \varepsilon^2 z - z^{\mu-1}$ in each sector S_k (Fig. 34).

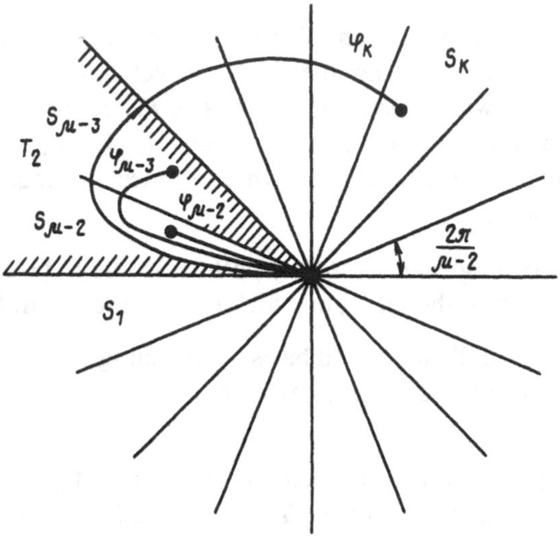

Fig. 34

One can show that the admissible system of paths shown in Fig. 34 defines the standard Dynkin diagram of the $A_{\mu-2}$-singularity $f|_{w=0} = -z^{\mu-1}$:

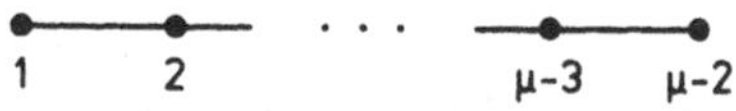

According to the preceding theorem, there exists a distinguished basis $\Delta_{1,1}$, $\Delta_{1,\mu-2}, \Delta_{2,\mu-3}, \Delta_{2,\mu-2}$ of the singularity D_μ, whose Dynkin diagram has the form

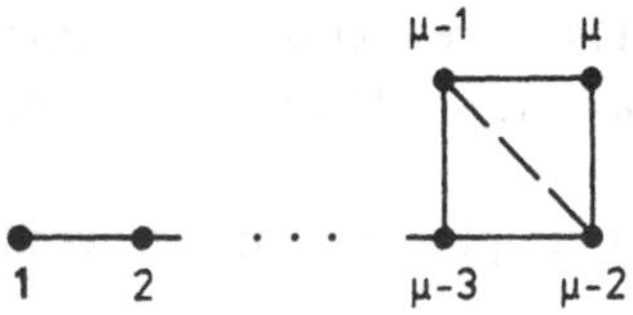

The operation $t_{\mu-2}$ defined in subsection 1.9 then yields the standard Dynkin diagram of the singularity D_μ:

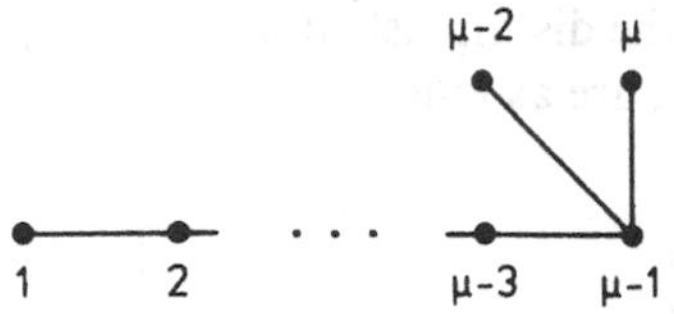

Gabrielov's theorem allows us to compute the intersection matrices for all the singularities in the list given in subsection 1.2.3, except the series **V**, in particular, for all bimodal singularities. The results of the computations are shown in the following table, in which **X** denotes the type of the singularity, f is the normal form (w is the singled-out variable), and $M_1, \ldots, M_{\mu'}$ are the upper bounds of the values of the parameter m in the distinguished basis Δ_{mj}.

X	f	$M_1, \ldots, M_{\mu'}$
$J_{k,i}$	$z^3 + z^2w^k + w^{3k+i}$	$3k+i-1, 3k-1$
E_{6k}	$z^3 + w^{3k+1}$	$3k, 3k$
E_{6k+1}	$z^3 + zw^{2k+1}$	$3k+1, 3k$
E_{6k+2}	$z^3 + w^{3k+2}$	$3k+1, 3k+1$
$X_{k,p}$	$z^4 - z^2w^{2k} + w^{4k+p}$	$4k-1, 4k-1, 4k+p-1$
$Y^k_{r,s}$	$z^2(z-w^k)^2 + z^2w^{2k+s} + (z-w^k)^2w^{2k+r}$	$4k+r-1, 4k+s-1, 4k-1$
$Z^k_{i,p}$	$z^4 - z^3w^k + z^2w^{2k+i} - w^{4k+3i+p}$	$4k-1, 4k+3i-p-1, 4k+3i-1$
$Z^k_{12k+6i-1}$	$z^4 - z^3w^k + w^{4k+3i+1}$	$4k+3i, 4k-1, 4k+3i$
Z^k_{12k+6i}	$z^4 - z^3w^k + zw^{3k+2i+1}$	$4k+3i+1, 4k-1, 4k+3i$
$Z^k_{12k+6i+1}$	$z^4 - z^3w^k + w^{4k+3i+2}$	$4k+3i+1, 4k-1, 4k+3i+1$
W_{12k}	$z^4 + w^{4k+1}$	$4k, 4k, 4k$
W_{12k+1}	$z^4 - zw^{3k+1}$	$4k+1, 4k, 4k$
$W_{k,i}$	$z^4 - z^2w^{2k+1} + w^{4k+2+i}$	$4k+1, 4k+1, 4k+i+1$
$W^\#_{k,2q-1}$	$(z^2 - w^{2k+1})^2 + zw^{3k+q+1}$	$4k+q+1, 4k+q, 4k+1$
$W^\#_{k,2q}$	$(z^2 - w^{2k+1})^2 + z^2w^{2k+q+1}$	$4k+q+1, 4k+q+1, 4k+1$
W_{12k+5}	$z^4 - zw^{3k+2}$	$4k+2, 4k+1, 4k+2$
W_{12k+6}	$z^4 + w^{4k+3}$	$4k+2, 4k+2, 4k+2$
$Q_{k,i}$	$z^3 + (w-z)u^2 + z^2w^k + w^{3k+i}$	$2, 2, 3k-1, 3k+i-1$
Q_{6k+4}	$z^3 + (w-z)u^2 + w^{3k+1}$	$2, 2, 3k, 3k$
Q_{6k+5}	$z^3 + (w-z)u^2 - zw^{2k+1}$	$2, 2, 3k+1, 3k$
Q_{6k+6}	$z^3 + (w-z)u^2 + w^{3k+2}$	$2, 2, 3k+1, 3k+1$
S_{12k-1}	$z^2u + (w-u)u^2 + w^{4k}$	$2, 4k-1, 4k-1, 4k-1$
S_{12k}	$z^2u + (w-u)u^2 + zw^{3k}$	$2, 4k, 4k-1, 4k-1$
$S_{k,i}$	$z^2u + (w-u)u^2 + z^2w^{2k} + w^{4k+i+1}$	$2, 4k, 4k, 4k+i$
$S_{k,2q-1}$	$z^2u + (w-u)u^2 - uw^{2k+1} + zw^{3k+q}$	$2, 4k+q, 4k+q-1, 4k$
$S_{k,2q}$	$z^2u + (w-u)u^2 - uw^{2k+1} + z^2w^{2k+q}$	$2, 4k+q, 4k+q, 4k$
S_{12k+4}	$z^2u + (w-u)u^2 + zw^{3k}$	$2, 4k+1, 4k, 4k+1$
S_{12k+5}	$z^2u + (w-u)u^2 + w^{4k+2}$	$2, 4k+1, 4k+1, 4k+1$
$T_{p,q,r}$	$zu(w-z-u) + z^p + u^q + (w-z-u)^r$	$p-1, q-1, r-1, 2$
U_{12k}	$z^3 + zu^2 + w^{3k+1}$	$3k, 3k, 3k, 3k$
$U_{k,2q-1}$	$z^3 + zu^2 - zw^{2k+1} + u^2w^{k+q}$	$3k+q, 3k+q, 3k, 3k+1$
$U_{k,2q(q>0)}$	$z^3 + zu^2 - zw^{2k+1} + uw^{2k+q+1}$	$3k+q+1, 3k+q, 3k, 3k+1$
U_{12k+4}	$z^3 + zu^2 + w^{3k+2}$	$3k+1, 3k+1, 3k+1, 3k+1$

The Dynkin diagrams of the distinguished basis of $f|_{w=0}$ defined by an admissible system of paths for T_m, are as follows:

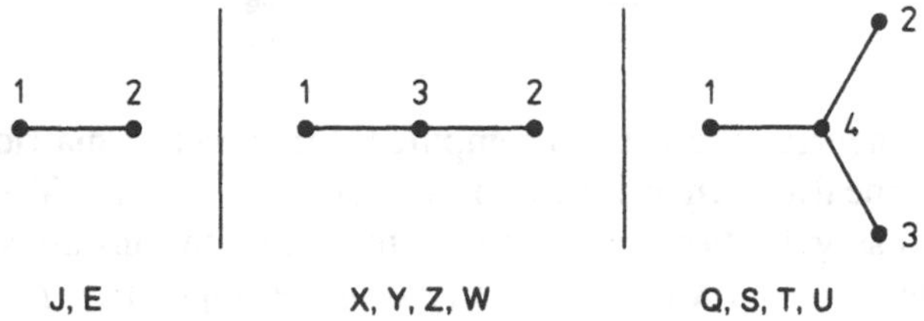

2.5. Modality and Quadratic Forms of Singularities. The methods of subsections 3.1 through 3.4 allow us to obtain the intersection forms of the initial segment of the classification of singularities. We give below the answer for functions of modality $m \leqslant 1$. As it turns out, the inertia index of the quadratic form of a singularity is related to its modality.

Definition. A function-germ with an isolated critical point is said to be *elliptic* if the quadratic form of its singularity (subsection 1.8) is negative definite.

Theorem ([336], [14]). *The elliptic singularities are precisely the simple singularities A_μ with $\mu \geqslant 1$, D_μ with $\mu \geqslant 4$, E_6, E_7, and E_8. The corresponding quadratic forms are described by the following Dynkin diagrams (the labeling of vertices is arbitrary – see subsection 1.9):*

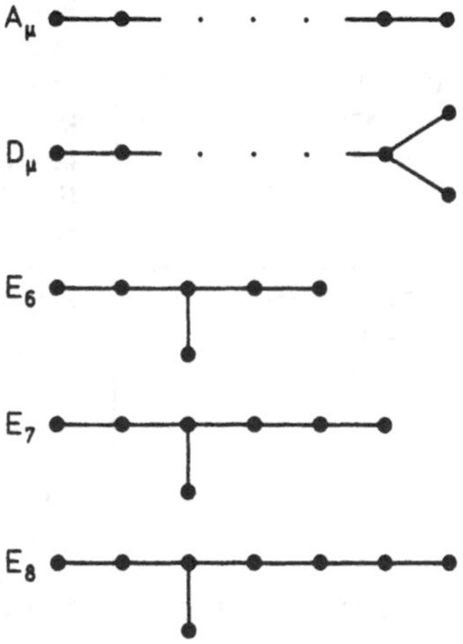

Definition. A function-germ with an isolated critical point is said to be *parabolic* if the corresponding quadratic form is semidefinite.

Theorem ([13]). *The list of parabolic singularities consists of three one-parameter families:*

$$T_{3,3,3} = P_8\colon x^3 + y^3 + z^3 + axyz,\ a^3 \neq -27,$$

$$T_{2,4,4} = X_9\colon x^2 + y^4 + z^4 + ay^2z^2,\ a^2 \neq 4,$$

$$T_{2,3,6} = J_{10}\colon x^2 + y^3 + z^6 + ay^2z^2,\ 4a^3 + 27 \neq 0.$$

The parabolic singularities are uniquely characterized by the following property: In a small enough neighborhood of a sufficient jet of the corresponding function there exist only finitely many orbits whose dimension is larger than the dimension of the orbit of that jet, and infinitely many orbits of exactly the same dimension.

Definition. A function-germ with an isolated critical point is said to be *hyperbolic* if the positive inertia index of the corresponding quadratic form is equal to one.

Theorem ([38]). *There is a three-index series of hyperbolic singularities:*

$$T_{p,q,r}\colon x^p + y^q + z^r + axyz, \quad a \neq 0,\ p^{-1} + q^{-1} + r^{-1} < 1,$$

with $\mu(T_{p,q,r}) = p + q + r - 1$.

It follows from the theorems given above that any unimodal singularity is either parabolic, or hyperbolic, or one of the 14 exceptional unimodal singularities (see subsection 1.2.3).

The proofs of these theorems available in the literature rely on the classification of singularities according to their modality, the list of adjacencies, and the following semicontinuity property of the inertia index of a quadratic form.

Suppose the singularity f of type X is adjacent to the singularity g of type Y: $Y \to X$. With this adjacency is associated an embedding $V_{*f} \to V_{*g}$ of non-singular level manifolds, and hence an embedding of the corresponding integral homology lattices,

$$L_f = H_{n-1}(V_{*f}) \subset L_g = H_{n-1}(V_{*g}).$$

The quadratic form of the singularity X is the restriction of the quadratic form of the singularity Y to the sublattice L_f. Consequently, the positive (or negative) inertia index of the quadratic form of X does not exceed the corresponding inertial index for Y.

The quadratic forms of the unimodal singularities are described in [132], [134]. Let $\tilde{T}_{p,q,r}$ denote the quadratic form with the Dynkin diagram shown in Fig. 35.

Theorem. *The singularity* $T_{p,q,r}$, $p^{-1} + q^{-1} + r^{-1} \leqslant 1$, *has a degenerate quadratic form that can be written as the direct sum of a null quadratic form and the form* $\tilde{T}_{p,q,r}$. *The form* $\tilde{T}_{p,q,r}$ *is nondegenerate and its positive inertia index is equal to one. The null form has rank two for parabolic singularities and rank one for hyperbolic singularities.*

Remark. For parabolic singularities the form $\tilde{T}_{p,q,r}$, $p^{-1} + q^{-1} + r^{-1} = 1$, is one of the forms E_6, E_7, E_8.

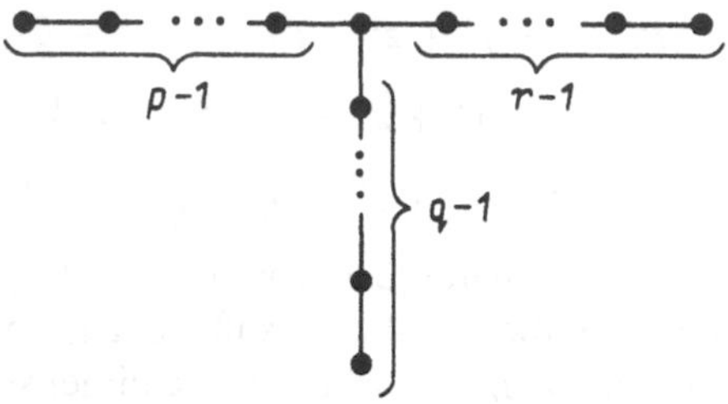

Fig. 35

Theorem. *Each of the 14 exceptional unimodal singularities has a nondegenerate quadratic form with positive inertia index two, which can be written as the direct sum of a form $\tilde{T}_{p,q,r}$ and a hyperbolic form U of rank two.*

The triplets (p, q, r) corresponding to these singularities are shown in the following table.

E_{12}	2, 3, 7	W_{12}	2, 5, 5
E_{13}	2, 3, 8	W_{13}	2, 5, 6
E_{14}	2, 3, 9	Q_{10}	3, 3, 4
Z_{11}	2, 4, 5	Q_{11}	3, 3, 5
Z_{12}	2, 4, 6	Q_{12}	3, 3, 6
Z_{13}	2, 4, 7	S_{11}	3, 4, 4
U_{12}	4, 4, 4	S_{12}	3, 4, 5

A comparison of the above table with the table given in subsection 1.2.6 reveals that the set of 14 exceptional unimodal singularities admits an involution that keeps the 6 singularities with $\mu = 12$ fixed: this involution maps each singularity, with a triplet of numbers defined by its quadratic form, into the singularity with the same triplet defined by the angles of a triangle in the Lobachevskiĭ plane, and vice versa (see subsection 1.2.6).

The Dynkin diagrams of unimodal singularities in some distinguished base were obtained in [116], [118] (Fig. 36).

The work [118] also gives all Dynkin diagrams of bimodal singularities.

2.6. The Monodromy Group and the Intersection Form

Definition. The *Milnor lattice L* of a singularity f of multiplicity μ is the μ-dimensional integer lattice defined in the homology group of the nonsingular fibre, $H_{n-1}(V_*) \simeq \mathbf{Z}^{\mu}$, and supplied with the intersection form.

The intersection form $\Phi(\cdot, \cdot) = (\cdot \circ \cdot)$ is symmetric [resp. skew-symmetric] for n odd [resp. even]. Let $O(L)$ denote the orthogonal [resp. symplectic] group of all isometries of the Milnor lattice L. The monodromy group Γ preserves the

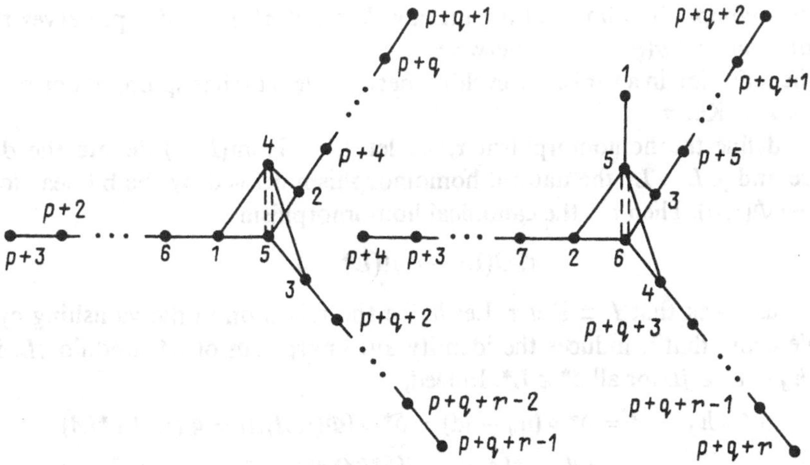

Fig. 36

intersection form, and hence is a subgroup of $O(L)$. Let us give a number of results that describe the monodromy group in terms of the Milnor lattice.

We begin with the symmetric case $n \equiv 3 \pmod 4$. Here the monodromy group is generated by the reflections h_Δ: $h_\Delta(\delta) = \delta + \Phi(\delta, \Delta)\Delta$, in the vanishing cycles, and the lattice L is even: $\Phi(\delta, \delta) \in 2\mathbb{Z}$ (for vanishing cycles $\Phi(\Delta, \Delta) = -2$).

Theorem ([288]). *There exists a unique even quadratic form on $\mathbb{Z}^\mu = H_{n-1}(V_*)$, invariant under the monodromy group Γ, such that its range contains -2.*

This form can be naturally identified with the quadratic form of the singularity. Thus, the monodromy group Γ determines uniquely the Milnor lattice L.

A converse of this result holds for "almost" all singularities: the monodromy group Γ can be described in terms of the lattice L.

Let us define the group $O^*(L) \subset O(L)$ as the intersection

$$O^*(L) = \operatorname{Ker} \sigma \cap \operatorname{Ker} \tau$$

of the kernels of the two natural homomorphisms of the group $O(L)$ that are defined as follows.

The homomorphism σ: $O(L) \to \{\pm 1\}$ is, in the case of a nondegenerate form Φ, the real spinorial norm. More precisely, each element $g \in O(L)$ can be written as a product of reflections $s_{\delta_i} = s_i$ in the realification $L_{\mathbb{R}} = L \otimes \mathbb{R}$ of the integer lattice L: $g = s_1 \cdot s_2 \cdot \ldots \cdot s_n$. We put $\sigma(g) = 1$ if the number of vectors δ_i of positive length is even, and $\sigma(g) = -1$ otherwise. For a degenerate lattice we define σ as the composition of the canonical homomorphism $L \to L/\operatorname{Ker} L$ and the spinorial norm on $L/\operatorname{Ker} L$.

Remark. The invariant definition of the spinorial norm goes as follows: All subspaces of $L_{\mathbb{R}}$ of maximal dimension on which the form Φ is positive definite admit a compatible choice of orientation. We put $\sigma(g) = 1$ if g preserves that orientation, and $\sigma(g) = -1$ otherwise.

The reflection in a vanishing cycle (of negative length) has spinorial norm one; hence, $\Gamma \subset \operatorname{Ker} \sigma$.

To define the homomorphism τ, we let $L^* = \operatorname{Hom}(L, \mathbb{Z})$ denote the dual lattice and $j: L \to L^*$ the natural homomorphism defined by the bilinear form $\Phi: \delta \mapsto \Phi(\cdot, \delta)$. Then τ is the canonical homomorphism

$$\tau: O(L) \to \operatorname{Aut}(L^*/jL).$$

Let us check that $\Gamma \subset \operatorname{Ker} \tau$. Let h_Δ be the reflection in the vanishing cycle Δ. We claim that it induces the identity automorphism of L^* modulo jL, i.e., $\delta^* \circ h_\Delta - \delta^* \in jL$ for all $\delta^* \in L^*$. Indeed,

$$\begin{aligned}\delta^* \circ h_\Delta - \delta^* &= \delta^* \circ (h_\Delta - \mathrm{id}) = \delta^* \circ (\Phi(\cdot, \Delta)\Delta) = \Phi(\cdot, \Delta)\delta^*(\Delta) \\ &= \Phi(\cdot, \delta^*(\Delta)\Delta) = j(\delta^*(\Delta)\Delta).\end{aligned}$$

Thus, the monodromy group Γ is a subgroup of $O^*(L) = \operatorname{Ker} \sigma \cap \operatorname{Ker} \tau$.

Remark. In the case of a nondegenerate lattice L, $O^*(L)$ is a subgroup of finite index in $O(L)$.

Theorem ([120]). *Let Γ be the monodromy group of an isolated singularity of a function f in an odd number of variables. Suppose that the singularity of f either is nonhyperbolic, or is hyperbolic of one of the types $T_{2,3,7}$, $T_{2,4,5}$, $T_{3,3,4}$. Then $\Gamma = O^*(L)$.*

This assertion is false for hyperbolic singularities of types different from those indicated. Let $\Gamma_s \subset O(L/\operatorname{Ker} L)$ denote the image of the monodromy group under its natural action on $L/\operatorname{Ker} L$. Then, for such hyperbolic singularities Γ_s is not a subgroup of finite index in $O^*(L/\operatorname{Ker} L)$.

The assertion of the theorem was first established for the exceptional unimodal singularities in [268] (see also [116], [117], [119], [199]).

The monodromy groups for which the assertion of the theorem is valid are completely determined by the integer lattice L. The set of vanishing cycles of such a singularity can also be described in terms of L.

Theorem (Looijenga, see [63]). *Suppose that for the isolated singularity f the monodromy group $\Gamma = O^*(L)$. Then $\Delta \in L$ is a vanishing cycle if and only if*

$$\Phi(\Delta, \Delta) = -2 \quad \text{and} \quad \Phi(\Delta, L) = \mathbb{Z}.$$

The situation is different for the hyperbolic singularities $T_{p,q,r}$. Although among these singularities there are "doubles" with isomorphic Milnor lattices (for example, $T_{2,7,7}$ and $T_{3,3,10}$), one has the following result.

Theorem ([64]). *The monodromy groups of distinct singularities $T_{p,q,r}$ are not isomorphic.*

Isomorphic quadratic forms also occur for certain pairs of bimodal singularities (Z_{17} and Q_{17}; E_{18} and Q_{18}; E_{19} and Z_{19}; Z_{18} and $J_{3,2}$; S_{17} and $W_{1,2}$; see

[116]). Thus, only the simple and unimodal nonequivalent singularities are distinguished from one another by their monodromy groups (see also [64]).

2.7. The Monodromy Group in the Skew-Symmetric Case. In the case of an even number of variables the intersection form is skew-symmetric and the monodromy group Γ is a subgroup of the symplectic group $O(L)$ of the isometries of the Milnor lattice L (see subsection 3.6).

A "skew-symmetric" monodromy group (unlike a "symmetric" one) is not uniquely determined by the Milnor lattice: for instance, the singularities A_6 and E_6, with isomorphic lattices, have nonisomorphic monodromy groups $\Gamma \subset O(L)$ (see [84]).

In the case in which the number of variables is even, the monodromy group can be described in terms of the Milnor lattice L and a finite $\mathbb{Z}_k$-module, defined by the lattice L and the set of vanishing cycles in L.

The group $O(L)$ acts on the finite lattice $L_{\mathbb{Z}_k} = L/kL$.

Definition. The *$\mathbb{Z}_k$-monodromy group*, denoted by $\Gamma_{\mathbb{Z}_k}$, is the image of the monodromy group Γ under the canonical homomorphism $i_k \colon O(L) \to O(L_{\mathbb{Z}_k})$.

The lattice embedding $\operatorname{Ker} L \to L$ induces the restriction homomorphism $j \colon O(L) \to O(\operatorname{Ker} L)$.

Theorem (Varchenko, Chmutov, A'Campo, Wajnryb [37], [5], [367]). *For simple singularities in an even number of variables the monodromy group $\Gamma \subset O(L)$ is uniquely determined by the conditions*

(1) $\Gamma \subset \operatorname{Ker} j$

and

(2) $i_2(\Gamma) = \Gamma_{\mathbb{Z}_2}$.

This result admits the following generalization for arbitrary singularities. By a theorem of Frobenius (see [232]), the lattice L has a basis in which the intersection form is given by a $\mu \times \mu$ matrix of the type

$$\left[\begin{array}{cc|c} 0 & A & \\ -A & 0 & 0 \\ \hline & 0 & 0 \end{array}\right],$$

where A is a diagonal matrix with eigenvalues $\lambda = \lambda_1 > \lambda_2 > \cdots > \lambda_n$ such that $\lambda_i \lambda_{i-1}^{-1} \in \mathbb{Z}$. The set of numbers $\lambda_1, \ldots, \lambda_n$ is uniquely determined.

Theorem ([82], [83]). *The monodromy group $\Gamma \subset O(L)$ of an isolated singularity in an even number of variables is uniquely determined by the conditions*

(1) $\Gamma \subset \operatorname{Ker} j$

and

(2) $i_{2\lambda}(\Gamma) = \Gamma_{\mathbb{Z}_{2\lambda}}$.

Remark. For singularities in two variables, $\lambda = 1$ (see [83], [367]).

The above theorems reduce the task of describing a skew-symmetric monodromy group to that of describing the finite monodromy group $\Gamma_{Z_{2\lambda}}$. Such a description has been obtained in [84] in terms of the set of vanishing cycles in L. The proofs of the theorems use the construction of bases of vanishing cycles with certain special properties. An algebraic proof of the indicated results can be found in [185], [186].

§ 3. Complex Monodromy and Period Maps

With an arbitrary locally trivial fibration are associated vector bundles of (co)homologies of its fibres. In the (co)homology bundle there is a canonically defined connection, the Gauss-Manin connection. In the case of the Milnor fibration the corresponding cohomology bundle with complex coefficients is naturally endowed with a structure of holomorphic bundle. The sections of the Milnor cohomology bundle are given by holomorphic forms, and the integrals of holomorphic forms over cycles that depend continuously on the parameters of the base of the fibration define period maps.

The analysis to which one is led in this way – in particular, the investigation of the asymptotics of period maps in the neighborhood of a critical point – allows us to obtain information on various invariants connected with the critical point. A description of the relevant notions and results is given in this section and the next.

3.1. The Cohomology Bundle and the Gauss-Manin Connection. Let $\pi: E \to B$ be an arbitrary locally trivial fibration with smooth base. For each $k \geqslant 0$ we define the complex vector bundle of k-dimensional cohomologies, with base B. Its fibre over a point $b \in B$ is the k-dimensional complex cohomology space $H^k(E_b; \mathbb{C})$ of the fibre E_b. The total space of the bundle, $\mathscr{H}_\pi^k = \bigcup_{b \in B} H^k(E_b; \mathbb{C})$, is the union of the cohomology spaces of the fibres, with the natural projection $\pi^*: \mathscr{H}_\pi^k \to B$.

For an arbitrary contractible open subset U of the base, the preimage $\pi^{-1}(U)$ is homeomorphic to the direct product of U and the fibre of the original fibration. The embedding $E_b \hookrightarrow \pi^{-1}(U)$ induces an isomorphism in cohomology $H^k(E_b; \mathbb{C}) \simeq H^k(\pi^{-1}(U); \mathbb{C})$. Such isomorphisms define trivializations of the projection π^* over contractible open subsets of the base. The local trivializations constructed in this manner provide $\pi^*: \mathscr{H}_\pi^k \to B$ with a structure of complex vector bundle.

Definition. The bundle $\pi^*: \mathscr{H}_\pi^k \to B$ is called the *k-dimensional cohomology bundle* associated with the fibration π.

The natural embeddings of the integer cohomology of a fibre of π (modulo torsion) in the cohomology of the fibre with real coefficients, and of the real cohomology of the fibre in the complex one, define in each fibre of the cohomology bundle π^* a real subspace together with an integer lattice in it. Thus, in the cohomology bundle subbundles of real and integer k-dimensional cohomologies arise canonically.

The cohomology bundle is not only locally trivial, but also locally trivialized. The transition functions of the local trivializations exhibited above are locally constant: the integer cohomology lattice present in each fibre is canonically transported to neighboring fibres. Such local trivializations define in each cohomology bundle an integrable connection ∇ that depends continuously on the point of the base; the integer cocycles are the horizontal sections of this connection.

Definition. The connection ∇ defined canonically in the cohomology bundle is called the *Gauss-Manin connection* (see [230], [231], [264]).

The existence of the connection ∇ means that in the cohomology bundle a parallel transport of fibres over curves in the base is defined for which the map of the fibre over the initial point into the fibre over the terminal point is a linear isomorphism of the fibres.

The integrability of the connection ∇ is equivalent to the following property: The parallel transport mapping of fibres depends only on the homotopy class of the paths that connect the corresponding points in the base.

3.2. Sections of the Cohomology Bundle

Definition. A *section* of the cohomology bundle over an open simply connected subset U of the base is said to be *covariantly constant* (or *horizontal*) if its values are invariant under parallel transport along paths in U.

Giving the value of a horizontal section over one point of the base determines the section uniquely over any simply connected subset of the base containing that point. In the case where the base is not simply connected, the continuation of a horizontal section along paths belonging to different homotopy classes may lead to different values of the section over the same point of the base. In that case we shall speak about a multi-valued covariantly constant section of the cohomology bundle.

Remark. Let $\tilde{B}$ be the universal covering of the base B. The canonical map $\tilde{B} \to B$ induces a bundle $\tilde{\mathcal{H}}_\pi^k \to \tilde{B}$ from the cohomology bundle over B. The connection ∇ in the cohomology bundle, in its turn, induces a connection in the bundle over $\tilde{B}$. Then the multi-valued covariantly constant sections of the cohomology bundle are precisely the images of the horizontal sections of the induced bundle $\tilde{\mathcal{H}}_\pi^k$.

The real and integral subbundles of the cohomology bundle are invariant under parallel transport with respect to the connection ∇ defined above.

Now suppose that the base of the original fibration is a complex manifold.

Definition. A *section* of the cohomology bundle is said to be *holomorphic* if its coordinates in an arbitrary frame of locally covariantly constant sections are holomorphic functions on the base.

This definition is correct, as the transition functions between covariantly constant frames are locally constant. In this way, the cohomology bundle is canonically endowed with a structure of complex vector bundle compatible with the connection ∇.

One analogously defines the *k-dimensional homology bundle* $\pi_*: \mathscr{H}_{k\pi} \to B$ and one extends to it all notions defined above.

The k-dimensional cohomology and homology bundles associated with a fibration π are dual to one another. Let s and δ be respective sections of these bundles over some subset U of the base. The pairing $s|\delta$ of these sections, defined by the fibrewise pairing of the cohomology and homology groups, is a complex-valued function on U. This duality is compatible with both the holomorphic structure and the connection present in each of the two bundles.

The parallel transport in the (co)homology bundle along a closed path starting at some selected point of the base defines a linear automorphism of the fibre over that point. Thus, the connection ∇ in the (co)homology bundle defines a homomorphism of the fundamental group of the base in the group of automorphisms of the fibre.

Definition. The image of the homomorphism constructed above is called the *complex-monodromy group.*

3.3. The Vanishing Cohomology Bundle. Let $f: (\mathbb{C}^n, 0) \to (\mathbb{C}, 0)$ be a germ with an isolated critical point of multiplicity μ at zero. With the singularity f are associated two fibrations: the fibration $V'_T \to T'$ of nonsingular fibres of f over a punctured disk T' and the Milnor fibration $V'_\Lambda \to \Lambda'$ over the complement to the level bifurcation set in the base of the miniversal deformation $F(z, \lambda)$ of the function $f(z)$ (see subsections 1.1 and 1.10). The reduced (co)homology groups of the fibres of these fibrations are nontrivial only in dimension $n - 1$.

Let $\mathscr{H}_f^* \to T'$ and $\mathscr{H}_F^* \to \Lambda'$ denote the bundles of $(n - 1)$-dimensional complex cohomologies associated with these fibrations.

Definition. The holomorphic bundles $\mathscr{H}_f^* \to T'$ and $\mathscr{H}_F^* \to \Lambda'$ are called the *vanishing cohomology bundles* of the singularity f.

The *vanishing homology bundles* $\mathscr{H}_{*f} \to T'$ and $\mathscr{H}_{*F} \to \Lambda'$ are defined in a similar manner.

Remark. By identifying the range of the function f with the λ_0-axis in the base Λ of the miniversal deformation via the mapping $t \mapsto (-\lambda_0, 0)$, we obtain natural embeddings

$$\begin{array}{ccc} V'_T & \hookrightarrow & V'_\Lambda \\ \downarrow & & \downarrow \\ T' & \hookrightarrow & \Lambda' \end{array}$$

Thus, the fibration of nonsingular fibres of f over the punctured disk T' is induced from the Milnor fibration. Correspondingly, the vanishing (co)homology bundles over T' are induced from the respective bundles over Λ'. This allows us to automatically carry over assertions about the (co)homology bundles over Λ' to the (co)homology bundles over T'.

3.4. The Period Map. The nonsingular fibre of the Milnor fibration $V'_\Lambda \to \Lambda'$ is a Stein manifold [169]. Hence, its cohomology can be computed using holomorphic differential forms [127], [298]. This allows us to give an analytic description for the Gauss-Manin connection in the vanishing cohomology bundles (see subsection 3.7).

Let ω be a holomorphic $(n-1)$-form defined in a neighborhood of zero in $\mathbb{C}^n \times \mathbb{C}^\mu$. The restriction ω_λ of ω to the nonsingular fibre V_λ is a closed form, since on the $(n-1)$-dimensional complex manifold V_λ there are no nonzero holomorphic n-forms. Consequently, ω_λ defines a cohomology class $[\omega]_\lambda \in H^{n-1}(V_\lambda; \mathbb{C})$ for each $\lambda \in \Lambda'$, i.e., a section of the vanishing cohomology bundle $\mathscr{H}_F^* \to \Lambda'$.

Definition. The global section $[\omega]$ of the vanishing cohomology bundle is called the *period map* of the form ω.

Theorem ([58], [229]). *The period map of a holomorphic form is a holomorphic section of the cohomology bundle.*

The theorem is equivalent to the assertion that the integral $\int_{\delta_\lambda} \omega_\lambda$ depends holomorphically on the parameter $\lambda \in \Lambda'$, where δ_λ is an integer cycle in the fibre V_λ of the Milnor fibration that depends continuously on λ (a covariantly constant section of the vanishing homology bundle).

Similarly, a holomorphic $(n-1)$-form ω on $\mathbb{C}^n$ yields a holomorphic period map of the bundle $\mathscr{H}_f^* \to T'$.

Example. Let $f = z_1^2 + \cdots + z_n^2$, a Morse singularity. The fibres of the fibration $V'_T \to T'$ are homotopy equivalent to the "vanishing" sphere $S_t = \sqrt{t}S^{n-1}$ of radius $r(t) = |t|^{1/2}$, where

$$S^{n-1} = \{z: z_1^2 + \cdots + z_n^2 = 1, \operatorname{Im} z_i = 0, i = 1, \ldots, n\}$$

(see subsection 1.3). The integral over S_t of a holomorphic form ω in general position is equal to the integral of $d\omega$ over the ball of radius $r(=r(t))$ and admits an expansion $c_n r^n + \cdots$, in which the first term is proportional to the volume of the ball and the value of $d\omega$ at zero. In the "coordinates" S_t the period map has the asymptotics

$$[\omega]_t | S_t = \int_{S_t} \omega = ct^{n/2} + \cdots$$

provided $d\omega|_0 \neq 0$.

3.5. The Residue Form. Let ω be a holomorphic n-form on $\mathbb{C}^n$ and let V_t be the nonsingular level manifold of the function $f: (\mathbb{C}^n, 0) \to (\mathbb{C}, 0)$.

Theorem ([213]). *There exists a holomorphic* $(n-1)$*-form* ψ*, defined in a neighborhood of* V_t*, such that* $\omega = df \wedge \psi$*. The restriction of* ψ *to* V_t *is a uniquely determined* $(n-1)$*-form.*

Definition. The restriction of the form ψ to the nonsingular level manifold V_t is called the *residue form* of ω and is denoted by ω/df_t.

The residue form is closed on the level manifold and defines a cohomology class $[\omega/df]_t \in H^{n-1}(V_t; \mathbb{C})$. Thus, the family ω/df of residue forms defines a section $[\omega/df]: T' \to \mathscr{H}_F^*$ of the vanishing cohomology bundle.

Definition. The global sections of the form $[\omega/df]$ of the vanishing cohomology bundle are called *geometric sections.*

The integral of the residue form ω/df_t over an integer cycle δ_t that depends continuously on t is a locally holomorphic function of the parameter t.

Theorem ([58], [229]). *Every geometric section* $[\omega/df]$ *of the vanishing cohomology bundle is holomorphic.*

Remark. A holomorphic $(n-1)$-form ω is the residue form of the n-form $df \wedge \omega$. Consequently, any result valid for residue forms also covers the case of ordinary $(n-1)$-forms.

Example. Let $f = z_1^2 + \cdots + z_n^2$ and let ω be a holomorphic n-form. Then the geometric section $[\omega/df]$ has the asymptotics

$$[\omega/df]_t|S_t = \int_{S_t} \omega_t/dt = \frac{d}{dt}\int_{B_t} \omega = \frac{d}{dt}[ct^{n/2} + \cdots] = \tilde{c}t^{n/2-1} + \cdots$$

(see the example in subsection 3.4).

Any holomorphic n-form ω on $\mathbb{C}^n \times \mathbb{C}^\mu$ defines a geometric section $[\omega/d_zF]$ of the bundle $\mathscr{H}_F^* \to \Lambda'$; this section is holomorphic.

Similarly, let ω be a holomorphic $(n+\mu)$-form on $\mathbb{C}^n \times \mathbb{C}^\mu$. In this case, on the fibres of the Milnor fibration there is a well defined holomorphic residue $(n-1)$-form $\omega/d_zF \wedge d\lambda$ $(d\lambda = d\lambda_1 \wedge \cdots \wedge d\lambda_\mu)$. This form, too, defines a holomorphic section of the vanishing cohomology bundle over Λ'.

3.6. Trivializations of the Cohomology Bundle. Here we show that the set of geometric sections defined by μ generic n-forms forms a basis in each fibre of the cohomology bundle, thereby defining a trivialization of this bundle.

The convolution of a section $[\omega]$ of the vanishing cohomology bundle with a multi-valued covariantly constant integer section δ of the homology bundle yields a multi-valued holomorphic function

$$[\omega]|\delta = \int_{\delta_\lambda} \omega_\lambda, \qquad \lambda \in \Lambda',$$

on the base Λ' of the bundle.

The branching of this function is described by the monodromy group. More precisely, let δ_* be a cycle lying in the fibre of the Milnor fibration over the point $\lambda_* \in \Lambda'$, γ a closed path in Λ' that starts and terminates at λ_*, and h_γ the corresponding monodromy transformation along γ. Then the analytic continuation along γ of the branch of the function, corresponding to the integral $\int_{\delta^*} \omega$, is equal to $\int_{h_\gamma \delta^*} \omega$.

Next, let us single out in the fibres of the Milnor fibration a basis of integer $(n-1)$-dimensional homology classes $\delta_1(\lambda), \ldots, \delta_\mu(\lambda)$ that depends continuously on the point $\lambda \in \Lambda'$. Any given set of n-forms $\omega_1, \ldots, \omega_\mu$ defines a family of residue forms $\omega_1/d_zF, \ldots, \omega_\mu/d_zF$ in the fibres of the Milnor fibration.

Let us consider the holomorphic function

$$\Delta: \lambda \mapsto \det{}^2\left(\left[\int_{\delta_j(\lambda)} \omega_i/d_zF\right]_{1 \leqslant i,j \leqslant n}\right)$$

on the base Λ'. The entries of the above matrix are multi-valued holomorphic functions on Λ'. Under analytic continuation along the closed path γ the value of Δ gets multiplied by the square of the determinant of the operator of integer monodromy along γ. Since the monodromy operator inverse to the latter (along the path $-\gamma$) is also integer, the square of the determinant equals one. Hence, Δ is a single-valued function on the base Λ'.

The level bifurcation set $\Sigma \subset \Lambda$ is an irreducible branched μ-sheeted covering over the hyperplane $\{\lambda_0 = 0\}$ in the base Λ of the versal deformation. Let $g(\lambda)$ be a holomorphic function that is a polynomial of degree μ in the variable λ_0 and vanishes on Σ.

Remark. The existence of a function g with the desired properties follows from the Weierstrass preparation theorem [169]. Any holomorphic function that vanishes on Σ lies in the ideal generated by g in the ring of germs of holomorphic functions at zero (this is a consequence of Hilbert's Nullstellensatz [178]).

Theorem ([346]). *The function Δ/g^{n-2} is holomorphic at the origin in the base Λ of the versal deformation.*

Remark. The proof of the theorem is essentially contained in the example given in subsection 3.5, since it suffices to verify that Δ/g^{n-2} is holomorphic at a nonsingular point of Σ.

Thus, the function Δ is meromorphic, and for $n > 1$, even holomorphic on Λ.

Let $\omega_1, \ldots, \omega_\mu$ be a set of holomorphic n-forms for which the restriction of Δ to the λ_0-axis has a zero of order $\mu(n-2)$ at the origin. From the above theorem it follows that the function Δ/g^{n-2} is invertible at zero, and consequently $\Delta \neq 0$ off Σ.

Definition. The set of forms $\omega_1, \ldots, \omega_\mu$ is called a *trivialization* if the function Δ has on the λ_0-axis a zero of order $\mu(n-2)$ at the origin.

Since, for a trivialization $\omega_1, \ldots, \omega_\mu$, the function Δ does not vanish in $\Lambda' = \Lambda \setminus \Sigma$, the set of sections $[\omega_i/d_zF]$ forms a basis in each fibre of the cohomology bundle $\mathcal{H}_F^* \to \Lambda'$ and provides a trivialization of the latter. The existence of a trivialization $\omega_1, \ldots, \omega_\mu$ is guaranteed by the following result.

Theorem ([346]). *The function Δ constructed from a set of generic forms $\omega_1, \ldots, \omega_\mu$ (i.e., such that the finite jets of $\omega_1, \ldots, \omega_\mu$ do not satisfy some complex-analytic relation) has on the λ_0-axis a zero of order $\mu(n-2)$.*

The proof of the existence of a trivialization was obtained for the first time in [58] using the technique of coherent analytic sheaves. The proof of the theorem stated above relies on asymptotic expansion of integrals.

One of the corollaries of the results discussed above is that the set of geometric sections $[\omega_i/d_zF]$, $i = 1, \ldots, \mu$ is a basis of the module of germs of geometric sections at the origin in the base Λ of the versal deformation over the ring of germs of holomorphic functions on Λ at the origin. In fact, let ω be an arbitrary n-form on $(\mathbb{C}^n \times \mathbb{C}^\mu, 0)$ and let $\delta_1(\lambda), \ldots, \delta_\mu(\lambda)$ be an integer homology basis in the fibres of the Milnor fibration that depends continuously on $\lambda \in \Lambda'$.

The system of equations

$$\int_{\delta_i(\lambda)} \omega/d_zF \equiv \sum_j p_j(\lambda) \int_{\delta_i(\lambda)} \omega_j/d_zF$$

for the unknown functions $p_j(\lambda)$ has a unique solution, since its principal determinant is different from zero for any $\lambda \in \Lambda'$. The auxiliary determinants have on Σ a zero of order larger than or equal to that of the principal determinant. Hence, by Cramer's rule, the functions $p_j(\lambda)$ admit a holomorphic extension to the point $0 \in \Lambda$. It follows that one can write the section $[\omega/d_zF]$ as

$$[\omega/d_zF] = \sum p_j[\omega_j/d_zF],$$

as needed.

This result can be alternatively formulated as follows. Let $\Omega^\bullet_{V_\Lambda/\Lambda} = \{\Omega^p_{V_\Lambda/\Lambda}, d^p\}$ be the complex of relative differential forms associated with the Whitney map $\pi\colon V_\Lambda \to \Lambda$: $\Omega^p_{V_\Lambda/\Lambda} = \Omega^p_V/d_zF \wedge \Omega^{p-1}_V$. Similarly, let $\Omega^\bullet_{V'_\Lambda/\Lambda'}$ denote the complex of relative forms on V'_Λ. The sheaf of sections of the vanishing cohomology bundle is canonically isomorphic to the $(n-1)$st cohomology group of the complex $\pi_*\Omega^\bullet_{V'_\Lambda/\Lambda'}$. The sheaf $H^{n-1}(\pi_*\Omega^\bullet_{V'_\Lambda/\Lambda'})$ admits the natural extension $\mathcal{H}(V_\Lambda/\Lambda) = H^{n-1}(\pi_*\Omega^\bullet_{V_\Lambda/\Lambda})$ to the entire base Λ.

Theorem ([58], [158], [295]). *$\mathcal{H}(V_\Lambda/\Lambda)$ is a coherent analytic sheaf on Λ. Its fibre over $0 \in \Lambda$ is a free module of rank μ over the ring $\mathcal{O}_\mu$ of holomorphic function-germs on $(\Lambda, 0)$.*

3.7. The Classical Complex Monodromy. The Gauss-Manin connection ∇ in the vanishing cohomology bundle $\mathcal{H}_f^* \to T'$, defined transcendentally in subsection 3.1, also admits an analytic description. In an arbitrary bundle with one-dimensional basis T' a connection is specified by the operator of covariant derivation ∇_t along the holomorphic vector field $\partial/\partial t$ on the base. Let ω be a holomorphic $(n-1)$-form on $\mathbb{C}^n$.

Theorem ([58]). *The covariant derivation ∇_t of the Gauss-Manin connection in the vanishing cohomology bundle $\mathscr{H}_f^* \to T'$ is specified by the formula*

$$\nabla_t[\omega] = [d\omega/df]$$

and is a singular first-order differential operator on $\mathscr{H}_f^*$. The monodromy of this singular operator can be canonically identified with the classical complex monodromy of the singularity f.

Example. Let $f(z)$ be a quasihomogeneous function of degree one with set of weights $\mathbf{v} = (v_1, \ldots, v_n)$. Consider the quasihomogeneous differential form

$$\omega_{\mathbf{k}} = \mathbf{z}^{\mathbf{k}} dz/df$$

of weight $l_{\mathbf{k}} = \langle \mathbf{k} + 1, \mathbf{v} \rangle - 1$, where $\mathbf{k} = (k_1, \ldots, k_n)$.

Then $\nabla_t[\omega_{\mathbf{k}}] = l_{\mathbf{k}}[\omega_{\mathbf{k}}]/t$. Thus, $\omega_{\mathbf{k}}$ is a joint eigenvector of the differential operator ∇_t and the classical-monodromy operator. The corresponding eigenvalue of the monodromy operator is equal to $\exp(2\pi i l_{\mathbf{k}})$.

Let $\mathbf{z}^{\mathbf{k}_1}, \ldots, \mathbf{z}^{\mathbf{k}_r}$ be a monomial basis of the local algebra Q_f (see subsection 1.3.3), and let $\omega_1, \ldots, \omega_\mu$ be the corresponding set of $(n-1)$-forms of degrees $l_1 \ldots, l_\mu$ ($\omega_s = \omega_{\mathbf{k}_s}$, $l_s = \deg \omega_s$).

Theorem ([318], [346]). *The set of forms $\omega_1, \ldots, \omega_\mu$ is a trivialization. In the basis $[\omega_i/df]$ the classical-monodromy operator is diagonal with eigenvalues* $\exp(2\pi i l_s)$, $s = 1, \ldots, \mu$.

For a description of the Gauss-Manin connection in the cohomology bundle $\mathscr{H}_F^* \to \Lambda'$ we refer the reader to [158].

3.8. Differential Equations and Asymptotics of Integrals. In this subsection it will be shown that the singular operator ∇_t defined by the Gauss-Manin connection in the cohomology bundle $\mathscr{H}_f^* \to T'$ has a regular singularity. This allows one to obtain a series of results on the asymptotics of integrals of holomorphic forms and on the structure of the classical-monodromy operator.

Let $\omega = (\omega_1, \ldots, \omega_\mu)$ be a trivialization of the cohomology bundle $\mathscr{H}_f^* \to T'$. The covariant derivative of the geometrical section defined by the form ω_i admits a unique expansion of the form

$$\nabla_t[\omega_i/df] = \sum p_{ij}(t)[\omega_j/df], \qquad t \neq 0,$$

where the functions p_{ij} are holomorphic in T' (this follows from the definition of a trivialization, subsection 3.6).

Remark. The functions p_{ij} do not necessarily admit a holomorphic extension to zero, since the covariant derivative $\nabla_t[\omega_i/df]$ is not a geometric section. However, the p_{ij} are meromorphic functions. This follows from the fact that the operator $f^k \nabla_t$ takes geometric sections again into geometric sections whenever f^k belongs to the gradient ideal $(f_1, \ldots, f_\mu)$; such a k does indeed exist because the singularity of f is isolated.

Let $P(t)$ denote the matrix $[p_{ij}(t)]_{i,j=1}^{\mu}$. Then the covariant derivative of the trivialization ω is determined from the system of ordinary differential equations

$$\nabla_t[\omega/df] = P[\omega/df].$$

To write this in coordinate form, let $I(t) = (\int_{\delta(t)} \omega_1/df, \ldots, \int_{\delta(t)} \omega_\mu/df)$ be the vector of coordinates of the covariantly constant section $\delta(t)$ of the homology bundle $\mathscr{H}_{*f} \to T'$. Then

$$\frac{dI}{dt} = PI.$$

Definition. The system $dI/dt = PI$ is called the *Picard-Fuchs equation* of the trivialization ω.

The mapping $I = (I_1, \ldots, I_\mu) \mapsto \sum_j I_j s_j^*$, where s_j^*, $j = 1, \ldots, \mu$, is the frame in the homology bundle dual to the frame $[\omega_j/df]$, $j = 1, \ldots, \mu$, establishes an isomorphism of the space of solutions of the Picard-Fuchs equation and the space of covariantly constant sections of the homology bundle. Under this isomorphism the monodromy of the system $dI/dt = PI$ is identified with the classical-monodromy operator h_{*f} of the singularity f in the homology of the nonsingular fibre.

Remark. The covariantly constant sections of the cohomology bundle in the basis $[\omega_i/df]$ and the monodromy operator h_f^* are described by the system $dI/dt = -P^*I$.

Now let $dI/dt = AI$ be a system of differential equations in which the entries of the matrix A are meromorphic at $t = 0$. The fundamental matrix (of solutions) of such a system has the form

$$\Phi = Qt^R,$$

where the entries of the matrix Q are holomorphic functions in T' and R is a constant matrix.

Remark. The monodromy of the system of equations $dI/dt = AI$ is expressible in terms of its fundamental matrix: $M = \exp(2\pi i R)$.

Definition. The singularity $(t = 0)$ of the system $dI/dt = AI$ is said to be *regular* if the entries of the matrix Q are meromorphic at the origin.

An equivalent condition is: in any sector $a < \arg t < b$, every solution $I(t)$ of the system grows no faster than a power of t as $t \to 0$: $\|I(t)\| = o(t^{-N})$ for some N.

The change of variables $I = QI'$ takes the given system with regular singularity into the system

$$\frac{dI'}{dt} = t^{-1}RI'$$

with a simple pole at $t = 0$. The converse is also true.

Theorem (Sauvage, see [176]). *A system of differential equations with a simple pole has a regular singularity.*

Thus, the system $dI/dt = AI$ is regular if and only if it can be transformed into a system $dI'/dt = t^{-1}RI'$ with a simple pole by means of a change of variables with a meromorphic matrix.

Remark. The fundamental matrix of the system $dI'/dt = t^{-1}RI'$ is $\Phi = \exp(R \ln t)$. Consequently, its solutions admit series expansions of the form $\sum a_{k,\alpha} t^{\alpha}(\ln t)^k$, where α are the eigenvalues of R (correspondingly, $\exp(2\pi i\alpha)$ are the eigenvalues of the monodromy operator M).

Let ω be a holomorphic $(n-1)$-form, $\delta(t)$ be a continuous family of integer homology classes of nonsingular fibres of the Milnor fibration, and $a < \arg t < b$ be a sector in T'.

Theorem ([58], [229], [252], [253]). *There exists a natural number N such that in the given sector,*

$$\left|\int_{\delta(t)} \omega\right| \leqslant \text{const}\,|t|^{-N}.$$

It follows from this result that the Picard-Fuchs equation has a regular singularity. The assertion of the theorem and a sharpened version (see [229]) allow us to derive the following result.

Theorem ([58], [229], [252], [104], [159], [191]). *Let ω be a holomorphic n-form. Then the function $\int_{\delta(t)} \omega/df$ admits in each sector $a < \arg t < b$ a series expansion $\sum a_{k,\alpha} t^{\alpha}(\ln t)^k/k!$. The series converges for sufficiently small values of the parameter. All numbers α are rational and larger than -1. For each α appearing in the series the number $\exp(2\pi i\alpha)$ is an eigenvalue of the classical-monodromy operator in homology. The coefficient $a_{k,\alpha}$ is equal to zero whenever the classical-monodromy operator has no Jordan blocks of dimension larger than or equal to $k+1$ with eigenvalue $\exp(2\pi i\alpha)$. In the case of the integral of a holomorphic $(n-1)$-form one has, in addition, that all α are nonnegative and $a_{k,0} = 0$ for $k > 0$.*

The information contained in the theorem on the eigenvalues of the classical-monodromy operator can be formulated as follows.

Theorem ([58]). *The characteristic polynomial $\det(h_* - \lambda E)$ of the classical-monodromy operator is a product of cyclotomic polynomials.*

In other words, the operator $h_*^N - E$ is nilpotent for some N.

The asymptotic series expansion of the integral of the residue form,

$$[\omega/df]|\delta = \int_{\delta(t)} \omega/df_t = \sum a_{k,\alpha} t^{\alpha}(\ln t)^k/k!,$$

defines a "coordinate-free" series expansion of the geometric section $[\omega/df]$ of the cohomology bundle $\mathscr{H}_f^* \to T'$ in covariantly constant sections. The coefficient $a_{k,\alpha}$ in the expansion of the integral depends linearly on the covariantly constant section δ of the homology bundle (for a fixed form ω). In other words, each coefficient $a_{k,\alpha}$ defines a linear function on the space of covariantly constant sections of the homology bundle, i.e., a covariantly constant multi-valued section $A^{\omega}_{k,\alpha}$ of the cohomology bundle:

$$A^{\omega}_{k,\alpha}|\delta = a_{k,\alpha}(\omega, \delta).$$

Thus, the series expansion of the integral $[\omega/df]\delta$ determines uniquely the expansion of the geometric section $[\omega/df]$:

$$[\omega/df] = \sum t^{\alpha}(\ln t)^k A^{\omega}_{k,\alpha}/k!.$$

This expansion will be used later to define the mixed Hodge structure in the vanishing cohomology (see § 4).

3.9. Nondegenerate Period Maps. Here and in the next subsections we give an account of the results of the papers [354], [346], which connect nondegenerate period maps of holomorphic forms of the vanishing cohomology bundle with the intersection form in the homology of the nonsingular fibre of a singularity f.

Consider the k-dimensional cohomology bundle $\pi^*\colon \mathcal{H}^k_\pi \to B$ associated with a locally trivial fibration π with smooth base (see subsection 3.1). The Gauss-Manin connection ∇ in the cohomology bundle defines, for each section s of this bundle, the map

$$\nabla s\colon T_*B \to \mathcal{H}^k_\pi$$

of the tangent bundle of the base into [the total space of] the cohomology bundle. This map sends any tangent vector at a point of the base into an element of the fibre of the cohomology bundle over the point, namely, the covariant derivative of the section s in the direction of the vector.

Remark. The tangent space to the fibre of the cohomology bundle at any of its points is identified with the fibre itself by means of the linear structure in the fibre.

Definition. The map $\nabla s\colon T_*B \to \mathcal{H}^k_\pi$ is called the *derivative of the section s.*

The derivative of a section is linear at each point of the base. The section s also defines the map $(\nabla s)^*\colon \mathcal{H}_{k\pi} \to T^*B$, the dual of the derivative.

The base and the fibre of the vanishing cohomology bundle $\mathcal{H}^*_F \to \Lambda'$ associated with the Milnor fibration of the singularity f have the same dimension (equal in turn with the multiplicity μ).

Definition. A *section* of the vanishing cohomology bundle $\mathcal{H}^*_F \to \Lambda'$ is said to be *nondegenerate* if its derivative is an isomorphism of bundles.

Thus, a nondegenerate section s yields an isomorphism $T_*\Lambda' \to \mathcal{H}^*_F$ and induces a local diffeomorphism of a neighborhood of an arbitrary point of the base into the fibre of $\mathcal{H}^*_F$ over the point. The map dual to the derivative of a nondegenerate section s defines an isomorphism $(\nabla s)^*\colon \mathcal{H}_{*F} \to T^*\Lambda'$ of the vanishing homology bundle onto the cotangent bundle of the base.

Let ω be a holomorphic $(n-1)$-form on $\mathbb{C}^n \times \mathbb{C}^\mu$ and let $[\omega]$ be its period map (see subsection 3.4). The covariant derivative of the period map $[\omega]$ along a vector field on the base Λ of the versal deformation is also a section of the cohomology bundle $\mathscr{H}_F^* \to \Lambda'$. Let us consider covariant derivates of period maps $[\omega]$ along the vector field $\partial/\partial\lambda_0$.

Remark. We are concerned here with a versal deformation $F(z, \lambda) = f(z) + \sum_{i=0}^{\mu-1} \lambda_i \varphi_i(z)$ with $\varphi_0(z) \equiv 1$.

Definition. The kth derivative $[\omega]^{(k)} = (\nabla_{\partial/\partial\lambda_0})^k [\omega]$ of the period map of the form ω is called the kth *associated period map of* ω.

Fix a frame $\delta_1(\lambda), \ldots, \delta_\mu(\lambda)$ of covariantly constant multi-valued sections of the vanishing homology bundle over Λ'. In coordinates relative to this frame,

$$[\omega]^{(k)} | \delta(\lambda) = \left(\frac{\partial}{\partial\lambda_0}\right)^k \left(\int_{\delta_1(\lambda)} \omega, \ldots, \int_{\delta_\mu(\lambda)} \omega\right).$$

Example. For the Morse function $f = z_1^2 + \cdots + z_n^2$, the asymptotics of the kth associated period map of a generic form ω on a vanishing cycle s_t has the form

$$[\omega]^{(k)} | s_t = ct^{n/2-k} + \cdots,$$

provided $n/2 - k$ is not a negative integer (see the example in subsection 3.4). For n even and $k > n/2$, the leading term of the asymptotics vanishes upon differentiation and the associated period map has a zero of higher order.

The vector fields $\partial/\partial\lambda_i$ provide a basis in each fibre of the tangent bundle. Consider the map

$$J^2 \colon \lambda \mapsto \det{}^2 \left[\frac{\partial}{\partial\lambda_i}([\omega]^{(k)} | \delta_j)\right],$$

i.e., the square of the Jacobian of the kth associated period map of the form ω, calculated in the bases $(\partial/\partial\lambda_i)$ and $(\delta_i(\lambda))$.

Example. For a Morse function, $J^2(t) \sim t^{n-2k-2}$ (except for the case when $n/2 - k$ is a negative integer).

An argument similar to the one used in subsection 3.6 shows that J^2 is a single-valued holomorphic function on Λ'. Let g be a holomorphic function that is a polynomial of degree μ in λ_0 and vanishes identically on Σ (see subsection 3.6).

Theorem. *The function J^2/g^{n-2k-2} is holomorphic at the origin in the base Λ of the versal deformation.*

This assertion follows from the fact that J^2/g^{n-2k-2} is holomorphic at the nonsingular points of Σ (see the preceding example) as well as off Σ, and consequently admits a holomorphic extension to the entire base Λ.

It follows from the preceding theorem that the order of the zero of the restriction of the function J^2 to the λ_0-axis is at least $\mu(n - 2k - 2)$.

Definition. The kth associated period map $[\omega]^{(k)}$ of the form ω is said to be *infinitesimally nondegenerate* if the order of the zero of J^2 on the λ_0-axis equals $\mu(n-2k-2)$.

Clearly, $[\omega]^{(k)}$ is nondegenerate whenever $J^2 \not\equiv 0$ off Σ, in particular whenever the germ of the holomorphic function J^2/g^{n-2k-2} at zero is invertible. From this one obtains the following:

Theorem. *Any infinitesimally nondegenerate section* $[\omega]^{(k)}$ *is nondegenerate.*

For a fixed k, the infinitesimal nondegeneracy property of kth associated period maps will be said to be *generic* if it holds in the complement of an analytic set of positive codimension in a space of jets of sufficiently high order.

Theorem. *The infinitesimal nondegeneracy property is generic for* $k = 0$. *If the intersection form in the homology of the nonsingular fibre of the function* f *is nondegenerate, then the infinitesimal nondegeneracy property is generic for all* k.

The example of Morse functions shows that the requirement that the intersection form be nondegenerate is essential: for n even the intersection form is degenerate and the kth associated period map is infinitesimally degenerate for all $k > n/2$.

In the case of quasihomogeneous functions the nondegeneracy of the period map of a generic form has been established in [220]; analogous results were announced in [287].

3.10. Stability of Period Maps. Let us define an equivalence relation on the set of all sections of the cohomology bundle $\mathscr{H}_F^* \to \Lambda'$.

To this end, let us pick a diffeomorphism in the component of identity in the group of the holomorphic diffeomorphisms $(\Lambda, 0) \to (\Lambda, 0)$ that preserve the bifurcation set Σ.

Next, let us connect this diffeomorphism with the identity map $\Lambda \to \Lambda$ by a smooth one-parameter family of diffeomorphisms. By integrating the connection along this family we obtain a lift of the initial diffeomorphism to a diffeomorphism of [the total space of] the cohomology bundle $\mathscr{H}_F^* \to \Lambda'$. The latter is uniquely determined by the initial diffeomorphism of the base and the homotopy class of the one-parameter family, thanks to the integrability of the Gauss-Manin connection.

The diffeomorphisms obtained in this manner form a subgroup in the group of diffeomorphisms of the cohomology bundle. This subgroup acts on the sections of the bundle.

Definition. Two sections of the vanishing cohomology bundle $\mathscr{H}_F^* \to \Lambda'$ are said to be *equivalent* if they lie in the same orbit of the subgroup constructed above.

Definition. The kth associated period map $[\omega]^{(k)}$ of the form ω is said to be *stable* if the kth associated period map of any holomorphic form sufficiently close to ω is equivalent to $[\omega]^{(k)}$.

Theorem. *Any infinitesimally nondegenerate* kth *associated period map is stable. For a quasihomogeneous function* f *all infinitesimally nondegenerate* kth *associated period maps of holomorphic forms are equivalent.*

3.11. Period Maps and Intersection Forms. The nondegenerate sections of the vanishing cohomology bundle $\mathscr{H}_F^* \to \Lambda'$ can be used to pull back to the base Λ' the structures that exist in the vanishing (co)homology bundle, in particular, the intersection form.

Specifically, any nondegenerate section s of the cohomology bundle $\mathscr{H}_F^* \to \Lambda'$ defines an isomorphism $(\nabla s)^*\colon \mathscr{H}_{*F} \to T^*\Lambda'$ and in this manner pulls-back the intersection form in the homology group $H_{n-1}(V_\lambda; \mathbb{C})$ of the fibre of the Milnor fibration to a 2-form on $T^*\Lambda'$.

Definition. The 2-form constructed above on the cotangent bundle $T^*\Lambda'$ is called the *intersection form of the section s.*

From the results of subsection 3.9 we derive the following assertions.

Theorem. *The intersection form of an infinitesimally nondegenerate* kth *associated period map is stable. If* f *is a quasihomogeneous function, than any two such forms are equivalent.*

Remark. The equivalence and stability of forms on $T^*\Lambda'$ should be understood in the sense of the definitions given in subsection 3.10, adapted to the cotangent bundle of Λ'.

Theorem. *The intersection form of an infinitesimally nondegenerate* kth *associated period map is holomorphic on* $T^*\Lambda'$, *and for* $k \geqslant (n-2)/2$ *it admits a holomorphic extension to* $T^*\Lambda$.

This last result permits us to extend the operation of convolution of invariants (see subsection 5.4) to the case of nonsimple singularities of functions.

In the remaining part of this subsection we shall assume that the intersection form in the homology of the fibre of the Milnor fibration is nondegenerate. In this case the intersection form defines a nondegenerate dual form in the fibres of the vanishing cohomology bundle; the latter will also be referred to as the intersection form (in cohomology).

Remark. For the generalization of the results presented below to the case of a degenerate intersection form we refer the reader to [38, p. 447].

In the case where the number n of variables is odd the intersection form is symmetric and defines a structure of complex pseudo-Riemannian space of curvature zero on the base Λ'. In the case where n is even the intersection form is skew-symmetric (i.e., a differential 2-form) and defines on the base Λ' a holomorphic symplectic structure.

Remark. The reason the Riemannian metric has zero curvature and the differential 2-form is closed, respectively, is that locally they are induced from a form with constant coefficients.

Theorem. *For $n = 2k + 2$ the intersection form defined on the tangent bundle by an infinitesimally nondegenerate* kth *associated period map of a holomorphic form extends to a holomorphic symplectic structure on the entire base Λ of the versal deformation.*

The restriction of the symplectic structure to the strata of the bifurcation set Σ carries information on the degeneracy types of the critical points over the strata (for instance, the stratum $(\mu/2)A_1$ is Lagrangian since the corresponding $\mu/2$ vanishing cycles "vanish" simultaneously on that stratum, and hence do not intersect).

Now suppose that the number of variables on which f depends is odd; the intersection form is then symmetric.

Definition. An infinitesimally nondegenerate kth associated period map of a holomorphic form for a function of $n = 2k + 1$ variables is called a *principal period map.*

In the case of a simple singularity, a principal period map identifies the base of the versal deformation with the bifurcation set Σ embedded in it and the orbit space of the corresponding Coxeter group with the hypersurface of the irregular orbits embedded in it (see subsection 5.6).

Theorem. *Suppose that the intersection form is nondegenerate. Then the isomorphism $T^*\Lambda' \to T_*\Lambda'$ given by a principal period map induces an isomorphism of the module of germs of holomorphic* 1*-forms on* $(\Lambda, 0)$ *onto the module of germs of holomorphic vector fields on* $(\Lambda, 0)$ *that are tangent to* Σ.

This theorem allows us to generalize the construction of vector fields tangent to the level bifurcation set Σ (see subsection 5.7) to the case of nonsimple singularities.

3.12. The Characteristic Polynomial and the Zeta Function of the Monodromy Operator. The characteristic polynomial of the classical-monodromy operator $h_f^{n-1}: H^{n-1}(V_t) \to H^{n-1}(V_t)$ can be calculated with the help of a resolution of the singularity of f at zero, or (in case that f satisfies a certain additional nondegeneracy condition) with the help of the Newton polyhedron of f in a suitable coordinate system. Note that the calculation of the characteristic polynomial includes that of its degree, i.e., the Milnor number of the function f.

Instead of the characteristic polynomial of the monodromy operator, it is convenient to look for its zeta function (defined below). In the case of isolated singularities the latter task is equivalent to the former.

Thus, let $f: (\mathbb{C}^n, 0) \to (\mathbb{C}, 0)$ be an arbitrary singularity. For each $i = 0, 1, \dots,$ the monodromy operator h_f^i acts on the space $H^i(V_t; \mathbb{C})$.

Definition. The *zeta function of the monodromy operator* h_f^* is the rational function

$$\zeta_f(z) = \prod_{i \geqslant 0} [\det(\mathrm{Id} - z h_f^i)]^{(-1)^q}.$$

When the singularity f is isolated and its Milnor number is μ, the characteristic polynomial $\chi_f(z)$ of the operator h_f^{n-1} is expressible through the zeta function as

$$\chi_f(z) = z^\mu[(z/(z-1))\zeta_f(1/z)]^{(-1)^{n-1}}.$$

This follows from observing that in the present case $H^0(V_t; \mathbb{C}) = \mathbb{C}$ and h_f^0 acts trivially on this space, whereas $H^i(V_t; \mathbb{C}) = 0$ for $i \neq 0, n-1$. Note also that the order of the function ζ_f is equal to the Euler characteristic of the nonsingular fibre V_t.

Definition. A *resolution of the singularity* of the function f at the point 0 is a pair (Y, π), where Y is a nonsingular n-dimensional complex manifold and $\pi: Y \to \mathbb{C}^n$ is a proper analytic map that takes Y into some neighborhood U of 0, such that

a) π defines a biholomorphism $Y \setminus V_0' \to U \setminus V_0$ in the complement of the hypersurface $V_0' = \pi^{-1}(V_0)$;

b) V_0' is the union of a finite number of nonsingular divisors in Y that have only normal mutual intersections; in particular, in suitable local coordinates in a neighborhood of any intersection point of k ($k \geqslant 1$) such divisors of the function $\tilde{f} = f \circ \pi$ can be written in the form $x_1^{m_1} \cdot \ldots \cdot x_k^{m_k}$.

According to [179], any singularity f admits a resolution.

Let $S = \pi^{-1}(0)$. For each positive integer m denote by S_m the set of points of S in the neighborhood of which $\tilde{f} = x_1^m$ in a suitable coordinate system.

Theorem (see [4]). *The zeta function of the classical-monodromy operator h_f^* of the singularity of the function $f: (\mathbb{C}^n, 0) \to (\mathbb{C}, 0)$ is given by the formula*

$$\zeta_f(z) = \prod_{m \geqslant 1} (1 - z^m)^{\chi(S_m)},$$

where $\chi(\cdot)$ designates the Euler characteristic.

Corollary. *The Euler characteristic of a nonsingular fibre V_t, $t \neq 0$, equals $\sum_{m \geqslant 1} \chi(S_m)$.*

In the case of an isolated singularity f the Euler characteristic $\chi(V_t)$ is equal to $1 - (-1)^n \mu(f)$, and we obtain an expression for the Milnor number of the singularity in terms of its resolution.

The theorem given above, in conjunction with the technique of toroidal resolutions of singularities developed in [192], [194], enables us to calculate the Milnor number and the zeta function of a singularity in terms of its Newton diagram. (For the definition of the Newton polyhedron $\Gamma_+(f)$ and the Newton diagram $\Gamma(f)$ of a power series f see subsection 1.3.1).

Definition. The *principal part* of a series $f = \sum f_\alpha x^\alpha$ is the polynomial $\sum_{\alpha \in \Gamma(f)} f_\alpha x^\alpha$. The Newton diagram $\Gamma(f)$ is said to be *suitable* if it intersects all coordinate axes in $\mathbb{R}^n$.

Any discrete topological characteristic of the singularity f (such as the Milnor number, the modality, or the Jordan form of the monodromy operator) takes the same value for almost all functions with a given Newton diagram. "Almost all" is given here the following precise meaning.

Definition. A series f is said to be *Γ-nondegenerate* if there is no face γ of the diagram $\Gamma(f)$ such that all n partial derivatives of the quasihomogeneous polynomial $\sum_{\alpha\in\gamma} f_\alpha x^\alpha$ vanish simultaneously at some point in the complement of the union of the coordinate planes.

The last condition is satisfied for almost all power series with the same Newton diagram.

Suppose the polyhedron $\Gamma(f)$ is suitable. Then the following quantities are meaningful: V_n, the n-dimensional volume of the set $\mathbb{R}^n_+ \setminus \Gamma_+(f)$; V_{n-1}, the sum of the $(n-1)$-dimensional volumes of the n sets $(\mathbb{R}^{n-1}_+)_i \setminus \Gamma_+(f)$, where $\mathbb{R}^{n-1}_i$ is the coordinate hyperplane $\{\alpha_i = 0\}$; V_{n-2}, the sum of the $(n-2)$-dimensional volumes of the $n(n-1)/2$ sets $(\mathbb{R}^{n-2}_+)_{ij} \setminus \Gamma_+(f)$, where $\mathbb{R}^{n-2}_{ij} = \{\alpha_i = \alpha_j = 0\}$, and so forth.

Theorem (see [203]). *Suppose the function f is Γ-nondegenerate and its Newton diagram $\Gamma(f)$ is suitable. Then the Milnor number μ of f is finite and equals*

$$n!V_n - (n-1)!V_{n-1} + (n-2)!V_{n-2} + \cdots + (-1)^n. \qquad (1)$$

For example, if $n = 2$, then $\mu(f) = 2S - a - b + 1$, where S denotes the area "under the diagram" and a, b are the coordinates of the intersection of $\Gamma(f)$ with the coordinate axes.

We next turn to the calculation of the zeta function of the monodromy in terms of the Newton polyhedron.

Theorem (see [343]). *Suppose the function $f\colon (\mathbb{C}^n, 0) \to (\mathbb{C}, 0)$ has a singularity at zero and is Γ-nondegenerate. Then the zeta function ζ_f of the classical-monodromy operator of the singularity f equals*

$$\prod_{l=1}^{n} (\zeta^l(z))^{(-1)^{l-1}},$$

where ζ^l are the polynomials defined in formula (2) *below.*

Corollary. *Under the assumptions of the theorem the Euler characteristic of the nonsingular fibre V_λ of the Milnor fibration equals $\sum_{l=1}^{n} (-1)^{l-1} \deg \zeta^l(z)$.*

If, in addition, the diagram $\Gamma(f)$ is suitable (and hence the singularity of f is isolated), the corollary is equivalent to Kushnirenko's formula (1) for the Milnor number $\mu(f)$.

The proof of the theorem is reduced to formula (1) with the help of a special resolution of the singularity of f that is constructed directly from its Newton diagram (see [194]).

Definition of the polynomials ζ^l. The polynomials $\zeta^l(z)$ are defined in terms of the faces of highest possible dimension (i.e., $l-1$) of the intersection of the diagram $\Gamma(f)$ with the l-dimensional coordinate planes in $\mathbb{R}^n$.

Let I be an arbitrary subset of $\{1, \dots, n\}$ with l elements ($|I| = l$), and let $L_I \simeq \mathbb{R}^l$ denote the coordinate plane spanned by the unit vectors e_i with $i \in I$. Let $j(I)$ be the number of $(l-1)$-dimensional faces of the diagram $\Gamma(f)$ that lie in the plane L_I. Denote by $L_I^1, \dots, L_I^{j(I)}$ the $(l-1)$-dimensional affine subspaces of L_I containing these faces. Now let $j \in \{1, \dots, j(I)\}$. In the l-dimensional lattice $L_I \cap \mathbb{Z}^n \simeq \mathbb{Z}^l$ consider the sublattice spanned by the vectors with integer components that lie in L_I^j. Let $m_j(I)$ denote the index of this sublattice. Next, define the $(l-1)$-dimensional volume in the affine space L_I^j in such a way that any cube spanned by an arbitrary basis of the lattice $L_I^j \cap \mathbb{Z}^n$ has volume one. Finally, let $V_j(I)$ denote the $(l-1)$-dimensional volume of the polyhedron $\Gamma(f) \cap L_I^j$. Then

$$\zeta^l(z) = \prod_{I, |I| = l} \prod_{j=1}^{j(I)} (1 - z^{m_j(I)})^{(l-1)! V_j(I)}. \tag{2}$$

§4. The Mixed Hodge Structure in the Vanishing Cohomology

In this section we define the mixed Hodge structure in the fibres of the vanishing cohomology bundle associated with the singularity of a function f and we discuss its connection with other invariants of the singularity.

The mixed Hodge structure in the vanishing cohomology was introduced for the first time by Steenbrink [316], [317], [319]. His definition is algebraic and is based on resolutions of singularities and the construction of a mixed Hodge structure in the cohomology of a quasiprojective variety due to Deligne [105].

An alternative, transcendental method for defining the mixed Hodge structure, which is the one we are going to follow here, uses the asymptotic expansion of integrals over vanishing cycles that we described in subsection 3.8. The mixed Hodge structure that arises in this manner is termed asymptotic and has been obtained in Varchenko's papers [344], [345], [346].

The mixed Hodge structures in the vanishing cohomology defined by Steenbrink and Varchenko possess identical weight filtrations, but their Hodge filtrations may differ. Despite this, the pure Hodge structures that they generate in the quotient spaces of the weight filtration coincide [344].

4.1. The Pure Hodge Structure. Let X be a compact complex manifold with a Hermitian metric $ds^2 = \sum \varphi_i \otimes \overline{\varphi}_i$, where φ_i are holomorphic 1-forms in some chart on X.

The manifold X is called a *Kähler manifold* if the (1, 1)-form

$$\omega = \frac{\sqrt{-1}}{2} \sum \varphi_i \wedge \overline{\varphi}_i$$

associated with the metric is closed. Let Ω^p denote the sheaf of holomorphic p-forms on X.

In the complex cohomology of the compact Kähler manifold X one has the Hodge decomposition [78], [161], [375]:

$$H^k(X;\mathbb{C}) \cong \bigoplus_{p+q=k} H^{p,q}(X), \qquad H^{p,q}(X) = \overline{H^{q,p}(X)},$$

where $H^{p,q}(X) = H^{p,q}_{\bar\partial}(X) = H^q(X;\Omega^p)$ is the Dolbeault cohomology of X; $H^{p,q}(X)$ may be regarded as the subspace of $H^{p+q}(X)$ generated by the closed differential forms of type (p, q). Let H be a finite-dimensional complex vector space that is the complexification of a real vector space $H_{\mathbb{R}}$ with a distinguished integer lattice $H_{\mathbb{Z}} \subset H_{\mathbb{R}}$: $H = H_{\mathbb{R}} \otimes_{\mathbb{R}} \mathbb{C}$.

Definition. A *pure Hodge structure of weight* k on H is a decomposition of H into a direct sum of subspaces $H^{p,q}$ with the property that $H^{p,q} = \overline{H^{q,p}}$ for all p, q with $p + q = k$.

Remark. Throughout this section we shall assume, as in this definition, that $p + q = k$.

Example. On the cohomology $H^k(X;\mathbb{C})$ of a compact Kähler manifold X there is the pure Hodge structure of weight k indicated above.

A pure Hodge structure of weight k on H defines a decreasing filtration

$$\{0\} = F^{k+1} \subset \cdots \subset F^{p+1} \subset F^p \subset \cdots \subset F^0 = H,$$

where $F^p = \bigoplus_{i \geqslant p} H^{i,k-i}$. The following relations hold:

$$F^p \cap \overline{F^q} = H^{p,q}, \qquad F^p \oplus \overline{F^{q+1}} = H.$$

Thus, the filtration determines the Hodge structure: the second of the above relations is necessary and sufficient in order for a decreasing filtration on the space H to define a pure Hodge structure on H.

Definition. A decreasing filtration F^p on H with the property that $F^p \oplus \overline{F^{q+1}} = H$ is called a *Hodge filtration.*

Let $\varphi\colon H \to H'$ be a linear map between spaces with pure Hodge structures that is rational with respect to the integer lattices singled out in H and H'.

Definition. The map φ is called a *morphism of type* (r, r) *of pure Hodge structures* if $\varphi(H^{p,q}) \subset H'^{p+r,q+r}$ for all pairs p, q.

In terms of Hodge filtrations the last condition says that $\varphi(F^p) \subset F'^{p+r}$.

Example. The map induced in cohomology by a holomorphic map of compact Kähler manifolds is a morphism of type (0, 0).

4.2. The Mixed Hodge Structure. Let H be as in subsection 4.1.

Definition. A *mixed Hodge structure* on H is specified by a pair of filtrations:

(1) an increasing filtration on H by subspaces W_k that are rational with respect to the integer lattice singled out in H; and

(2) a decreasing filtration F^p on H, such that the following relations hold for all integers p:

$$F^p \cap W_k + \overline{F^{q+1}} \cap W_k + W_{k-1} = W_k,$$

$$F^p \cap \overline{F^{q+1}} + W_{k-1} = W_{k-1}.$$

W and F are called the *weight filtration* and the *Hodge filtration*, respectively.

The Hodge filtration F defines a pure Hodge structure of weight k on the subspace W_k "modulo" W_{k-1}. More precisely, let

$$H^{(k)} = \mathrm{gr}_k W = W_k/W_{k-1}, \qquad F^p_{(k)} = (F^p \cap W_k)/W_{k-1} \subset H^{(k)},$$

and

$$H^{p,q} = F^p_{(k)} \cap \overline{F^q_{(k)}}.$$

Then one has the direct sum decomposition

$$H^{(k)} = \bigoplus_{p+q=k} H^{p,q}.$$

Let $\varphi\colon H \to H'$ be a linear map between spaces with mixed Hodge structures W, F and W', F', respectively, that is rational with respect to the integer lattices singled out in H and H'.

Definition. The map φ is called a *morphism of type* (r, r) *of mixed Hodge structures* if

$$\varphi(W_k) \subset W'_{k+2r}, \qquad \varphi(F^p) \subset F'^{p+r}.$$

A morphism of type (r, r) of mixed Hodge structures induces a map $H^{(k)} \to H'^{k+2r}$ that is a morphism of type (r, r) of pure Hodge structures of weights k and $k + 2r$, respectively.

Example. Let $\overline{X}$ be a nonsingular complex subvariety in projective space and $D \subset \overline{X}$ a nonsingular subvariety of codimension one in $\overline{X}$. Let us define a mixed Hodge structure on the cohomology of the complement $X = \overline{X} \setminus D$ (see [114]).

The cohomology of X can be calculated by using the de Rham complex of C^∞-differential forms on X. Let us first define the Hodge filtration: $F^p \subset H^k(X; \mathbb{C})$ is the subspace consisting of the cohomology classes defined by forms with no less than p holomorphic differentials dz_i.

To define the weight filtration let us consider the exact Gysin cohomology sequence

$$H^{k-2}(D) \xrightarrow{\gamma_k} H^k(\overline{X}) \xrightarrow{i^*} H^k(X) \xrightarrow{\mathrm{Res}} H^{k-1}(D) \xrightarrow{\gamma_{k+1}} \cdots$$

induced by the embedding $i\colon X \to \overline{X}$; the homomorphism γ is the dual, in the sense of Poincaré, of the Gysin homomorphism $\delta \mapsto \delta \cap D$, $\delta \in H_k(\overline{X})$.

The cohomology groups $H^{k-2}(D)$, $H^{k-1}(D)$, and $H^k(\overline{X})$ of Kähler manifolds are endowed with pure Hodge structures of respective weights $k-2$, $k-1$, and k. The homomorphism γ maps forms of type (p, q) into forms of type $(p+1, q+1)$ and is a morphism of type $(1, 1)$ of pure Hodge structures; the residue homomorphism Res takes (p, q)-forms into $(p-1, q)$-forms. We define the weight filtration on $H^k(X)$ by setting

$$W_{k-1} = 0, \qquad W_k = \operatorname{Im} i^*, \qquad W_{k+1} = H^k(X).$$

Note that $W_k/W_{k-1} \cong H^k(\bar{X})/\operatorname{Im}\gamma$ inherits the pure Hodge structure of weight k present in $H^k(\bar{X})$. The quotient space W_{k+1}/W_k is endowed with a Hodge structure of weight $k+1$ with respect to which the isomorphism $W_{k+1}/W_k \overset{\text{Res}}{\rightleftarrows} \operatorname{Ker}\gamma_{k+1}$ is a morphism of type $(-1, -1)$ of pure Hodge structures.

In order to define a mixed Hodge structure on X when D is a union of nonsingular subvarieties of codimension one, one proceeds analogously, using instead of the Gysin sequence the corresponding spectral sequence.

4.3. The Asymptotic Hodge Filtration in the Fibres of the Cohomology Bundle. The asymptotics of sections of the vanishing cohomology bundle associated with the singularity of a function allows us to define a Hodge filtration in the fibres of that bundle.

Let $\mathscr{H}_f^* \to T'$ be the bundle of the cohomology spaces of the nonsingular fibres of the singularity f over the punctured disk T' (see subsection 3.3). The geometric section of this bundle defined by a holomorphic n-form ω admits a series expansion

$$[\omega/df] = \sum_k t^\alpha (\ln t)^k A^\omega_{k,\alpha}/k!$$

in the variable $t \in T'$ whose coefficients are covariantly constant sections of the cohomology bundle (see subsection 3.8).

Let $\alpha(\omega)$ denote the smallest value of α for which at least one of the sections $A^\omega_{0,\alpha}, A^\omega_{1,\alpha}, \ldots$ is different from zero.

Definition. The number $\alpha(\omega)$ is called the *order of the geometric section* $[\omega/df]$. The section

$$[\omega/df]_{\max} = \sum_k t^{\alpha(\omega)} (\ln t)^k A^\omega_{k,\alpha(\omega)}/k!$$

is called the *principal part of the geometric section* $[\omega/df]$.

Let us define the *asymptotic Hodge filtration* F^p of the cohomology bundle $\mathscr{H}_f^*$ as follows: in each fibre $H^{n-1}(V_t; \mathbb{C})$ we single out the subspace F_t^p spanned by the values of the principal parts of the geometric sections of order less than or equal to $n - p - 1$.

The properties of the Hodge filtration are described by the following:

Theorem ([346]). (i) *F^p is an analytic subbundle of the cohomology bundle $\mathscr{H}_f^*$;*
(ii) *the asymptotic Hodge filtration is decreasing*: $F^{p+1} \subset F^p$;
(iii) *the subspace $F_t^p \subset H^{n-1}(V_t; \mathbb{C})$ is the direct sum of its intersections with the root subspaces of the monodromy operator, i.e., the Hodge filtration is invariant under the semisimple part of the monodromy operator*;
(iv) $F^n = \{0\}$, $F^0 = H_f^*$.

4.4. The Weight Filtration. The weight filtration in the cohomology bundle is defined as the standard filtration connected with the monodromy operator (see [160], [162], [293]).

For any nilpotent operator N acting on a finite-dimensional vector space H there exist one, and only one increasing filtration

$$0 \subset \cdots \subset W_k \subset W_{k+1} \subset \cdots \subset H$$

for which:

(i) $N(W_k) \subset W_{k-2}$;

(ii) $N^k\colon W_{s+k}/W_{s+k+1} \to W_{s-k}/W_{s-k-1}$ is an isomorphism for any integer k (see [293]).

Definition. The filtration W_k is called the *weight filtration of index s* of the nilpotent operator N.

A linear operator is said to be *semisimple* if the underlying space admits a basis of eigenvectors of the operator. An operator is said to be *unipotent* if all its eigenvalues are equal to one. Any invertible operator M decomposes uniquely into a product $M_u M_s$ of a unipotent operator M_u and a semisimple operator M_s such that M_u and M_s commute; M_u and M_s are called the *unipotent* and respectively the *semisimple parts* of M [299].

Now let M be the classical-monodromy operator and let $N = \sum(-1)^{i+1}(M_u - E)^i/i$ be the logarithm of its unipotent part. We use the nilpotent operator N to define the weight filtration W_k of the cohomology bundle $\mathscr{H}_f^*$ as follows.

For each eigenvalue λ of the monodromy operator M let $H_\lambda^{n-1}(V_t; \mathbb{C})$ denote the root subspace corresponding to λ in the fibre $H^{n-1}(V_t; \mathbb{C})$ of $\mathscr{H}_f^*$. Let ${}_\lambda W_{k,t}$ be the weight filtration of the nilpotent operator N, acting on $H_\lambda^{n-1}(V_t; \mathbb{C})$. We set the index s of the weight filtration ${}_\lambda W_{k,t}$ equal to n if $\lambda = 1$ and to $n-1$ otherwise. The collection of filtrations ${}_\lambda W_{k,t}$ on each root subspace $H_\lambda^{n-1}(V_t; \mathbb{C})$ defines a filtration

$$W_{k,t} = \bigoplus_\lambda {}_\lambda W_{k,t}$$

in any fibre $H^{n-1}(V_t; \mathbb{C})$ of the cohomology bundle $\mathscr{H}_f^*$, and hence a weight filtration $W_k \subset \mathscr{H}_f^*$ of that bundle.

Remark. The rule according to which the filtration index depends on the eigenvalue of the monodromy operator in the definition of the weight filtration is chosen so as to ensure that the weight and Hodge filtrations of the cohomology bundle will form together a mixed Hodge structure.

Theorem ([346]). (i) *W_k is an analytic subbundle of the cohomology bundle $\mathscr{H}_f^*$;*

(ii) *the weight filtration is increasing*: $W_{k+1} \supset W_k$;

(iii) *the weight filtration is invariant under the Gauss-Manin connection, i.e., the covariant derivative of a section of any of the subbundles W_k belongs again to W_k;*

(iv) *the weight filtration is invariant under the action of the semisimple part of the monodromy operator;*

(v) $W_{2n-2} = \mathscr{H}_f^*$.

Assertions (i) to (iv) follow immediately from the definitions, while (v) is a consequence of Steenbrink's theorem (see the end of subsection 4.5 below).

4.5. The Asymptotic Mixed Hodge Structure. Two filtrations were introduced above in the vanishing cohomology bundle H_f^*. The asymptotic Hodge filtration F^p is defined by the asymptotic behavior of integrals over vanishing cycles as the base parameter t tends to a critical value. The weight filtration W_k is described by the Jordan structure of the monodromy operator and reflects the behavior of integrals over vanishing cycles under analytic continuation of such integrals along paths that go around the critical value $t = 0$ of the parameter.

Theorem ([344], [345], [346]). *The weight and Hodge filtrations constitute a mixed Hodge filtration in the fibres of the vanishing cohomology bundle* $\mathscr{H}_f^*$ *associated with the singularity of the function* f (see subsection 4.2).

Example. Let $f = z_1^2 + \cdots + z_n^2$ be a Morse singularity. The cohomology bundle $\mathscr{H}_f^* \to T'$ is one-dimensional. From the example given in subsection 3.4 it follows that

$$\{0\} = F^{[n/2]+1} \subset F^{[n/2]} = \mathscr{H}_f^*,$$

and the subbundle $F^{[n/2]}$ is spanned by the principal part of a generic n-form ω.

The eigenvalue of the monodromy operator is $(-1)^n$. Thus, for a Morse singularity,

$$\{0\} = W_{2[n/2]-1} \subset W_{2[n/2]} = \mathscr{H}_f^*,$$

and the unique nontrivial quotient space of the weight filtration coincides with the fibre $H^{n-1}(V_t; \mathbb{C})$ of $\mathscr{H}_f^*$. The Hodge filtration induces on each fibre of $\mathscr{H}_f^*$ a pure Hodge structure of weight $2[n/2]$:

$$H^{n-1}(V_t; \mathbb{C}) = H_t^{[n/2],[n/2]} = F_t^{[n/2]} \cap \overline{F_t^{[n/2]}}.$$

Let $f\colon (\mathbb{C}^n, 0) \to (\mathbb{C}, 0)$ be a germ with an isolated singularity, and let $g\colon (\mathbb{C}^n, 0) \to (\mathbb{C}^n, 0)$ be a map-germ of finite multiplicity at zero. Assume that the singularity of the function $f \circ g$ is isolated. The germ g induces a mapping $g^*\colon \mathscr{H}_f^* \to \mathscr{H}_{f\circ g}^*$ of cohomology bundles.

Theorem ([346]). The map g^* *is a morphism of type* (0, 0) *of mixed Hodge structures in the vanishing cohomology.*

Another manifestation of the functoriality of the mixed Hodge structure is a set of formulas that express the spectrum of a direct sum of singularities $f \oplus g$ in terms of the spectra of the "components" f and g (see subsection 4.7).

From the theorem on the mixed Hodge structure one derives the following bound on the dimension of the Jordan blocks of the monodromy operator.

Theorem ([317]). *The dimension of any Jordan block of the monodromy operator is at most* n; *the dimension of any block with eigenvalue* 1 *is at most* $n - 1$.

4.6. The Hodge Numbers and the Spectrum of a Singularity. It follows from the theorem on the mixed Hodge structure that the asymptotic Hodge filtration F^p induces a decomposition of each quotient bundle $H^{(k)} = W_k/W_{k-1}$ into a direct sum of subbundles:

$$H^{(k)} = \bigoplus_{p+q=k} H^{p,q},$$

where $H^{p,q} = F^p \cap W_k / F^{p+1} \cap W_k + W_{k-1}$.

Next, from the theorems of subsections 4.3 and 4.4 we conclude that the semisimple part M_s of the monodromy operator acts on the bundles $H^{p,q}$. Let $h^{p,q}$ and $h^{p,q}_\lambda$ denote the dimensions of the fibre of $H^{p,q}$ and of the root subspace of M_s associated with the eigenvalue λ in that fiber, respectively.

Definition. The numbers $h^{p,q}$ and $h^{p,q}_\lambda$ are called the *Hodge numbers* of the mixed Hodge structure defined by the filtrations F^p, W_k.

The Hodge numbers enjoy the following symmetry properties:

(i) $h^{p,q}_\lambda = h^{q,p}_{\bar\lambda}$;

(ii) $h^{p,q}_\lambda = h^{n-1-p,n-1-q}_\lambda$, $\lambda \neq 1$;

(iii) $h^{p,q}_1 = h^{n-p,n-q}_1$.

To each eigenvalue λ of the operator M_s, acting in the fibre of the bundle $H^{p,q}$, one associates the pair of numbers

$$(n-1-l_p(\lambda), p+q), \qquad \text{if } \lambda \neq 1,$$

$$(n-1-l_p(\lambda), p+q-1), \qquad \text{if } \lambda = 1,$$

where $l_p(\lambda) = \ln(\lambda/2\pi i)$ [with the choice of branch [Re $l_p(\lambda)$] = p].

Definition. The unordered set of μ pairs of numbers obtained in the indicated manner is called the *set of spectral pairs* of the singularity f. The set of first numbers in the μ spectral pairs is called the *spectrum* of the singularity f.

The next theorem enables us to describe the spectrum of a singularity without resorting to the theorem on the mixed Hodge structure.

Theorem ([346], [348]). *Let $\omega_1, \ldots, \omega_\mu$ be a set of μ n-forms that gives a trivialization of the cohomology bundle $\mathcal{H}_f^* \to T'$* (see subsection 3.6), *such that sum of the orders $\alpha(\omega_i)$ of these forms $\sum \alpha(\omega_i) = \mu(n/2 - 1)$. Then the spectrum of the singularity f coincides with the set $\alpha(\omega_1), \ldots, \alpha(\omega_\mu)$.*

Thus, the spectrum of a singularity is completely determined by the Hodge filtration in the vanishing cohomology bundle and does not depend on the weight filtration.

As a corollary of the theorem we note that the sum of the numbers in the spectrum of the singularity equals $\mu(n/2 - 1)$.

Example. For a Morse singularity $f = z_1^2 + \cdots + z_n^2$ the Hodge numbers are $h^{[n/2],[n/2]} = h^{[n/2],[n/2]}_{(-1)^n} = 1$, and the spectral pair is $(n/2 - 1, n - 1)$.

The symmetry properties of the Hodge numbers can be formulated in terms of spectral pairs as follows: *the set of spectral pairs, regarded as a subset of* $\mathbb{R}^2$, *is centrally symmetric with respect to the point* $(n/2 - 1, n - 1)$. Correspondingly, *the spectrum of the singularity is symmetric about the point* $n/2 - 1$.

4.7. Computing the Spectrum. Let $f\colon (\mathbb{C}^n, 0) \to \mathbb{C}, 0)$ be a quasihomogeneous function of degree one with weights $\mathbf{v} = (v_1, \ldots, v_n)$, and let $\mathbf{z}^{\mathbf{k}}$, $\mathbf{k} \in I$, be a set of monomials that form a basis of the local algebra Q_f (see subsection 1.3.2).

Theorem ([318], [346]). *The spectrum of the singularity* f *coincides with the set of numbers* $l_{\mathbf{k}} = \langle \mathbf{k} + 1, \mathbf{v} \rangle - 1$. *The second numbers in the spectral pairs are all equal to* $n - 1$ (see the example in subsection 3.7).

The spectrum of a critical point of a semi-quasihomogeneous function is identical to the spectrum of its quasihomogeneous part.

For functions that are nondegenerate with respect to their Newton polyhedron formulas are available expressing the spectrum through the distribution of integer points in cones connected with the faces of the polyhedron [22], [102], [291], [317], [355]. The Hodge numbers $h^{p,q}$ can also be expressed in terms of the geometry of the Newton polyhedron [103].

The spectrum of a function of two variables whose zero level set is an irreducible curve-germ is expressible in terms of its Puiseux expansion [289]; see also the paper by Varchenko in [87].

Let $f \oplus g$ be a direct sum of singularities $f\colon (\mathbb{C}^n, 0) \to (\mathbb{C}, 0)$ and $g\colon (\mathbb{C}^m, 0) \to (\mathbb{C}, 0)$ (see subsection 2.2). The cohomology bundle $\mathscr{H}^*_{f \oplus g}$ is isomorphic to the tensor product of the cohomology bundles of the "components", $\mathscr{H}^*_f \otimes \mathscr{H}^*_g$. One can show that if the sets of forms $\omega_1, \ldots, \omega_{\mu(f)}$ and $\varphi_1, \ldots, \varphi_{\mu(g)}$ are trivializations of $\mathscr{H}^*_f$ and $\mathscr{H}^*_g$, respectively, then the set of forms $\{\omega_i \wedge \varphi_j\}$ is a trivialization of $\mathscr{H}^*_{f \oplus g}$. The order of the exterior product is expressible in terms of the orders of the factors as $\alpha(\omega \wedge \varphi) = \alpha(\omega) + \alpha(\varphi) + 1$. The theorem of subsection 4.6 therefore admits the following consequence.

Theorem ([346]). *Let* $\{\alpha_i\}_{i=1}^{\mu(f)}$ *and* $\{\beta_j\}_{j=1}^{\mu(g)}$ *be the spectra of the critical points of* f *and* g, *respectively. Then the spectrum of the critical point of the direct sum* $f \oplus g$ *is* $\{\alpha_i + \beta_j + 1\}_{i=1, j=1}^{\mu(f), \mu(g)}$.

This theorem describes, in particular, how the spectrum behaves under stabilization: the spectrum of the critical point of the function $f + z^2$ is obtained from the spectrum of the singularity f through a shift to the right by 1/2.

The spectra of simple, unimodal and bimodal singularities are given in the three tables below (see [155]). Here the number of variables is $n = 3$, and for each critical point the numbers L_s and N, in terms of which the spectrum $\{\alpha_s\}$ is given by the formula $\alpha_s = (L_s/N) - 1$, are indicated. In view of the symmetry of the spectrum about the point $n/2 - 1 = 1/2$, only half of each spectrum is shown (i.e., for $s \leqslant \mu/2$), except for the case of the singularities A_μ, D_μ, and $T_{p,q,r}$.

	N	L_r		N	L_r
A_μ	$\mu + 1$	$\mu + 1 + k,\ 1 \leqslant k \leqslant \mu$	E_6	12	13, 16, 17
D_μ	$2\mu - 1$	$3\mu - 3,\ 2\mu - 1 + 2k$, $0 \leqslant k \leqslant \mu - 2$	E_7	18	19, 23, 25
			E_8	30	31, 37, 41, 43
P_8	3	3, 4, 4, 4	Z_{12}	22	21, 25, 27, 29, 31, 33
X_9	4	4, 5, 5, 6	Z_{13}	18	17, 20, 22, 23, 25, 29
J_{10}	6	6, 7, 8, 8, 9	W_{12}	20	19, 23, 24, 27, 28, 29
Q_{10}	24	23, 29, 31, 32, 35	W_{13}	16	15, 18, 19, 21, 22, 23
Q_{11}	18	17, 21, 23, 24, 25	E_{12}	42	41, 47, 53, 55, 59, 61
Q_{12}	15	14, 17, 19, 20, 20, 22	E_{13}	30	29, 33, 37, 39, 41, 43
S_{11}	16	15, 19, 20, 21, 23	E_{14}	24	23, 26, 29, 31, 32, 34, 35
S_{12}	13	12, 15, 16, 17, 18, 19	$T_{p,q,t}$	pqt	$pqt,\ 2pqt,\ (p + k_1)qt$,
U_{12}	12	11, 14, 15, 15, 17, 18			$p(q + k_2)t,\ pq(t + k_3)$,
Z_{11}	30	29, 35, 37, 41, 45			$0 < k_1 < p,\ 0 < k_2 < q$, $0 < k_3 < t$

$J_{3,p}$	$18(p + 9)$	$17(p + 9),\ 23(p + 9),\ 25(p + 9),\ 9(2p + 17 + 2k)$, $1 \leqslant k \leqslant (p + 10)/2$
$Z_{1,p}$	$14(p + 7)$	$13(p + 7),\ 17(p + 7),\ 19(p + 7),\ 21(p + 7)$, $7(2p + 13 + 2k),\ 1 \leqslant k \leqslant (p + 7)/2$
$W_{1,p}$	$12(p + 6)$	$11(p + 6),\ 14(p + 6),\ 16(p + 6),\ 17(p + 6)$, $6(2p + 11 + 2k),\ 1 \leqslant k \leqslant (p + 7)/2$
$W^{\#}_{1,p}$	$12(p + 12)$	$11(p + 12),\ 17(p + 12),\ 12(p + 12 + k)$, $1 \leqslant k \leqslant (p + 11)/2$
$Q_{2,p}$	$12(p + 6)$	$11(p + 6),\ 15(p + 6),\ 16(p + 6),\ 17(p + 6)$, $6(2p + 11 + 2k),\ 1 \leqslant k \leqslant (p + 6)/2$
$S_{1,p}$	$10(p + 5)$	$9(p + 5),\ 12(p + 5),\ 13(p + 5),\ 14(p + 5)$, $5(2p + 9 + 2k),\ 1 \leqslant k \leqslant (p + 6)/2$
$S^{\#}_{1,p}$	$10(p + 10)$	$9(p + 10),\ 13(p + 10),\ 10(p + 10 + k)$, $1 \leqslant k \leqslant (p + 10)/2$
$U_{1,p}$	$9(p + 9)$	$8(p + 9),\ 11(p + 9),\ 13(p + 9),\ 9(p + 9 + k)$, $1 \leqslant k \leqslant (p + 8)/2$

Q_{16}	21	19 22 25 26 28 28 29 31
Q_{17}	30	27 31 35 37 39 40 41 43
Q_{18}	48	43 49 55 59 61 64 65 67 71
S_{16}	17	15 18 20 21 22 23 24 25
S_{17}	24	21 25 28 29 31 32 33 35
U_{16}	15	13 16 18 18 19 21 21 22
W_{17}	20	18 21 23 24 26 27 28 29
W_{18}	28	25 29 32 33 36 37 39 40 41
Z_{17}	24	22 25 28 29 31 32 34 35
Z_{18}	34	31 35 39 41 43 45 47 49 51
Z_{19}	54	49 55 61 65 67 71 73 77 79
E_{18}	30	28 31 34 37 38 40 41 43 44
E_{19}	42	39 43 47 51 53 55 57 59 61
E_{20}	66	61 67 73 79 83 85 89 91 95 97

4.8. Semicontinuity of the Spectrum. Here we give results describing the behavior of the spectrum of a critical point under deformations.

Let $f(z)$ be a germ with an isolated singularity at zero, and let $F(z, \lambda)$, $\lambda \in \Lambda$, be a deformation of f. With no loss of generality we may assume that $F(z, \lambda)$ is versal. The fibration of nonsingular fibres of f over the punctured disk, $V'_T \to T'$, is a subfibration of the Milnor fibration $V'_\Lambda \to \Lambda'$ over the complement $\Lambda' = \Lambda \setminus \Sigma$ of the level bifurcation set in the base of the deformation (see subsections 1.10 and 3.3).

Denote by $\mathrm{sp}(\lambda)$ the union of the spectra of the singularities of the fiber V_λ, $\lambda \in \Lambda$ (for a nonsingular fiber V_λ, $\lambda \in \Lambda'$, we put $\mathrm{sp}(\lambda) = \varnothing$). Thus, $\mathrm{sp}(0)$ is exactly the spectrum of the singularity f.

Definition ([350]). A subset $M \subset \mathbb{R}$ is called a *semicontinuity set of the spectrum* if

$$\#(\mathrm{sp}(0) \cap M) \geqslant \#(\mathrm{sp}(\lambda) \cap M).$$

The semicontinuity of the spectrum of a singularity with respect to deformations is described by the following

Theorem ([320] [350], [87]). *Any semi-interval* $]\alpha, \alpha + 1]$ *is a semicontinuity set.*

The semicontinuity of the spectrum was first conjectured by Arnol'd [25] for the sets $]-\infty, \alpha]$, and follows from the above theorem.

The assertion on the semicontinuity of the spectrum on a half-line can be recast in the following form.

Order the spectrum of the critical point: $\alpha_1 \leqslant \alpha_2 \leqslant \cdots \leqslant \alpha_\mu$. Suppose that a critical point of type X with the spectrum $\alpha_1 \leqslant \cdots \leqslant \alpha_\mu$ is adjacent (see subsection 1.2.7) to a critical point of type X' with the spectrum $\alpha'_1 \leqslant \cdots \leqslant \alpha'_{\mu'}$, where $\mu' \leqslant \mu$.

Theorem. *A necessary condition for the adjacency* $X \to X'$ *is that the spectra be adjacent in the sense that* $\alpha_i \leqslant \alpha'_i$ (see [25]).

From the symmetry of the spectrum about the point $n/2 - 1$ and the necessary condition of adjacency of the spectra one derives the two-sided inequalities

$$\alpha_i \leqslant \alpha'_i \leqslant \alpha_i + (\mu - \mu').$$

For example, if $\mu' = \mu - 1$, then the spectrum of the singularity X' interlaces that of the singularity X.

The relationship between the spectra of the singularities X and X' is identical to that between the semiaxes of an ellipsoid in $\mathbb{R}^\mu$ and the semiaxes of its section by the subspace $\mathbb{R}^{\mu'}$.

For adjacencies of the initial segment of the classification of singularities the condition of adjacency of the spectra is "almost" sufficient (see subsection 2.7, and also [155]).

A particular consequence of the preceding theorem is the following assertion concerning the behavior of the spectrum under multiplicity-preserving deformations of a critical point.

Theorem ([345], [348]). *The spectrum is constant on the* $\mu =$ const *stratum.*

Thus, for example, under multiplicity-preserving deformations, the largest possible order of the integral of a holomorphic form over the classes of a covariantly constant family of vanishing cycles does not change. This means that if for some deformation the largest possible order does change, then the critical point decomposes under that deformation. This reasoning leads to the following lower bound for the codimension of the $\mu =$ const stratum.

Theorem ([347], [353]). *The codimension of the* $\mu =$ const *stratum in the base of the versal deformation is not smaller than the number of spectral numbers* α_i *such that* $\alpha_i < \alpha_1 + 1$, *where* α_1 *is the left endpoint of the spectrum.*

In the case of a critical point of a semi-quasihomogeneous function this lower bound for the codimension of the $\mu =$ const stratum coincides with the upper bound [203]; this enables us to specify the modality and the stratum explicitly.

Theorem ([347]). *The modality of a semi-quasihomogeneous function coincides with its intrinsic modality. The* $\mu =$ const *stratum in the base of the monomial versal deformation* $F = f + \sum_{i=1}^{\mu} \lambda_i e_i$ *of a quasihomogeneous function* f *is given by the equations* $\{\lambda_i = 0 \colon e_i$ *is a lower monomial*$\}$ (*see subsections* 1.3.6 *and* 1.3.7).

For homogeneous germs this theorem is proved in [136].

Steenbrink's theorem on the semicontinuity of the spectrum is a somewhat weaker version of Varchenko's conjecture [350] that the spectrum is semicontinuous on the interval $]\alpha, \alpha + 1[$. That conjecture has been confirmed for $\alpha < -1$ [349], for singularities of functions in two variables [350], and for deformations of quasihomogeneous functions by lower monomials [350].

4.9. The Spectrum and the Geometric Genus. Let V be the zero level set of a germ $f \colon (\mathbb{C}^n, 0) \to (\mathbb{C}, 0)$. Let Ω_V^p denote the sheaf of germs of holomorphic p-forms on V. Further, let $\pi \colon W \to V$ be a resolution of singularities at the point $0 \in V$, i.e., a proper holomorphic mapping of a nonsingular manifold W that is an isomorphism in the complement of the divisor with normal intersections $\pi^{-1}(0) \subset W$. Denote by $\pi_* \Omega_W^p$ the direct image of the sheaf Ω_W^p under π.

Definition. The $\mathbb{C}$-dimension of the space $H_{V,0} = \Omega_{V,0}^{n-1} / \pi_* \Omega_{W,\pi^{-1}(0)}^{n-1}$ is called the *geometric genus* of the singular point $0 \in V$.

Theorem ([290]). *The geometric genus of a singular point is equal to the number of its spectral numbers lying in the semi-interval* $]0, 1]$.

From the theorem on the semicontinuity of the spectrum given in subsection 4.8 one derives the following result.

Theorem ([125]). *The geometric genus of a singularity is semicontinuous with respect to deformations.*

4.10. The Mixed Hodge Structure and the Intersection Form. Here we give a description of the indices of inertia of the quadratic form of a singularity in terms of the mixed Hodge structure.

Let Φ be the intersection form in the homology $H_{n-1}(V_*; \mathbb{R})$ of the nonsingular level V_* of the singularity of the germ f (see subsection 1.6). Let μ_0 denote the dimension of the kernel of Φ. If n is even, then the intersection form Φ is skew-symmetric and μ_0 is its unique real invariant. If n is odd, then Φ is symmetric and hence can be diagonalized by means of a real linear transformation. Let μ_+ and μ_- be the positive and respectively the negative index of inertia of the corresponding quadratic form. The numbers μ_0, μ_+, μ_- form a complete set of real invariants of the form Φ.

These invariants can be expressed in terms of the mixed Hodge structure associated with the singularity.

Theorem ([317]).

$$\mu_0 = \sum_{p+q\leqslant n} h_1^{p,q} - \sum_{p+q\geqslant n+2} h_1^{p,q}.$$

For n odd

$$\mu_\pm = \sum_{p+q=n+1} h_1^{p,q} + 2 \sum_{p+q\geqslant n+2} h_1^{p,q} + \sum_{\lambda\neq 1} \sum_{q} h_\lambda^{p,q},$$

where in the formula for μ_+ [resp. μ_-] q assumes only even [resp. odd] values.

In particular, if n is even then $\mu - \mu_0$ is also even. For $n \equiv 3 \pmod 4$, $\mu - \mu_-$ is even, whereas for $n \equiv 1 \pmod 4$, $\mu - \mu_+$ is even.

Another corollary of the theorem is the following nondegeneracy condition for the intersection form.

Theorem. *The intersection form Φ is nondegenerate if and only if 1 is not an eigenvalue of the monodromy operator.*

In the case of a quasihomogeneous function the Hodge numbers $h_\lambda^{p,q}$ can be expressed in terms of the quasihomogeneous structure of the local algebra of the singularity (see subsection 4.7). In this case the preceding theorem can be recast in the following form.

Theorem ([318]). *The dimension μ_0 of the kernel of the intersection form is equal to the number of integers in the spectrum of the singularity. For n odd, μ_+ [resp. μ_-] is equal to the number of nonintegers in the spectrum whose integer part is odd [resp. even].*

Example. For a Pham singularity $f = z_1^{\alpha_1} + \cdots + z_n^{\alpha_n}$ the intersection form is nondegenerate if the numbers $\alpha_1, \ldots, \alpha_n$ are pairwise coprime.

The intersection form of a quasihomogeneous singularity is expressible through the multiplication operation in its local algebra [147], [351].

4.11. The Number of Singular Points of a Complex Projective Hypersurface. Let $V \subset \mathbb{C}P^n$ be an algebraic hypersurface of degree d whose singular points are all isolated.

What is the maximal number $N_n(d)$ of nondegenerate singular points that a hypersurface of degree d may have?

A complete answer to this question is known only for $n = 1$ and $n = 2$, namely $N_1(d) = [d/2]$ and $N_2(d) = d(d-1)/2$.

The first nontrivial case is $n = 3$.

Theorem (Basset [46]).

$$N_3(d) \leqslant [d(d-1)^2 - 5 - \{d(d-1)(3d-14) + 25\}^{1/2}]/2.$$

This yields an estimate with the asymptotics $d^3/2$ as $d \to \infty$. Subsequently the estimate has been improved and generalized to the case $n > 3$; however, in all the works the asymptotics of the estimate as $d \to \infty$ was $d^n/2$ [47], [67], [315], [148].

The theorem on the semicontinuity of the spectrum enables us to obtain an upper estimate with a new asymptotics.

Let $A_n(d)$ be the number of integer points $(k_1, \ldots, k_n)$ that lie strictly inside the interior of the cube $[0, d]^n$ and satisfy the inequalities $(n-2)d/2 + 1 \leqslant \sum k_i \leqslant nd/2$. For example, $A_3(d) = 23d^3/48 +$ (terms of lower order in d).

Theorem ([350]). *Let $V \subset \mathbb{CP}^n$ be an algebraic hypersurface of degree d with isolated singular points. Then the number of singular points of V does not exceed $A_n(d)$.*

For fixed n the number $A_n(d)$ is a polynomial of degree n in d. The coefficient of the highest power d^n in this polynomial has the asymptotics $\sqrt{6/\pi n}$ for $n \to \infty$.

Examples of hypersurfaces with a large number of singular points, which provide a lower bound for the number $N_n(d)$, were given by Chmutov (see [38, p. 419]). His hypersurface is defined by the affine equation

$$\sum_{i=1}^{n} T_d(x_i) = [1 + (-1)^{n+1}]/2,$$

where $T_d(x_i)$ is the Chebyshev polynomial of degree d with critical values ± 1. The number of singular points of this hypersurface is a polynomial of degree n in d; the coefficient of d^n has the asymptotics $\sqrt{2/\pi n}$ for $n \to \infty$.

Thus, the upper and lower bounds available presently for the number $N_n(d)$ have the same order and differ by the coefficient $\sqrt{3}$.

For $d = 3$ and small values of n the results of the theorem and the known examples are shown in the table below.

n	1	2	3	4	5	6	7	8	9
$A_n(3)$	1	3	4	10	15	35	56	126	196
Examples	1	3	4	10	15	33	54	118	189

Note that the equality $N_5(3) = 15$ is a theorem of Togliatti [331].

4.12. The Generalized Petrovskiĭ-Oleĭnik Inequalities. Let us consider a nonsingular real algebraic curve of degree m in the real projective plane. The (connected) components of this curve (each of which is diffeomorphic to a circle) are called *ovals*.

Finding the relative disposition of the ovals of a real algebraic curve is one of the classical problems in geometry (cf. Hilbert's 16th problem). A complete solution is known only for $m \leqslant 6$ (in the case $m = 6$ this is a result obtained in 1969), whereas even today no description is known of all possible dispositions of ovals of a curve of degree 8 [168], [365], [36].

Along with descriptions of the disposition of ovals of low-degree curves, general estimates of various numerical characteristics of algebraic curves and hypersurfaces of a given degree are known. The Petrovskiĭ-Oleĭnik inequalities are one set of such estimates.

Let $V \subset \mathbb{RP}^{n-1}$ be a nonsingular projective hypersurface of degree d, given by a homogeneous polynomial f in n variables. If d is even, denote by V_+ and V_- the subsets of $\mathbb{RP}^{n-1}$ defined by the condition $f \geqslant 0$ and $f \leqslant 0$, respectively.

Definition. The *Petrovskiĭ number* $\Pi_n(d)$ is the number of integer points that lie strictly inside the interior of the cube $[0, d]^n$ in the hyperplane passing through its center and orthogonal to its principal diagonal:

$$\Pi_n(d) = \#\{\mathbf{k} = (k_1, \dots, k_n): 0 < k_i < d, \textstyle\sum k_i = nd/2\}.$$

In the notations introduced above, the Petrovskiĭ-Oleĭnik inequalities are written as

$$|\chi(V) - 1| \leqslant \Pi_n(d), \qquad \text{if } n \text{ is even,}$$

and

$$|\chi(V_+) - \chi(V_-)| \leqslant \Pi_n(d), \qquad \text{if } n \text{ is odd and } d \text{ is even;}$$

here $\chi(\cdot)$ denotes the Euler characteristic. They generalize an inequality of Petrovskiĭ obtained for curves of even degree [260], i.e., for $n = 3$.

The following unified form of the Petrovskiĭ-Oleĭnik inequalities was proposed in [22].

Let ind denote the index of the singular point 0 of the gradient, in $\mathbb{R}^n$, of the polynomial f that defines the hypersurface V.

Theorem ([22]). *The number appearing in the left-hand side of the Petrovskiĭ-Oleĭnik inequalities is equal to* |ind| *for any value of* n.

The Petrovskiĭ number $\Pi_n(d)$, which appears in the right-hand side of the inequalities, is expressible in terms of the mixed Hodge structure of the critical point 0 of the polynomial f, regarded as a function on $\mathbb{C}^n$.

Theorem ([22]). *The Petrovskiĭ number $\Pi_n(d)$ coincides with the number of spectral pairs of the critical point of the germ f that are equal to* $(n/2 - 1, n - 1)$.

The resulting estimate for the index of the gradient vector field in terms of the number of spectral pairs of the type indicated also remains valid in the general case. The following assertion, formulated as a conjecture in [22], was proved in [351].

Theorem. *Let f be an arbitrary germ of real-analytic function on $\mathbb{R}^n$ having an isolated (in the complex sense) singular point at zero. Then the absolute value of the index of the gradient vector field ∇f does not exceed the number of spectral pairs of the germ f that are equal to $(n/2 - 1, n - 1)$.*

The number ind coincides, up to sign, with the reduced (i.e., reduced by 1) Euler characteristic of the local nonsingular complex level manifolds $V_\pm = f^{-1}(\pm\varepsilon)$, $0 < \varepsilon \ll 1$ (see [22]).

Let σ_+ and σ_- denote the actions of the complex-conjugation involution in the homology of the manifolds V_ε and $V_{-\varepsilon}$, respectively. If $(\cdot, \cdot)$ denotes the intersection form in the mid-dimensional homology of the corresponding manifold, the bilinear forms $(\sigma_\pm \cdot, \cdot)$ are symmetric for any n. The following local analogue of the Atiyah-Singer formula [40] for involutions on $4k$-dimensional manifolds holds for the number ind:

Theorem ([173], [351]; conjectured in [22]). *For $n = 2k + 1$,*

$$\mathrm{ind} = (-1)^k(\mathrm{sign}(\sigma_+ \cdot, \cdot) - \mathrm{sign}(\sigma_- \cdot, \cdot))/2.$$

Here sign denotes the difference of the positive and negative indices of inertia of a quadratic form.

Theorem ([173]). *For $n = 2k$,*

$$\mathrm{sign}(\sigma_+ \cdot, \cdot) = -\mathrm{sign}(\sigma_- \cdot, \cdot).$$

Concerning the index of a nongradient vector field see subsection 3.1.12.

§ 5. Simple Singularities

In the case of a simple singularity the quadratic form is negative definite, and the monodromy group Γ is a finite group that is generated by reflections in the Euclidean space $H_{n-1}(V_*; \mathbb{R})$ and preserves the integer lattice $H_{n-1}(V_*; \mathbb{Z})$, i.e., a Weyl group. In view of this connection, let us begin the present section by reviewing the needed information about reflection groups.

5.1. Reflection Groups ([56]). Consider the Euclidean space $\mathbb{R}^\mu$, with the inner product denoted by $\langle \cdot, \cdot \rangle$. The *reflection* s_v, $v \in \mathbb{R}^\mu$, in the hyperplane $H = \{x: \langle x, v \rangle = 0\}$, is defined as the orthogonal transformation

$$s_v: x \mapsto x - 2\langle x, v\rangle/\langle v, v\rangle$$

of $\mathbb{R}^\mu$. The reflection s_v is an involution whose fixed-point set coincides with the hyperplane H; the latter will be referred to as the *mirror* of s_v.

Consider a finite irreducible group W generated by reflections in $\mathbb{R}^\mu$ (such groups will be referred to as *Coxeter groups*). Each reflection $s_i \in W$ determines a *mirror* H_i. The set of all such mirrors divides $\mathbb{R}^\mu$ into domains, called the *chambers* of the group W. Fix one of the chambers, C, and for each mirror H_i choose from the two half-spaces into which H_i divides $\mathbb{R}^\mu$ the one that contains C. Let $H_1, \ldots, H_k$ be the minimal set of mirrors such that the intersection of the chosen half-spaces coincides with C. Then $H_1, \ldots, H_k$ are called the *walls of the chamber* C.

Theorem. (i) *The chamber C is a simplicial cone.*
(ii) *The group W is generated by the reflections in the walls $H_1, \ldots, H_k$ of C.*
(iii) *C is a fundamental domain of W.*
(iv) *The angles made by the walls of C can take only the values π/k, $k \geqslant 2$.*
(v) *The group W is defined by the following set of relations:*

$$s_i^2 = 1, \qquad (s_i s_j)^{m_{ij}} = 1,$$

where s_i is the reflection in the wall H_i of C and the angle between the walls H_i and H_j is π/m_{ij}.

Remark. The angles between walls are measured inside the chamber C.

The geometry of a chamber of the group W is completely determined by the *Coxeter graph*, whose vertices correspond to the walls of the chamber; two vertices are connected by an edge of multiplicity k if the angle made by the corresponding walls equals π/k. Groups with identical Coxeter graphs are isomorphic.

Remark. According to a widely accepted convention, one does not draw edges of multiplicity 2, and for edges of multiplicity 3 one does not indicate the multiplicity.

Example. In the plane $\mathbb{R}^2$ consider the symmetry group W of a regular polygon with p sides. W is generated by the reflections in the lines passing through the vertices and the midpoints of the sides of the p-gon.

The Coxeter graph of W consists of two vertices connected by an edge of multiplicity p (Fig. 37).

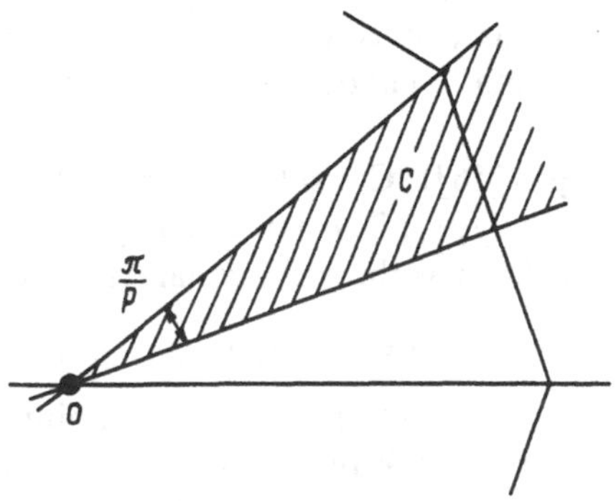

Fig. 37

Example. Consider the group $W = S_{\mu+1}$ of permutations of $\mu + 1$ elements, which acts in $\mathbb{R}^{\mu+1}$ by permuting the coordinates:

$$\sigma\colon (x_1, \ldots, x_{\mu+1}) \mapsto (x_{\sigma(1)}, \ldots, {}_{\sigma(\mu+1)}).$$

This representation is reducible and decomposes into the direct sum of the trivial one-dimensional representation in the line $\{x\colon x_1 = \cdots = x_{\mu+1}\}$ and the irreducible μ-dimensional representation in the subspace $\mathbb{R}^\mu = \{x\colon x_1 + \cdots + x_{\mu+1} = 0\}$. That μ-dimensional representation is generated by the reflections s_{ij} in the planes $\{x\colon x_i = x_j\}$. Consider the chamber $C = \{x\colon x_1 \leqslant x_2 \leqslant \cdots \leqslant x_{\mu+1}\}$ with walls $H_i = \{x\colon x_i = x_{i+1}\}$. The angles between walls are $\widehat{H_i, H_j} = \pi/2$ if $i - j > 1$, and $\widehat{H_i, H_{i+1}} = \pi/3$. Correspondingly, the Coxeter diagram of W has the form

(μ vertices).

Theorem. *W is a finite irreducible reflection group if and only if its Coxeter graph is isomorphic to one of the following graphs:*

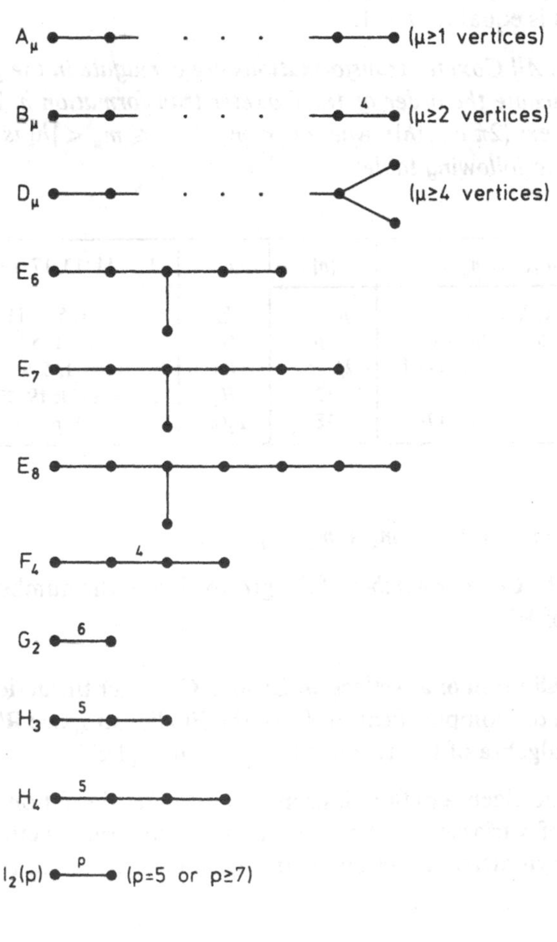

The composition $h = s_1 \cdot s_2 \cdot \ldots \cdot s_\mu$ of the reflections s_i in the walls of one of the chambers of the group W is called a *Coxeter transformation.*

Example. For a group W of type A_μ the Coxeter transformation h is the composition of the reflections s_i that correspond to the transpositions $(i, i+1)$ in the group of permutations. Hence, the composition $s_1 \cdot s_2 \cdot \ldots \cdot s_\mu$ corresponds to the cyclic permutation $(1, 2, 3, \ldots, \mu + 1) \mapsto (\mu + 1, 1, 2, \ldots, \mu)$ and is given in $\mathbb{R}^{\mu+1}$ by the matrix

$$\begin{bmatrix} 0 & 0 & \cdots & 0 & 1 \\ 1 & 0 & \cdots & 0 & 0 \\ 0 & 1 & \cdots & 0 & 0 \\ \cdots & \cdots & \cdots & \cdots & \cdots \\ 0 & 0 & \cdots & 1 & 0 \end{bmatrix}$$

The characteristic polynomial of this matrix is $t^{\mu+1} - 1$, and so the eigenvalues of the Coxeter transformation are $\exp(2\pi ik/(\mu + 1))$, $k = 1, \ldots, \mu$ (the eigenvalue 1 of the matrix should be disregarded, as it corresponds to the invariant line $\{x: x_1 = x_2 = \cdots = x_{\mu+1}\}$ in $\mathbb{R}^{\mu+1}$). The order of the Coxeter transformation is equal to $\mu + 1$.

Theorem. (i) *All Coxeter transformations are conjugate in the group W.*

(ii) *Let $|h|$ denote the order of the Coxeter transformation h. Then the eigenvalues of h are* $\exp(2\pi\, im_j/|h|)$, *where $0 < m_1 \leqslant \cdots \leqslant m_\mu < |h|$ is the set of integers shown in the following table:*

W	$m_1, \ldots, m_\mu$	$\|h\|$	W	$m_1, \ldots, m_\mu$	$\|h\|$
			E_8	1, 7, 11, 13, 17, 19, 23, 29	30
A_μ	$1, 2, \ldots, \mu$	$\mu + 1$	F_4	1, 5, 7, 11	12
B_μ	$1, 3, 5, \ldots, 2\mu - 1$	2μ	G_2	1, 5	6
D_μ	$1, 3, 5, \ldots, 2\mu - 3, \mu - 1$	$2\mu - 2$	H_3	1, 5, 9	10
E_6	1, 4, 5, 7, 8, 11	12	H_4	1, 11, 19, 29	30
E_7	1, 5, 7, 9, 11, 13, 17	18	$I_2(p)$	$1, p - 1$	p

(iii) $m_1 = 1$, $m_\mu = |h| - 1$, $m_k + m_{\mu+1-k} = |h|$.

$|h|$ is called the *Coxeter number* of the group W, and the numbers m_j are called the *exponents* of W.

5.2. The Swallowtail of a Reflection Group. Consider the action of the Coxeter group W on the complexification $\mathbb{C}^\mu$ of the Euclidean space $\mathbb{R}^\mu$. Let $\mathbb{C}[[x]]^W$ denote the subalgebra of W-invariant functions in $\mathbb{C}[[x]]$.

Example. The algebra of functions invariant under the action of the group of permutations of variables is the algebra of symmetric functions. As is well known, every symmetric function of the variables $x_1, \ldots, x_{\mu+1}$ is expressible through the generators λ_i, $i = 0, \dots, \mu$:

$$\lambda_i(x) = (-1)^k \sum_{j_1 \neq j_2 \neq \dots \neq j_k} x_{j_1} x_{j_2} \dots x_{j_k}, \qquad k = \mu - i + 1,$$

among which there are no relations. It follows that the homogeneous polynomials λ_i, $i = 0, 1, \dots, \mu - 1$, of respective degrees $\mu + 1, \mu, \dots, 2$, generate the free algebra $\mathbb{C}[[x]]^W$ for a group of type A_μ (the generator $\lambda_\mu \equiv 0$ on the hyperplane $\mathbb{C}^\mu \subset \mathbb{C}^{\mu+1}$).

Theorem ([314]). *$\mathbb{C}[[x]]^W$ is a free algebra with μ generators, which are homogeneous polynomials of degrees $m_i + 1$, where m_i are the exponents of W.*

A consequence of this theorem is that the orbit space $\mathbb{B}^\mu = \mathbb{C}^\mu/W$ of the action of W on $\mathbb{C}^\mu$ is isomorphic to $\mathbb{C}^\mu$. The canonical projection $\mathbb{C}^\mu \to \mathbb{B}^\mu$ will be referred to as the *Vieta map*. It is a branched covering whose number of sheets equals the order $|W|$ of the group W. The set $\Sigma(W) \subset \mathbb{B}^\mu$ of branching points of this covering is a hypersurface, called the *(generalized) swallowtail of* W. It coincides with the image of the set of irregular orbits (i.e., the union H of all mirrors of W) under the Vieta map. The functions on $\mathbb{B}^\mu$ will be called the *invariants* of the Coxeter group W.

Example. The orbit of a point $x \in \mathbb{C}^{k+1}$ such that $x_1 + \dots + x_{\mu+1} = 0$ under the action of the group A_μ is described by the unordered set of $\mu + 1$ points (counting multiplicities) $x_1, \dots, x_{\mu+1}$ on $\mathbb{C}$ and is given by the polynomial

$$\prod_{i=1}^{\mu+1} (t - x_i) = t^{\mu+1} + \lambda_{\mu-1} t^{\mu-1} + \dots + \lambda_1 t + \lambda_0,$$

with the coefficients $\lambda_i(x)$ specified in the preceding example. The swallowtail $\Sigma(A_\mu)$ is the set of zeros of the discriminant of the polynomial. The real parts of the hypersurfaces $\Sigma(A_2) = \{4\lambda_1^3 + 27\lambda_0^2 = 0\}$ and $\Sigma(A_3)$ are shown in Fig. 38.

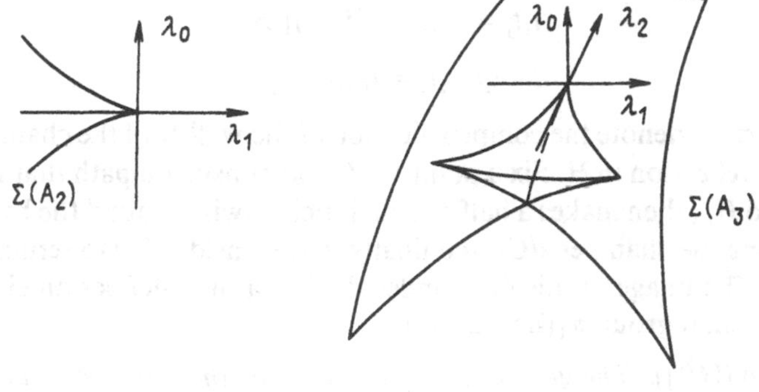

Fig. 38

Example. The mapping $\mathbb{C}^{\mu+1} \to \mathbb{C}^{\mu}$,

$$(x_1, \dots, x_{\mu+1}) \mapsto (x_1 - \bar{x}, \dots, x_\mu - \bar{x}), \qquad \bar{x} = \frac{1}{\mu}(x_1 + \cdots + x_\mu),$$

defines a locally trivial fibration of the complement $\{x: x_1 + \cdots + x_{\mu+1}v = 0, x_i \neq x_j\}$ of the set of mirrors of the Coxeter group A_μ, with base $\{x: x_1 + \cdots + x_\mu = 0, x_i \neq x_j\}$, the complement of the set of mirrors of the group $A_{\mu-1}$, and with fibre {$\mathbb{C}$ with μ points removed}. From the exact homotopy sequence of this fibration it follows that the homotopy groups π_i, $i \geqslant 2$, of the complement of the mirrors of the group A_μ are equal to those of the base, that is, of the complement of the mirrors of $A_{\mu-1}$. By induction, one concludes that all the groups π_i with $i \geqslant 2$ vanish. Therefore, the complement of the mirrors of A_μ, and consequently the complement of the swallowtail, $\mathbb{B}^\mu \setminus \Sigma(A_\mu)$, has the homotopy type of an Eilenberg-Mac Lane $K(\pi, 1)$-space.

This result (but not the argument that established it!) admits the following generalization.

Theorem ([106]). *Let $\mathscr{H} = \{H_i\}$ be a finite set of hyperplanes in the space $\mathbb{R}^\mu$ such that $\mathscr{H}$ divides $\mathbb{R}^\mu$ into simplicial cones. Let $\mathscr{H}_\mathbb{C}$ denote the set of complexified hyperplanes in the complexification $\mathbb{C}^\mu$ of $\mathbb{R}^\mu$. Then the complement $\mathbb{C}^\mu \setminus \mathscr{H}_\mathbb{C}$ is a $K(\pi, 1)$-space.*

An immediate consequence of this result is the following:

Theorem. *The complement $\mathbb{B}^\mu \setminus \Sigma(W)$ of the swallowtail of a Coxeter group W is an Eilenberg-Mac Lane space.*

5.3. The Artin-Brieskorn Braid Group. The fundamental group of the complement $\mathbb{B}^\mu \setminus \Sigma(W)$ of the swallowtail of the Coxeter group W is called the *Artin-Brieskorn braid group* of W (or a *generalized braid group*).

Example. The complement of the swallowtail of the group A_μ coincides with the space of polynomials of degree $\mu + 1$ without multiple zeroes, the fundamental group of which is the group $\mathrm{Br}(\mu + 1)$ of braids of $\mu + 1$ strands (subsection 1.9), with generators t_i, $i = 1, \dots, \mu$, and the set of defining relations

$$t_i t_j = t_j t_i, \qquad |i - j| > 1,$$

$$t_i t_{i+1} t_i = t_{i+1} t_i t_{i+1}.$$

Let $H_\mathbb{C} \subset \mathbb{C}^\mu$ denote the complexification of the wall H of the chamber C, and let s be the reflection in H. Fix a point $* \in C$ and consider a path that goes inside C from $*$ to $H_\mathbb{C}$, then makes a half-circuit anticlockwise around the hypersurface $H_\mathbb{C}$, reaching the chamber $s(C)$, and finally moves inside $s(C)$ to terminate at the point $s(*)$. The image of this path under the Vieta map defines an element t_s of the fundamental group $\pi_1(\mathbb{B}^\mu \setminus \Sigma(W))$.

Theorem ([60]). *The generalized braid group of the group W is generated by the elements $t_1, \dots, t_\mu$ (corresponding to the reflections $s_1, \dots, s_\mu$ in the walls of the* *chamber C), subject to the set of defining relations*

$$\underbrace{t_i t_j \dots}_{m_{ij}} = \underbrace{t_j t_i \dots}_{m_{ij}},$$

where m_{ij} is the multiplicity of the edge that connects the vertices i and j in the Coxeter graph.

Remark. According to the convention made above, for vertices i, j of the graph that are not connected by an edge, $m_{ij} = 2$ and the corresponding generators commute: $t_i t_j = t_j t_i$, and for unlabeled edges, $m_{ij} = 3$ and the corresponding generators satisfy the usual relation of the braid group: $t_i t_j t_i = t_j t_i t_j$.

A description of the cohomology ring of generalized braid groups can be found in [128], [85], [338], [156]; see also §6.

5.4. Convolution of Invariants of a Coxeter Group. Let W be a Coxeter group acting on the complexification $\mathbb{C}^\mu$ of the Euclidean space $\mathbb{R}^\mu$. The inner product $\langle\cdot,\cdot\rangle$ on $\mathbb{C}^\mu$, being W-invariant, defines an isomorphism $i\colon T^*\mathbb{C}^\mu \to T_*\mathbb{C}^\mu$ between the cotangent and tangent bundles of $\mathbb{C}^\mu$, by virtue of which to each holomorphic vector field ξ there corresponds the holomorphic 1-form $i^{-1}(\xi) = \langle\cdot,\xi\rangle$.

The Vieta map $\pi\colon \mathbb{C}^\mu \to \mathbb{B}^\mu = \mathbb{C}^\mu/W$ for W enables us to define the map

$$\pi_* i\colon T^*\mathbb{B}^\mu \to T_*\mathbb{B}^\mu, \qquad \omega \mapsto \pi_* \circ i \circ \pi^*\omega,$$

where ω is a 1-form on $\mathbb{B}^\mu$. This map is well defined, since the vector field $i(\pi^*\omega)$ is W-invariant. Moreover, since a W-invariant vector field is tangent to the mirrors, its image under the projection π_* is tangent to the swallowtail $\Sigma(W) \subset \mathbb{B}^\mu$. Thus, the map $\pi_* i$ takes 1-forms into vector fields tangent to $\Sigma(W)$.

Theorem ([23]). *The map $\pi_* i\colon T^*\mathbb{B}^\mu \to T_*\mathbb{B}^\mu$ establishes an isomorphism between the module of holomorphic 1-forms on $\mathbb{B}^\mu$ and the module of holomorphic vector fields that are tangent to the swallowtail $\Sigma(W)$.*

To each invariant $\varphi\colon \mathbb{B}^\mu \to \mathbb{C}$ there corresponds a vector field ξ_φ on the orbit manifold, namely, the image of the differential $d\varphi$ under the map $\pi_* i$. We call ξ_φ a *potential vector field with potential* φ.

Let us define a symmetric bilinear operation Φ on the set of invariants by assigning to each pair of invariants φ, ψ the inner product of their Euclidean gradients in $\mathbb{C}^\mu$:

$$\Phi(\varphi, \psi) = \pi_* \langle i\pi^* d\varphi, i\pi^* d\psi \rangle$$

(note that the inner product of two W-invariant fields in $\mathbb{C}^\mu$ is a W-invariant function, and hence gives a function on the orbit space).

Example. Consider the action of the Coxeter group A_2 on $\mathbb{C}^2 = \{(x_1, x_2, x_3)\colon x_1 + x_2 + x_3 = 0\}$. Calculating the convolution of the basis invariants $\lambda_1 = x_1x_2 + x_2x_3 + x_1x_3$ and $\lambda_0 = -x_1x_2x_3$ one obtains

$$\Phi(\lambda_1, \lambda_1) = -2\lambda_1, \Phi(\lambda_1, \lambda_0) = -3\lambda_0, \Phi(\lambda_0, \lambda_0) = 2\lambda_1^2/3.$$

For example,

$$\pi^* d\lambda_0 = -x_1x_2dx_3 - x_1x_3dx_2 - x_2x_3dx_1 + \tfrac{1}{3}(x_1x_2 + x_2x_3 + x_1x_3)(dx_1 + dx_2 + dx_3)$$
$$= \tfrac{1}{3}\sum (x_ix_j + x_kx_j - 2x_ix_k)\, dx_j,$$

since the gradient in $\mathbb{C}^3$ must be subsequently projected on the plane $x_1 + x_2 + x_3 = 0$. This gives

$$\langle i\pi^* d\lambda_0, i\pi^* d\lambda_0\rangle = \tfrac{1}{9}[(x_1x_3 + x_2x_3 - 2x_1x_2)^2 + (x_1x_2 + x_3x_2 - x_1x_3)^2 + (x_3x_1 + x_2x_1 - 2x_2x_3)^2]$$
$$= \tfrac{2}{3}(x_1x_3 + x_2x_3 + x_1x_2)^2,$$

and so

$$\pi_*\langle i\pi^* d\lambda_0, i\pi^* d\lambda_0\rangle = \tfrac{2}{3}\lambda_1^2.$$

It is readily seen that $\Phi(\varphi, \psi)$ is the derivative of the invariant ψ along the vector field ξ_φ, and consequently the Leibniz identity holds for Φ:

$$\Phi(\varphi, \psi_1\psi_2) = \psi_1\Phi(\varphi, \psi_2) + \psi_2\Phi(\varphi, \psi_1).$$

Let $\lambda_0, \ldots, \lambda_{\mu-1}$ be an arbitrary basis of invariants of W, i.e., a coordinate system on $\mathbb{B}^\mu$. Correspondingly, we have μ vector fields $\xi_{\lambda_0}, \ldots, \xi_{\lambda_{\mu-1}}$ tangent to the swallowtail $\Sigma(W)$; these will be referred to as basis vector fields.

Theorem ([313], [19], [386]). *The μ basis fields are linearly independent at each point of $\mathbb{B}^\mu \setminus \Sigma(W)$, whereas on the variety Σ of irregular orbits the determinant formed by their components has a zero of multiplicity one. At each point $\lambda \in \Sigma(W)$ the basis fields span the tangent space to the stratum of the swallowtail to which λ belongs.*

Remark. $\mathbb{C}^\mu$ admits a natural stratification for which the strata are intersections of mirrors. The images of these strata under the projection $\pi\colon \mathbb{C}^\mu \to \mathbb{B}^\mu$ define the stratification of $\mathbb{B}^\mu$ alluded to in the theorem.

The components of the basis vector fields are the entries $\Phi(\lambda_i, \lambda_j)$ of the matrix of convolution of basis invariants:

$$\xi_{\lambda_i} = \sum_j \Phi(\lambda_i, \lambda_j)\, \partial/\partial\lambda_j.$$

It follows that the task of describing the module of vector fields tangent to $\Sigma(W)$ reduces to that of computing the convolution of the basis invariants.

Example. For the basis λ_1, λ_0 of invariants of A_2 the basis vector fields have the form

$$\xi_{\lambda_1} = -2\lambda_1\partial/\partial\lambda_1 - 3\lambda_0\partial/\partial\lambda_0, \qquad \xi_{\lambda_0} = -3\lambda_0\partial/\partial\lambda_1 + \tfrac{2}{3}\lambda_1^2\partial/\partial\lambda_0.$$

In the present case the swallowtail Σ is the semicubic parabola $4\lambda_1^3 + 27\lambda_0^2 = 0$. The field ξ_{λ_1} is Eulerian, whereas the field ξ_{λ_0} is Hamiltonian with Hamilton function $4\lambda_1^3 + 27\lambda_0^2$. Any vector field tangent to $\Sigma(A_2)$ is uniquely representable in the form $\varphi_1\xi_{\lambda_1} + \varphi_2\xi_{\lambda_2}$, where φ_1 and φ_2 are holomorphic functions.

The explicit form of the convolution of invariants of Coxeter groups was obtained in [19], [23], [145].

The vector fields tangent to the swallowtail $\Sigma(W)$ form the Lie algebra of the group of diffeomorphisms of the orbit space $\mathbb{B}^\mu$ that preserve $\Sigma(W)$. By integrating such vector fields one can obtain any diffeomorphism in the component of identity of this group. The number of connected components of the group of diffeomorphisms that preserve the swallowtail is equal to the number of elements in $A(W)/W$, where $A(W)$ is the group of the linear automorphisms of $\mathbb{R}^\mu$ that preserve the set of mirrors of the group W [225].

The Lie algebra of the vector fields tangent to $\Sigma(W)$ is infinite-dimensional. However, from it one can produce a finite-dimensional Lie algebra upon replacing each vector field by its linear part at zero. The resulting algebra replaces in most computations the full algebra of vector fields tangent to the swallowtail.

The indicated finite-dimensional algebra of linear vector fields is described in terms of the operation of linearized convolution of invariants, Φ_0, in exactly the same way as the full algebra of vector fields tangent to $\Sigma(W)$ is described by the convolution of invariants of the Coxeter group; Φ_0: $T^* \times T^* \to T^*$ is defined as

$$\Phi_0(d\varphi, d\psi) = d\Phi(\varphi, \psi),$$

where $d\varphi$, $d\psi$ are the differentials of the invariants φ, ψ in the cotangent space T^* at zero to the orbit space.

Example. Let λ_1, λ_0 be again the basis invariants of the group A_2. Then the matrix of convolution of basis invariants, $[\Phi(\lambda_i, \lambda_j)]$, determines the matrix of linearized convolution of invariants, $[\Phi_0(d\lambda_i, d\lambda_j)]$:

$$[\Phi(\lambda_i, \lambda_j)] = \begin{bmatrix} \frac{2}{3}\lambda_1^2 & -3\lambda_0 \\ -3\lambda_0 & -2\lambda_1 \end{bmatrix} \Rightarrow [\Phi_0(d\lambda_i, d\lambda_j)] = \begin{bmatrix} 0 & -3\lambda_0 \\ -3\lambda_0 & -2\lambda_1 \end{bmatrix}.$$

5.5. Root Systems and Weyl Groups ([56]). Let R be a finite set of vectors in the Euclidean space $\mathbb{R}^\mu$, such that $0 \notin R$ and the linear span of R is the entire $\mathbb{R}^\mu$. Such a set R is called a *root system* if

(i) the reflection s_α, $\alpha \in R$, maps R into itself, and

(ii) for each pair $\alpha, \beta \in R$ the map s_α takes the integer lattice spanned by the vectors α and β into itself.

The elements of R are called *roots*. The reflection s_α takes the root α into $-\alpha$, and any root β into

$$s_\alpha(\beta) = \beta - 2(\langle\alpha, \beta\rangle/\langle\alpha, \alpha\rangle)\alpha.$$

We see that condition (ii) is equivalent to the requirement that $n(\alpha, \beta) = 2\langle\alpha, \beta\rangle/\langle\alpha, \alpha\rangle \in \mathbf{Z}$. A root system R is said to be *reduced* if for every root α, R contains no roots collinear to α except for $-\alpha$. The reflections s_α in the [hyperplanes orthogonal to the] roots α generate a group $W(R)$, called the *Weyl group* of the root system R. The Weyl group acts transitively on any of the sets consisting of all roots of an equal length. If $W(R)$ is irreducible, then the root system R is said to be *irreducible*. From now on we shall be concerned only with reduced irreducible root systems.

Remark. Any Weyl group is a Coxeter group, but the converse is false.

Example. Let $e_1, \ldots, e_{\mu+1}$ be an orthonormal basis of the Euclidean space $\mathbb{R}^{\mu+1}$. The vectors $e_i - e_j$, $i \neq j$, form a root system R in the hyperplane $\mathbb{R}^\mu = \{x\colon x_1 + \cdots + x_{\mu+1} = 0\}$. The reflection corresponding to the root $e_i - e_j$ is simply the permutation of the ith and jth coordinates; the Weyl group $W(R)$ coincides with the Coxeter group A_μ.

For each mirror H of the group $W(R)$ there is a pair of roots $\pm\alpha$ orthogonal to H, i.e., such that s_α is the reflection in H. Let C be a chamber of the Weyl group $W(R)$. To each wall H_i of C we associate the root α_i orthogonal to H_i that lies on the same side of H_i as C. The resulting set of roots $\alpha_1, \ldots, \alpha_\mu$ is called a *basis of the root system R* (relative to the chamber C). The inner products of the basis elements are negative: $\langle \alpha_i, \alpha_j \rangle < 0$. All bases of the root system R are conjugate under the Weyl group $W(R)$.

Example. The set of vectors $e_1 - e_2, e_2 - e_3, \ldots, e_\mu - e_{\mu+1}$ is a basis of the root system R of the preceding example.

Theorem. *Let $\alpha_1, \ldots, \alpha_\mu$ be a basis of the root system R. Then every root $\beta \in R$ is a linear combination with integer coefficients (simultaneously $\geqslant 0$ or $\leqslant 0$) of $\alpha_1, \ldots, \alpha_\mu$. The linear combinations of roots with integer coefficients form a discrete integer lattice $L \subset \mathbb{R}^\mu$ that is invariant under the Weyl group $W(R)$.*

For each basis $\alpha_1, \ldots, \alpha_\mu$ of the root system R there is a root $\bar{\alpha} = \bar{n}_1\alpha_1 + \cdots + \bar{n}_\mu\alpha_\mu$, called a *maximal root*, such that for any other root $\beta \in R$, $\beta = n_1\alpha_1 + \cdots + n_\mu\alpha_\mu$, one has $\bar{n}_i \geqslant n_i$, $i = 1, \ldots, \mu$.

A pair of roots α_i, α_j in a basis is necessarily (up to a permutation of the places of α_i and α_j) in one of the following relative positions:

(i) $\widehat{\alpha_i, \alpha_j} = \pi/2$;

(ii) $\widehat{\alpha_i, \alpha_j} = 2\pi/3$, $\|\alpha_i\| = \|\alpha_j\|$;

(iii) $\widehat{\alpha_i, \alpha_j} = 3\pi/4$, $\|\alpha_i\| = \sqrt{2}\,\|\alpha_j\|$;

(iv) $\widehat{\alpha_i, \alpha_j} = 5\pi/6$, $\|\alpha_i\| = \sqrt{3}\,\|\alpha_j\|$.

Accordingly, given a basis $\alpha_1, \ldots, \alpha_\mu$ of the root system R, we construct a graph called the *Dynkin diagram* of R. The vertices of the graph correspond to the elements of the basis; a pair of vertices $\langle i \rangle$, $\langle j \rangle$ is connected by an edge of multiplicity k if the angle between the corresponding roots is $\widehat{\alpha_i, \alpha_j} = \pi(k+1)/(k+2)$. If $\|\alpha_i\| > \|\alpha_j\|$, then the edge connecting the vertices $\langle i \rangle$ and $\langle j \rangle$ (if present) is given the orientation from $\langle i \rangle$ to $\langle j \rangle$.

Example. The Dynkin diagram of the root system $\{e_i - e_j,\ i \neq j\}$ is as follows:

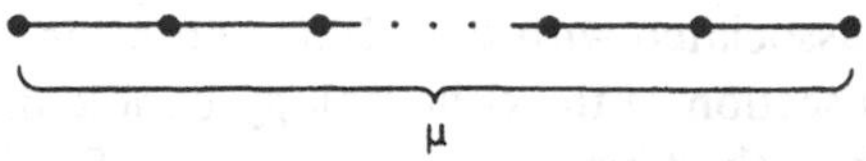

The Dynkin diagram of a root system R uniquely determines R up to an isomorphism.

Theorem. *The complete classification of root systems is given by the following list of their Dynkin diagrams:*

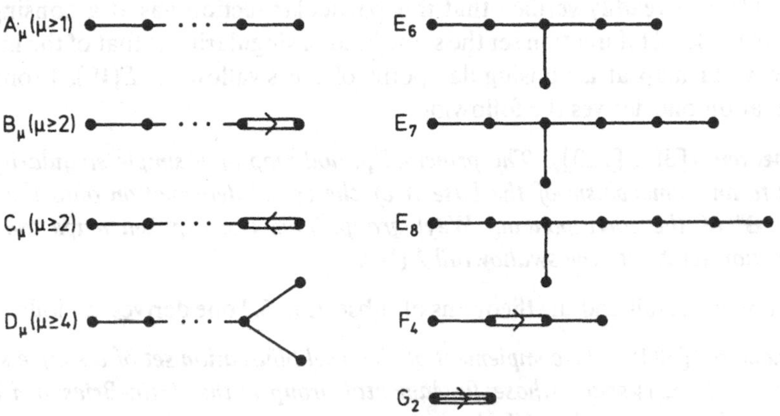

The above diagrams are pairwise nonisomorphic (except for $B_2 = C_2$).

5.6. Simple Singularities and Weyl Groups. Let f be an elliptic singularity (subsection 3.5). Then the intersection index (taken with the minus sign) defines a Euclidean space structure on $H_{n-1}(V_*; \mathbb{R}) = \mathbb{R}^\mu$ (we assume again that the number of variables $n \equiv 3 \pmod 4$). The monodromy group Γ is generated by the reflections in the vanishing cycles and preserves the integer lattice $L = H_{n-1}(V_*)$ that these cycles span. Thus, the set of vanishing cycles is a root system R in $\mathbb{R}^\mu$ and the monodromy group Γ coincides with the Weyl group $W(R)$. Since all vanishing cycles have the same length, the root system R is of the type A_μ, D_μ, E_6, E_7, or E_8 (subsection 4.4). A comparison of the Dynkin diagrams of elliptic singularities (subsection 3.5) and the Dynkin diagrams of the root systems shows that all the above possibilities are realized.

Theorem ([12]). *Let f be one of the simple singularities of type A_μ, D_μ, E_μ. Then the set of vanishing cycles in the homology $H_{n-1}(V_*; \mathbb{R})$ of the nonsingular fibre is a root system R of the same type, and the monodromy group Γ coincides with the Weyl group $W(R)$. The classical-monodromy operator h is a Coxeter transformation in the Weyl group $W(R)$ and its eigenvalues* $\exp(2\pi i m_j/|h|)$ *are uniquely determined by the exponents m_j of $W(R)$.*

Let $\Lambda \subset \mathbb{C}^\mu$ be the base of the miniversal deformation of the simple singularity f and let $\Sigma \subset \Lambda$ be its level bifurcation set (subsection 1.10). Consider the cohomology bundle associated with the Milnor fibration $V'_\Lambda \to \Lambda' = \Lambda \setminus \Sigma$ (see subsection 3.3). Each section of the cohomology bundle defines a multi-valued map of the base Λ' into the fibre $H^{n-1}(V_*; \mathbb{C})$ over a fixed point $*$, defined up to the action of the monodromy group Γ. In the case under consideration, $H^{n-1}(V_*; \mathbb{C}) \simeq \mathbb{C}^\mu$, $\Gamma = W(R)$, and the given section defines a single-valued map of the base into the orbit space of the group $W(R)$:

$$\Lambda' \to \mathbb{B}^\mu = \mathbb{C}^\mu / W(R).$$

Now take as a section the principal period map of a holomorphic form (subsection 3.11). It is readily verified that this particular section has at a nonsingular point of the level bifurcation set the same kind of singularity as that of the inverse of the Vieta map at a nonsingular point of the swallowtail $\Sigma(W)$. From this observation one derives the following

Theorem ([38], [220]). *The principal period map of a simple singularity extends to an isomorphism of the base Λ of the versal deformation onto the orbit space $\mathbb{B}^\mu$ of the corresponding Weyl group. This isomorphism maps the level bifurcation set Σ into the swallowtail $\Sigma(W)$.*

From this result and the theorems of subsection 5.2 one derives the following

Theorem ([38]). *The complement of the level bifurcation set of a simple singularity is a $K(\pi, 1)$-space whose fundamental group is the Artin-Brieskorn braid group of the corresponding Weyl group.*

The set of all distinguished bases of vanishing cycles for a simple singularity also admits a description in terms of the corresponding Weyl group.

Theorem (Deligne). *The set of elements $\Delta_1, \ldots, \Delta_\mu$ in the root system R is a distinguished basis of vanishing cycles for the corresponding singularity if and only if the product of the corresponding reflections $s_1, \ldots, s_\mu$ in the Weyl group $W(R)$ is the Coxeter element h given by the classical-monodromy operator.*

5.7. Vector Fields Tangent to the Level Bifurcation Set. The above identification of the pair (Λ, Σ), consisting of the base of the versal deformation of a simple singularity and the level bifurcation set Σ embedded in Λ, with the pair $(\mathbb{B}^\mu, \Sigma(W))$ defined by the corresponding Coxeter group allows to describe the Lie algebra of the vector fields tangent to Σ and the finite-dimensional Lie algebra of linearizations of such fields in terms of convolution of invariants of the group W (subsection 5.4).

As it turns out, the operation of linearized convolution of invariants of a Coxeter group can be simply expressed through the multiplication operation in the local graded algebra Q_f of the corresponding singularity.

The normal forms f of simple singularities are quasihomogeneous of order 1 with positive weights $v_1, \ldots, v_n$ of the individual variables. Let $\mathscr{D} = \sum v_i x_i \partial/\partial x_i$ denote the quasihomogeneous Euler operator. Then $\mathscr{D}f = f$. The operator $\mathscr{D}$ maps the gradient ideal $(\partial f/\partial x_i)$ into itself, and consequently it acts on the local algebra Q_f as a derivation.

Theorem ([145]). *The linearized convolution of invariants $T^* \times T^* \to T^*$ is equivalent, as a bilinear operation, with the map $Q_f \times Q_f \to Q_f$ given by the formula*

$$(p, q) \mapsto (\mathscr{D} + 2|h|^{-1})pq,$$

where $|h|$ is the Coxeter number of the Coxeter group associated with f.

The identification of T^* with Q_f that implements the equivalence in the theorem is not canonical, as we presently explain.

The tangent space at zero to the base of the versal deformation, $T_* \Lambda_0$, is canonically isomorphic to Q_f (the isomorphism takes the velocity vector of a deformation into its class in Q_f). We say that an element $\alpha \in Q_f^*$ of the dual space is *admissible* if the linear function $\alpha: Q_f \to \mathbb{C}$ does not vanish on the annihilator of the maximal ideal in Q_f. This annihilator is one-dimensional and is generated by the class of the Hessian of f. Consequently, the nonadmissible elements form a hyperplane in Q_f^*.

With each element $\alpha \in Q_f^*$ one can associate a bilinear form ψ_α on Q_f by the rule $\psi_\alpha(p, q) = \alpha(p \cdot q)$; ψ_α is nondegenerate whenever α is admissible. Denote by $N_\alpha: Q_f \to Q_f^*$ the linear operator defined by the form ψ_α. Consider the family of operations $\Psi_\alpha: Q_f^* \times Q_f^* \to Q_f^*$,

$$\Psi_\alpha(\beta, \gamma) = \mathscr{R}^* N_\alpha(N_\alpha^{-1}\beta \cdot N_\alpha^{-1}\gamma),$$

where α is admissible and $\mathscr{R} = E - \mathscr{D}$.

The identification of the orbit space $\mathbb{B}^\mu$ with the base Λ of the versal deformation is not unique and is defined up to a diffeomorphism $\mathbb{B}^\mu \to \mathbb{B}^\mu$ that preserves the swallowtail Σ. The tangent space to the group of all such diffeomorphisms generates from the operation of linearized convolution of invariants $\Phi_0: T^* \times T^* \to T^*$ a family of bilinear operations $\Phi_a: T^* \times T^* \to T^*$, parametrized by elements $a \in T^*$, in the following manner. Let $a \in T^*$ and let $\varphi: \mathbb{B}^\mu \to \mathbb{C}$ be an invariant such that $d\varphi = a$. The linearization of the potential vector field ξ_φ with potential φ (subsection 5.4) defines a linear operator $A_a: T_* \to T_*$ (namely, the linear part at the origin of the diffeomorphism $\exp \xi_\varphi$ obtained by integrating ξ_φ "to time 1"), and also the operation

$$\Phi_a(b, c) = A_a^{*-1}\Phi_0(A_a^* b, A_a^* c),$$

where $A_a^*: T^* \to T^*$ denotes the dual of the operator A_a.

Any identification $i: (\mathbb{B}^\mu, \Sigma(W)) \to (\Lambda, \Sigma)$ takes the family of operations $\Phi_a: T^* \times T^* \to T^*$ into a family of operations $i_* \Phi_a: Q_f^* \times Q_f^* \to Q_f^*$.

Theorem ([145], [23]). *An identification $i: \mathbb{B}^\mu \to \Lambda$ takes the family of operations Φ_a on T^* into the family of operations Ψ_α on Q_f^*. The relationship between the parameters $i_* a$ and α can be linearly expressed through the inversion operation in the local algebra Q_f.*

This theorem can be also formulated in terms of the Lie algebra of the linear parts of the vector fields tangent to $\Sigma(W)$. Specifically, let v_a denote the linear part of the vector field ξ_φ, where $a = d\varphi$. Let L_W denote the μ-dimensional Lie algebra spanned by the fields v_a. Consider the μ-dimensional Lie algebra L of the derivations v_q of the tangent space Q_f at zero to the base of the versal deformation; here $v_q = M_q(E - \mathscr{D})$, where $M_q\colon Q_f \to Q_f$ is the operator of multiplication by q and $\mathscr{D}$ is the Euler operator.

Theorem. *An identification $i\colon \mathbb{B}^\mu \to Q_f$ maps the Lie algebra L_W into the Lie algebra L. Moreover, the relationship between the parameters a and q of the vector fields v_a and $v_q = i_*(v_a)$ is linear.*

5.8. The Complement of the Function Bifurcation Set. In the base Λ of the versal deformation $F(x, \lambda)$ of a simple singularity f consider the hypersurface $\hat{\Xi}$ consisting of the values of the parameter λ for which $x \mapsto F(x, \lambda)$ is not a Morse function, i.e., it has degenerate critical points or distinct critical points with equal critical values. The pair $(\Lambda, \hat{\Xi})$ is diffeomorphic to the direct product of the base of the truncated versal deformation, with the function bifurcation set Ξ embedded in it, and the complex line $\mathbb{C}$.

Let us assign to each point $\lambda \in \Lambda$ the polynomial $t^\mu + \cdots$ of degree μ whose roots are exactly the μ critical values of the function $F(\cdot, \lambda)$, counting multiplicities. In this manner we define a map P of Λ into $\mathbb{C}^\mu$ which takes the hypersurface $\hat{\Xi}$ into the discriminant $\Delta \subset \mathbb{C}^\mu$ (the set of the polynomials with multiple roots). The complement $\mathbb{C}^\mu \setminus \Delta$ is a $K(\pi, 1)$-space whose fundamental group is isomorphic to the braid group $\mathrm{Br}(\mu)$ of μ strands (see the example in subsection 5.4).

Theorem ([219], [226]). *The map P defines a regular covering*

$$\Lambda \setminus \hat{\Xi} \to \mathbb{C}^\mu \setminus \Delta$$

of order $\nu = \mu!|h|^\mu|W|^{-1}$, where $|h|$ and W are the Coxeter number and respectively the order of the corresponding Weyl group.

Theorem. *The complement of the function bifurcation set for a simple singularity is a $K(\pi, 1)$-space whose fundamental group is isomorphic to a subgroup of index ν of the braid group* $\mathrm{Br}(\mu)$ *of μ strands.*

5.9. Adjacency and Decomposition of Simple Singularities. The stratifications of the level and function bifurcation sets are described by the Dynkin diagrams of the corresponding root systems.

Theorem ([37], [304]). *A simple singularity of type X is adjacent to a simple singularity of type Y if and only if the Dynkin diagram of the root system of Y embeds in the Dynkin diagram of the root system of X.*

The next result describes the stratification of the level bifurcation set $\Sigma(X)$ of a simple singularity of type X. Let us define the stratum $\Sigma_X(X_1, \ldots, X_k) \subset \Sigma(X)$ as the set of points $\lambda \in \Sigma(X)$ in the base of the deformation for which the function $F(\cdot, \lambda)$ takes the critical value zero at critical points of types $X_1, \ldots, X_k$.

Theorem (Grothendieck, see [107]). *The stratum $\Sigma_X(X_1, \ldots, X_k)$ is not empty if and only if, after a number of vertices together with all the edges that terminate* *at them are removed, the Dynkin diagram of the root system X decomposes into the disconnected union of the Dynkin diagrams of the root systems of* $X_1, \ldots, X_k$.

Example. The Dynkin diagram of the root system D_4 decomposes into three copies of the diagram of the root system A_1:

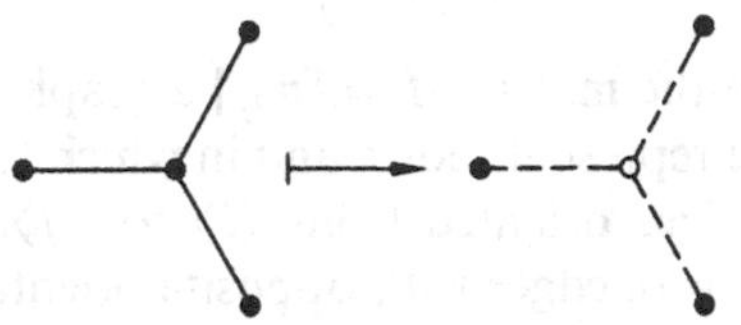

Consequently, there is a deformation f_ε of the function f with a D_4-singularity at zero such that to one of the critical values of f_ε there correspond three Morse critical points.

An analogous stratification can be defined for the function bifurcation set $\Xi(X)$. To this end, we define the stratum $\Xi_X[\mathscr{X}_1, \ldots, \mathscr{X}_k]$, where $\mathscr{X}_i = (X_{i1}, \ldots, X_{ij_i})$, as the set of points $\lambda \in \Xi(X)$ for which the function $F(\cdot, \lambda)$ has exactly k distinct critical values and the ith critical value is attained at j_i critical points of the types $X_{i1}, \ldots, X_{ij_i}$. The next theorem describes the one-dimensional strata $\Xi_X[\mathscr{X}_1, \mathscr{X}_2]$, corresponding to a pair of critical values of $F(\cdot, \lambda)$.

The description of strata with a higher number of critical points also reduces to this case.

Theorem ([224]). *If the stratum* $\Xi_X[\mathscr{X}_1, \mathscr{X}_2] \neq \varnothing$, *then:*

(1) $\sum_{1 \leqslant j \leqslant j_1} \mu(X_{1j}) + \sum_{1 \leqslant j \leqslant j_2} \mu(X_{2j}) = \mu(X)$;

(2) *the strata* $\Xi_X[\mathscr{X}_1]$ *and* $\Xi_X[\mathscr{X}_2]$ *are not empty;*

(3) *if* $X = D_\mu$, *then the stratum* $\Xi_{A_{\mu-2}}[\mathscr{X}_i] \neq \varnothing$ *for some index* $i \in \{1, 2\}$.

Conversely, all strata $\Xi_X[\mathscr{X}_1, \mathscr{X}_2]$ *for which conditions* (1)–(3) *are satisfied, are not empty, with one exception:* $\Xi_{E_6}[(A_1, A_1), (A_2, A_2)] = \varnothing$.

Remark. Conditions (2) and (3) are described by the preceding theorem in terms of Dynkin diagrams.

Example. The stratum $\Xi_{D_4}[(A_1, A_1), (A_1, A_1)] = \varnothing$, since it does not satisfy condition (3). Consequently, there is no deformation of the critical point D_4 under which it decomposes into two pairs of Morse critical points with the same critical values.

An analogous description of the function bifurcation set of a real simple singularity was obtained by Chislenko [80].

5.10. Finite Subgroups of SU_2, Simple Singularities, and Weyl Groups. Any finite subgroup of SU_2 defines a simple singularity (as the quotient space of $\mathbb{C}^2$ by its action; see subsection 1.2.4), which in turn defines a reflection group (the corresponding monodromy group). Thus, to each symmetry group of a regular polyhedron there corresponds one of the Coxeter groups A_μ, D_μ, E_μ. The next result establishes this correspondence directly (without recourse to singularities).

Let $\{R_i\}$ be a complete set of pairwise nonisomorphic irreducible representations of the finite group G, and let R be a fixed representation of G. Then

$$R \otimes R_i = \bigoplus_j m_{ij} R_j.$$

Let us associate with the matrix $M = [m_{ij}]$ a graph $\hat{\Gamma}$, whose vertices correspond to the irreducible representations, and in which two vertices $\langle i \rangle$ and $\langle j \rangle$ are connected by m_{ij} edges oriented from $\langle i \rangle$ to $\langle j \rangle$. Let us also make the convention that each pair of edges with opposite orientations is represented by a single nonoriented edge.

Theorem ([245]). *Let G be a finite subgroup of SU_2 and R be its canonical two-dimensional representation. Then the graph $\hat{\Gamma}$ associated with R is the completion of the Dynkin graph of the corresponding Weyl group; its Cartan matrix equals $C = 2E - M$ and the eigenvalues of C are the values of the character of the representation R.*

Remark. One obtains the completion of the Dynkin graph (the graph of the affine Weyl group) from the ordinary Dynkin diagram of the root system by adjoining to the basis of the root system the opposite of the maximal root.

5.11. Parabolic Singularities. After the simple singularities, in increasing order of complexity, there follow the parabolic singularities. They form three families that depend on a complex parameter γ, the modulus of the singularity:

$$P_8\colon x^3 + y^3 + z^3 - 3\gamma xyz, \qquad \gamma^3 \neq 1;$$

$$X_9\colon x^4 + 2\gamma x^2 y^2 + y^4 + z^2, \qquad \gamma^2 \neq 1;$$

$$J_{10}\colon x(x^2 + 2\gamma xy + y^4) + z^2, \qquad \gamma^2 \neq 1.$$

The reason these singularities are termed parabolic is that their intersection forms in the mid-dimensional homology have a nontrivial (two-dimensional) kernel, but nevertheless do not change sign. For more complex singularities with even-dimensional Milnor fibre the intersection form necessarily changes sign.

The Dynkin diagrams of the parabolic functions in some distinguished bases are shown below.

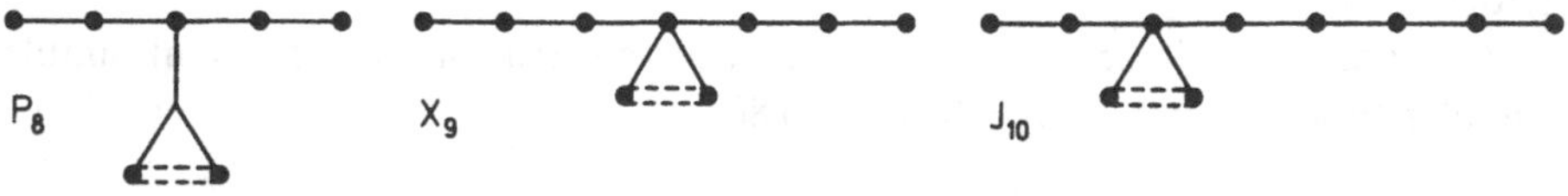

Functions belonging to the same parabolic family are equivalent if and only their j-invariants coincide (i.e., the j-invariants of the exceptional elliptic curves of the minimal resolution of the singularity of the surface $f = 0$; see [286], [177, Chapter 4, §4]). One can show that

$$j(P_8) = \frac{\gamma^3(\gamma^3+8)^3}{64(\gamma^3-1)^3}; \qquad j(X_9) = \frac{(\gamma^4+3)^3}{27(\gamma^4-1)^2};$$

$$j(J_{10}) = \frac{64(4\gamma^2-3)^3}{\gamma^2-1}.$$

All functions in the parabolic families are quasihomogeneous. Moreover, the derivative $\partial f/\partial\gamma$ is a highest-weight element in the local ring of f (f and $\partial f/\partial\gamma$ have the same weight).

Consider a quasihomogeneous deformation of the parabolic singularity f: $F = f + \lambda_1 e_1 + \cdots + \lambda_{\mu-1} e_{\mu-1}$, where, in the cases P_8 and X_9, $e_1, \ldots, e_{\mu-1}$, $\partial f/\partial\gamma$ is a monomial basis of the local ring of the function $f|_{\gamma=0}$ and for J_{10}, $\{e_1, \ldots, e_{\mu-1}\} = \{x^a y^b\colon a + b \leqslant 3, a \leqslant 2\}$. The germ of F at the point $x = y = z = 0$, $\gamma = \gamma_0$, $\lambda = 0$ gives, for any value of γ_0 except those excluded above, a miniversal deformation of the corresponding parabolic singularity. Here one takes γ and λ as deformation parameters.

Theorem ([221, Part I]). *The versal deformation F is topologically trivial along the parameter γ.*

(For the definition of the topological triviality see § III.3.)

Consider the global parameter space $\Lambda = \Gamma \times \mathbb{C}^{\mu-1}$ of the deformation F, where Γ is the complex line with the exceptional values of the modulus γ removed. Λ contains the level bifurcation set Σ of the corresponding parabolic family. The stratification of Σ is described by the following Grothendieck-type result.

Theorem ([189]). *The disconnected union of the standard Dynkin diagrams of the critical points corresponding to a single critical value of the deformation of any of the parabolic singularities can be obtained by deleting certain vertices from the diagrams*

1°. $\tilde{E}_6$ *for the singularity* P_8;
2°. $\tilde{E}_7$ *or* $(\tilde{D}_6, A_1)$ *for the singularity* X_9;
3°. $\tilde{E}_8, \tilde{D}_8, (\tilde{E}_7, A_1)$, or $(\tilde{E}_6, A_2)$ *for the singularity* J_{10}.

The converse holds true provided that one deletes at least one vertex from the components $\tilde{E}_*$ or $\tilde{D}_*$.

In this theorem $\tilde{E}_*$ and $\tilde{D}_*$ designate the standard completions of the diagrams of the corresponding root systems:

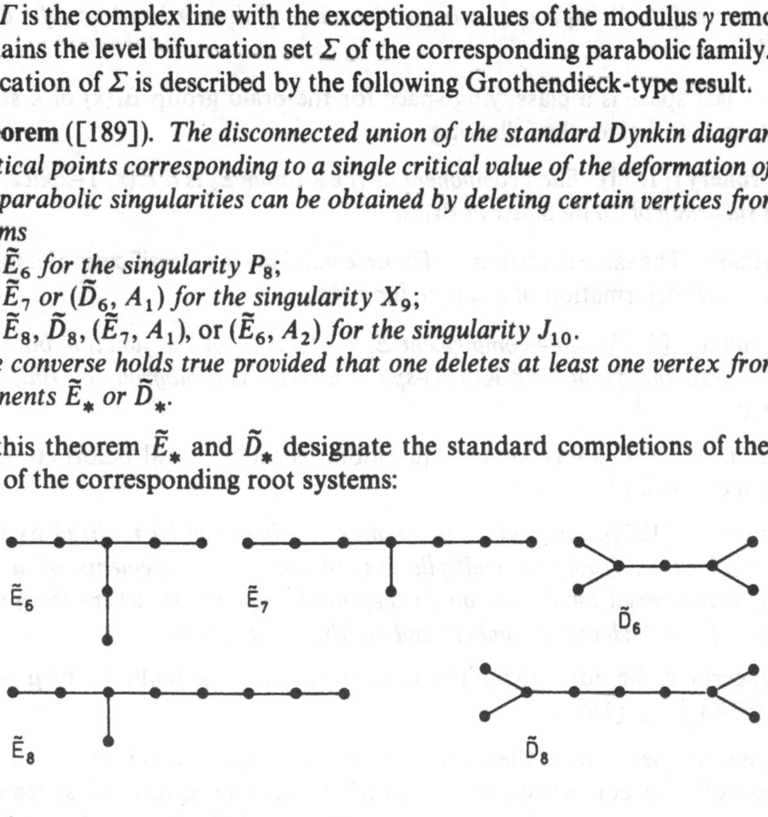

They determine quadratic forms with a single zero eigenvalue.

If $(\gamma, \lambda) \in \Lambda \setminus \Sigma$, then the surface $F_{\gamma,\lambda} = 0$ in $\mathbb{C}^3$ is homeomorphic to the nonsingular fibre of the parabolic function in question. On fixing such a point (γ, λ), we obtain an action of the group $\pi_1(\Lambda \setminus \Sigma)$ in the homology of the nonsingular fibre, i.e., a monodromy group. However, since here the parameter space Λ is regarded as a global object rather than a germ, the resulting monodromy is also global. If in the case of simple singularities there are no differences between the local and global situations, this is no longer true for parabolic families.

Theorem ([189]). *The local monodromy group of a parabolic singularity is a subgroup of infinite index in the global monodromy group. Moreover, the global monodromy acts in a nontrivial manner in the subspace isotropic with respect to the intersection form in the homology of the nonsingular fibre.*

This result is completely natural, for $\pi_1(\Lambda \setminus \Sigma)$ includes the [homotopy classes of the] loops that go around the exceptional hyperplanes $\gamma = \text{const}$.

The space Λ is also stratified into sets Ξ_k, $k = 1, \ldots, \mu$, each of which corresponds to functions with the same number k of distinct critical values. For example, the closure of $\Xi_{\mu-1}$ is the function bifurcation set of the parabolic family. Let φ_k be the map that sends each point $(\gamma, \lambda) \in \Xi_k$ into the unordered set of critical values of the function $F_{\gamma,\lambda}$. The following analogue of the Lyashko-Looijenga theorem for simple singularities holds true:

Theorem ([187]). *φ_k is a covering of the space of all unordered sets of k distinct complex numbers.*

This last space is a classifying space for the braid group $\mathrm{Br}(k)$ of k strands. One therefore derives the following

Corollary ([187]). *Each component of the stratum Ξ_k is a $K(\pi, 1)$-space, where π is a subgroup of finite index in* $\mathrm{Br}(k)$.

Remark. The same assertion is of course valid for the stratification of the base of the versal deformation of a simple function.

Corollary ([187]). *The complement $\Xi_\mu = \Lambda \setminus \bar{\Xi}_{\mu-1}$ of the function bifurcation set of a parabolic family is a $K(\pi, 1)$-space, where π is a subgroup of finite index in* $\mathrm{Br}(\mu)$.

The next result holds for any quasihomogeneous versal deformation of a parabolic singularity.

Theorem ([187]). *Any set of μ complex numbers can be realized as the set of critical values (counting multiplicities) of one of the elements of a quasihomogeneous versal family, the only exception being the case where the given set consists of $\mu - 1$ identical numbers and another, distinct one.*

Concerning the nonrealizability of critical values of multiplicity $\mu - 1$ see [73], [368], [70], [188].

Remark. The covering theorem given above remains valid if one takes for φ_k the map of some connected component of the stratum Ξ_k into the space of sets of k complex numbers, in which critical values corresponding to points of distinct type are regarded as distinct. For instance, the stratum of the functions that have singularities of the types D_4, A_1, and $2A_1$ on three critical levels covers the space of ordered triples of complex numbers.

Let us describe the decompositions

$$P \to \mathscr{X} = (\mathscr{X}_1; \ldots; \mathscr{X}_k), \qquad \mathscr{X}_k = (X_{i1}, \ldots, X_{ij_i})$$

of the parabolic singularity P into functions with k distinct critical values (see subsection 5.9). For P_8 this problem was solved first by Wall [368]; later, Jaworski proved the following result.

Theorem ([189]). 1°. *If there exists a decomposition*

$$P \to \mathscr{X} = (\mathscr{X}_1; \ldots; \mathscr{X}_k)$$

of the parabolic singularity P, then

a) $\sum \mu(X_i) = \mu(P)$;

b) *there exist the decompositions*

$$P \to \mathscr{X}_i, \qquad i = 1, \ldots, k,$$

into critical points on a single level.

2°. *For $k \geqslant 3$ every decomposition possessing properties* a) *and* b) *is realized for all values of the modulus of singularity γ, except for the following cases:*

i) *there are no decompositions*

$$P_8 \to (4A_1; A_2, A_1; A_1),$$
$$P_8 \to (2A_2; A_2, A_1; A_1),$$
$$P_8 \to (4A_1; 2A_1; 2A_1),$$
$$P_8 \to (2A_2; 2A_1; 2A_1),$$
$$X_9 \to (D_4; D_4; A_1);$$

ii) *the decompositions*

$$P_8(\gamma) \to (D_4; A_2; A_2),$$
$$P_8(\gamma) \to (4A_1; A_2; A_2),$$
$$J_{10}(\gamma) \to (D_4; D_4; A_2)$$

exist only for finitely many values of the modulus γ.

3°. *All decompositions of parabolic singularities, except for $X_9 \to (4A_1; 4A_1; A_1)$, can be "factored" through decompositions into functions with two critical levels.*

Remarks. 1°. Decompositions into two critical levels exist only for a finite number of values of the modulus γ. The complete list of such decompositions is given in [189].

2°. In the case of the family P_8 conditions a) and b) of part 1° of the theorem are satisfied for exactly 42 collections of singularities. Of these, only 23 are realized as a result of decomposition. For the family X_9 the corresponding numbers are 84 and 59, and for the family J_{10} they are 221 and 196 [189].

§ 6. Topology of Complements of Discriminants of Singularities

6.1. Complements of Discriminants and Braid Groups. Let $f = x^m$ be the standard A_{m-1}-singularity,

$$F(x, \lambda) = x^m + \lambda_1 x^{m-1} + \cdots + \lambda_{m-1} x + \lambda_m \qquad (1)$$

a versal deformation of f, and $\Sigma \subset \mathbb{C}^m$ the *discriminant* of F, i.e., the set of values of the parameter λ for which the polynomial $F(\cdot, \lambda)$ has multiple roots. The complement $\mathbb{C}^m \setminus \Sigma$ is biholomorphically equivalent to the space of all m-element subsets of $\mathbb{C}$: the biholomorphism implementing the equivalence assigns to each polynomial the set of its roots. The results given below concerning the topology of the space $\mathbb{C}^m \setminus \Sigma$ are formulated in terms of classifying spaces; for the reader's convenience we briefly review this terminology (see e.g., [183], [281]).

With every topological group G a topological space BG, the *classifying space of* G, is uniquely associated (up to homotopy equivalence). Also, any continuous homomorphism $G_1 \to G_2$ defines a unique (up to homotopy) map $BG_1 \to BG_2$, and hence a ring homomorphism $H^*(BG_2) \to H^*(BG_1)$. If G is discrete, then BG is an Eilenberg-Mac Lane $K(G, 1)$-space, i.e., a space characterized by the properties $\pi_1(BG) = G$, $\pi_i(BG) = 0$ for $i \geqslant 2$; in this case the ring $H^*(BG)$ coincides with the cohomology ring of the group G and can be denoted simply by $H^*(G)$.

Now, $\mathbb{C}^m \setminus \Sigma$ is a $K(G, 1)$-space, where $G = \mathrm{Br}(m)$ is the Artin braid group of m strands (see subsection 5.2). Consequently, $H^*(\mathbb{C}^m \setminus \Sigma)$ coincides with the cohomology ring of the group $\mathrm{Br}(m)$, the study of which is our next objective.

Proposition. *For any coefficient domain A the group $H^i(\mathrm{Br}(m), A)$ is trivial for all $i \geqslant m$.*

In fact, the space $\mathbb{C}^m \setminus \Sigma$ is homotopy equivalent to its intersection with the hyperplane $\{\lambda_1 = 0\}$, which is again the base of a versal deformation of the function x^m and corresponds to polynomials (1) the sum of whose roots is zero. But that intersection is a Stein manifold, and as such it is homotopy equivalent to an $(m - 1)$-dimensional cellular complex.

6.2. The mod-2 Cohomology of Braid Groups. For each m one has the obvious homomorphisms

$$\mathrm{Br}(m) \to S(m) \to O(m), \qquad (2)$$

where $S(m)$ is the group of permutations of m elements and the second arrow is given by the action of $S(m)$ through coordinate relabeling. These homomorphisms induce ring homomorphisms $H^*(BO(m)) \to H^*(S(m)) \to H^*(\mathrm{Br}(m))$.

Theorem (see [128]). *The composite homomorphism* $H^*(BO(m); \mathbb{Z}_2) \to H^*(\mathrm{Br}(m); \mathbb{Z}_2)$ *is an epimorphism for any* m.

We recall the structure of the ring $H^*(BO(m); \mathbb{Z}_2)$. As $BO(m)$ one can take the *stable Grassmann manifold* G_∞^m, i.e., the set of all m-dimensional subspaces of $\mathbb{R}^\infty$. The ring $H^*(BO(m); \mathbb{Z}_2)$ is the algebra of polynomials in m generators $w_1, \ldots, w_m$, $\deg w_i = i$, called *universal Stiefel-Whitney classes*. Over G_∞^m one has the "*tautological*" m-dimensional vector bundle: its fiber over a point $p \in G_\infty^m$ (corresponding to an m-dimensional subspace $L(p) \subset \mathbb{R}^\infty$) is the subspace $L(p)$ itself. This bundle is universal in the following precise sense.

Proposition (see [249]). *For any m-dimensional real vector bundle E over an arbitrary paracompact space B there exists a continuous map* $\mathrm{cl}(E)\colon B \to G_\infty^m$, *unique up to homotopy, such that E is isomorphic to the bundle induced from the tautological bundle by* $\mathrm{cl}(E)$.

Consequently, any m-dimensional vector bundle E over B determines m elements $w_i(E) \in H^i(B; \mathbb{Z}_2)$, the images of the universal classes w_i under the homomorphism in cohomology induced by the map $\mathrm{cl}(E)$. The $w_i(E)$ are called the *Stiefel-Whitney classes of the bundle E*.

Over the space $\mathbb{C}^m \setminus \Sigma = K(\mathrm{Br}(m), 1)$ one has an m-dimensional real vector bundle E whose fibre over a point λ is the vector space of the functions on the set of roots of the polynomial $F(\cdot, \lambda)$. E is induced from the tautological bundle over G_∞^m by the map $B\,\mathrm{Br}(m) \to BO(m)$ that corresponds to the composite homomorphism (2). The preceding theorem can therefore be rephrased as follows.

Theorem. *The Stiefel-Whitney classes of the bundle E generate the ring* $H^*(\mathrm{Br}(m); \mathbb{Z}_2)$ *multiplicatively.*

For each m one has the obvious homomorphism $\mathrm{Br}(m) \to \mathrm{Br}(m+1)$ given by the operation of drawing an additional trivial (rectilinear) $(m+1)$st strand. The direct limit $\varinjlim B(m)$ of these groups is called the *stable braid group* and is denoted by Br.

Theorem ([128]). 1. *The ring $H^*(\mathrm{Br}; \mathbb{Z}_2)$ is generated by a certain set of elements $a_{n,k}$, $k \geqslant 0$, $n > 0$, of dimensions $2^k(2^n - 1)$, which satisfy, aside from the usual (anti)commutation relations, only the relations $a_{n,k}^2 = 0$.*

2. *The Bockstein homomorphism $\beta_2 = \mathrm{Sq}^1$ is given by the formula*

$$\beta_2 a_{n,k} = a_{n+1,0} a_{n,1} a_{n,2} \cdots a_{n,k-1}.$$

3. *The limit map $H^*(\mathrm{Br}; \mathbb{Z}_2) \to H^*(\mathrm{Br}(m); \mathbb{Z}_2)$ is an epimorphism for all m. The kernel of this homomorphism is generated by all possible products $a_{n_1,k_1} \cdots a_{n_s,k_s}$ such that $2^{n_1+k_1+\cdots+n_s+k_s} > m$.*

4. *The i-th Stiefel-Whitney class of the bundle E over $\mathbb{C}^m \setminus \Sigma$ is equal to the intersection index with the set of all points λ with the property that at most $m - i$ of the real parts of the roots of the polynomial $F(\cdot, \lambda)$ are distinct.*

Corollary. *For any m, the group* $H^i(\mathrm{Br}(m); \mathbb{Z}_2)$ *is nontrivial for* $i = m - d(m)$ *and trivial for* $i > m - d(m)$, *where* $d(m)$ *is the number of units in the binary representation of m.*

6.3. An Application: Superposition of Algebraic Functions. The study of the cohomology of braid groups was initiated in [9], [128] in connection with the problem of the representability of algebraic functions as superpositions of functions in a smaller number of variables.

Theorem ([10]). *If* $k < m - d(m)$, *then a universal m-valued function* (1) *in variables* $\lambda_1, \ldots, \lambda_m$ *cannot be decomposed into a complete superposition of functions in k variables.*

This theorem is based on the following result.

Lemma ([10]). *Suppose that the algebraic function* (1) *decomposes into a complete superposition of algebraic functions in k variables. Then there are a k-dimensional Stein manifold K, an m-dimensional vector bundle F over K, and a map* $\varkappa\colon \mathbb{C}^m \setminus \Sigma \to K$ *such that the bundle* $E \to \mathbb{C}^m \setminus \Sigma$ *is isomorphic to the bundle induced from F by* $\varkappa$.

In view of the naturalness of the Stiefel-Whitney classes, this implies that all classes $w_i(E)$ and their products lie in the subring $\varkappa^*(H^*(K; \mathbb{Z}_2))$. But for $k < m - d(m)$ this conclusion contradicts the results of subsection 6.2, because $H^i(K) = 0$ for $i > k$.

The definitive result in this problem was obtained by V. Ya. Lin by a different method. Namely, he established the following

Theorem ([216], [217]). *A universal algebraic function* (1) *does not admit a decomposition into a complete superposition of algebraic functions in k variables if* $k < m - 1$.

6.4. The Integer Cohomology of Braid Groups

Theorem ([115], [85], [338]). *For* $p \neq 2$ *a prime number, the ring* $H^*(\mathrm{Br}; \mathbb{Z}_p)$ *is the tensor product of the truncated polynomial algebras* $\mathbb{Z}_p[x_i]/\{x_i^p\}$ *with generators* x_i, $i \geqslant 0$, dim $x_i = 2p^{i+1} - 2$, *and the exterior (Grassmann) algebra with generators* y_j, $j \geqslant 0$, dim $y_j = 2p^j - 1$. *The canonical map* $H^*(\mathrm{Br}; \mathbb{Z}_p) \to H^*(\mathrm{Br}(m); \mathbb{Z}_p)$ *is an epimorphism and its kernel is generated by the monomials* $x_{r_1} \cdot \ldots \cdot x_{r_s} \cdot y_{l_1} \cdot \ldots \cdot y_{l_t}$ *for which* $2(p^{r_1+1} + \cdots + p^{r_s+1} + p^{l_1} + \cdots + p^{l_t}) > m$. *The Bockstein operator* β_p *corresponding to the exact sequence* $0 \to \mathbb{Z}_p \to \mathbb{Z}_{p^2} \to \mathbb{Z}_p \to 0$ *is defined by the formulas* $\beta_p x_i = y_{i+1}$, $\beta_p y_j = 0$.

Theorem ([85], [338]). *For any* $q \geqslant 2$ *and any m,* $H^q(\mathrm{Br}(m); \mathbb{Z}) = \bigoplus_p \beta_p H^q(\mathrm{Br}(m); \mathbb{Z}_q)$, *where the sum is taken over all prime p (including* $p = 2$). *In particular, the groups* $H^q(\mathrm{Br}(m); \mathbb{Z})$ *have no* p^2*-torsion.*

Corollary ([9]). 1. (Finiteness theorem): *for any m, all groups* $H^q(\mathrm{Br}(m))$ *with* $q \geqslant 2$ *are finite (and* $H^0 = H^1 = \mathbb{Z}$).

2. (Stabilization theorem): $H^q(\mathrm{Br}(m)) = H^q(\mathrm{Br}(2q-2))$ *for all* $m \geqslant 2q-2$.
3. *The first few stable groups* $H^q(\mathrm{Br})$ *are*: $H^0 = H^1 = \mathbb{Z}$, $H^2 = 0$, $H^3 = H^4 = \mathbb{Z}_2$, $H^5 = H^6 = \mathbb{Z}_6$.

Theorem (G.B. Segal [297]). $H^*(\mathrm{Br}) \cong H^*(\Omega^2 S^3)$, *where* $\Omega^i(\cdot)$ *denotes the i-th loop space.*

6.5. The Cohomology of Braid Groups with Twisted Coefficients. Let $\pm\mathbb{Z}$ denote the system of local coefficients (groups) on the space $K(\mathrm{Br}(m), 1)$ which is locally isomorphic to $\mathbb{Z}$, but turns over under transport along loops that correspond to braids defining odd permutations of the strands' endpoints. By definition, $H^*(K(\mathrm{Br}(m), 1); \pm\mathbb{Z}) \cong H^*(\mathrm{Br}(m); T)$, where T is the unique nontrivial representation $\mathrm{Br}(m) \to \mathrm{Aut}(\mathbb{Z})$. One can think of the group $H^*(\mathrm{Br}(m); T)$ as follows. Let $L \to K(\mathrm{Br}(m), 1)$ be a one-dimensional vector bundle with the property that the orientation of its fiber changes under transport over loops in the base that define the odd permutations. Let L_0 denote the complement of the zero section in L. Then $H^i(\mathrm{Br}(m); T) \cong H^{i+1}(L, L_0; \mathbb{Z})$.

Theorem ([360]). *The group* $H^{m-1}(\mathrm{Br}(m); T)$ *is trivial if m is not a power of a prime number, and is isomorphic to* $\mathbb{Z}_p$ *if* $m = p^l$. *For each prime p, if m is a sum of k powers of p, then* $H^{m-k}(\mathrm{Br}(m); T \otimes \mathbb{Z}_p)$ *is nontrivial.*

Theorem ([360]). *The homomorphism* $H^*(S(m); T) \to H^*(\mathrm{Br}(m); T)$ *induced by the first homomorphism in* (2) *is an epimorphism.*

The bundle $E \to \mathbb{C}^m \setminus \Sigma$, defined in subsection 6.2, possesses a nonzero section that sends each $\lambda \in \mathbb{C}^m \setminus \Sigma$ into the function identically equal to 1 on the set of roots of the polynomial $F(\cdot, \lambda)$. Let E' denote the quotient of E by the subbundle spanned by this section. Then E', like E, is not orientable, and hence it does not admit an Euler class defined as an element in the integer cohomology of $\mathbb{C}^m \setminus \Sigma$. However, E' does admit an Euler class defined as an element in the cohomology with coefficients in $\pm\mathbb{Z}$, this being connected with the fact that the system $\pm\mathbb{Z}$ turns over precisely on transport along the loops that destroy the orientation of the bundle E'.

Theorem ([360]). *For* $m = p^k$ *the Euler class* $e(E')$ *is a generator of the group* $H^{m-1}(\mathbb{C}^m \setminus \Sigma; \pm\mathbb{Z})$. $e(E')$ *is equal to the intersection index with the (suitably oriented) set of all points* $\lambda \in \mathbb{C}^m \setminus \Sigma$ *for which all the roots of the polynomial* $F(\cdot, \lambda)$ *have the same real part.*

We define the *Coxeter representation* X_m of the group $S(m)$ (and hence also of $\mathrm{Br}(m)$) as the representation $S(m) \to \mathrm{Aut}(\mathbb{Z}^m)$ that acts by permutations of the basis vectors. The restriction $\tilde{X}_m$ of the Coxeter representation to the sublattice $\{c \in \mathbb{Z}^m: c_1 + \cdots + c_m = 0\}$ is called the *reduced Coxeter representation*.

Theorem ([360]). *The group* $H^i(\mathrm{Br}(m); X_m)$ *is trivial for* $i > m-1$, *and is isomorphic to* $\mathbb{Z}$ *for* $i = 0, m-1$, *to* $\mathbb{Z}^2$ *for* $i = 1$, *and to* $\mathbb{Z}^2 \oplus$ (torsion) *for* $i = 2, 3, \ldots, m-2$. *The group* $H^i(\mathrm{Br}(m); \tilde{X}_m)$ *is trivial for* $i = 0$ *and* $i > m-1$, *and is isomorphic to* $\mathbb{Z}$ *for* $i = 1, m-1$, *and to* $\mathbb{Z}^2 \oplus$ (torsion) *for* $i = 2, 3, \ldots, m-2$.

6.6. Genus of Coverings Associated with an Algebraic Function, and Complexity of Algorithms for Computing Roots of Polynomials. The *topological complexity of an algorithm* was defined in [312] as a branching number (of the IF operators). The topological complexity of a problem is accordingly defined as the minimal topological complexity of the algorithms that solve the problem.

Consider the following problem $P(m, \varepsilon)$: given an arbitrary set of complex numbers $\lambda = (\lambda_1, \ldots, \lambda_m)$ belonging to the polydisk $B^m = \{\lambda \in \mathbb{C}^m: |\lambda_i| \leqslant 1, i = 1, \ldots, m\}$, compute the roots of the polynomial (1) with accuracy ε (i.e., guess a set of numbers $z_1, \ldots, z_m$ such that, for some ordering of the true roots $\xi_1, \ldots, \xi_m$ of the polynomial (1), $|z_i - \xi_i| \leqslant \varepsilon$ for $i = 1, \ldots, m$). The set $B^m \setminus \Sigma$ is again a $K(\mathrm{Br}(m), 1)$-space.

Smale [312] applied to the estimation of the topological complexity the notion of genus introduced by Shvarts in [301] (and rediscovered in [312]). Specifically, let X, Y be normal Hausdorff topological spaces (for example, manifolds) and let $p: X \to Y$ be a continuous map such that $Y = p(X)$.

Definition ([301]). The *genus of the map p* is defined to be the minimal cardinality of the covers of Y by open sets U_i with the property that the restriction $p|U_i$ admits a continuous section for all i.

Consider the $(m!)$-sheeted covering $p_m: M^m \to B^m\setminus\Sigma$, whose fiber over the point λ is the set of all possible ordered collections of roots of the polynomial $F(\cdot, \lambda)$. The covering p_m is a principal $S(m)$-bundle (see [183]).

Theorem ([312]). *For sufficiently small $\varepsilon > 0$ the topological complexity of the problem $P(m, \varepsilon)$ is not smaller than the genus of the covering p_m minus* 1.

From general results [301] on the genus of principal bundles one derives the following estimate.

Theorem. *Suppose that for some i and some $S(m)$-module A there is an element of the group $H^i(S(m); A)$ that is taken by the canonical homomorphism c: $\mathrm{Br}(m) \to S(m)$ into a nonzero element of $H^i(\mathrm{Br}(m); c^*A)$. Then the genus of the covering p_m is not smaller than $i + 1$.*

Combining this with the results of subsection 6.5 one immediately obtains the following:

Theorem. *For sufficiently small $\varepsilon > 0$ the topological complexity of the problem $P(m, \varepsilon)$ is not smaller than $m - \min D_p(m)$, where $D_p(m)$ is the sum of the digits in the base-p representation of the number m and the minimum is taken over all prime p.*

On the other hand, one has the following assertion.

Theorem ([360]). *For any $\varepsilon > 0$ and any m there exists an algorithm of topological complexity $m - 1$ that solves problem $P(m, \varepsilon)$.*

Corollary. *If m is a power of a prime number, then for any sufficiently small ε the topological complexity of problem $P(m, \varepsilon)$ equals $m - 1$.*

Now let $P_1(m, \varepsilon)$ be a problem that differs from $P(m, \varepsilon)$ in that one is required to compute approximately only one of the roots of the polynomial. With the algebraic function (1) is associated yet another covering of $B^m \setminus \Sigma$, this time m-sheeted, whose fiber over the point λ consists of all roots of the polynomial $F(\cdot, \lambda)$; we denote this covering by φ_m. By analogy with Smale's theorem one can establish the following result.

Theorem. *For sufficiently small $\varepsilon > 0$ the topological complexity of problem $P_1(m, \varepsilon)$ is not smaller than the genus of the covering φ_m minus 1.*

Theorem ([360]). *If m is a power of a prime number, then the genus of the covering φ_m equals m.*

Thus, from a "topological" point of view, the problems of computing all roots or only one root have the same complexity for $m = p^k$, with p prime.

6.7. The Cohomology of Brieskorn Braid Groups and Complements of the Discriminants of Singularities of the Series C and D. The definition of the Brieskorn braid groups was given in subsection 5.3. The cohomology of these groups for the series C and D was computed in [154], [156].

Theorem C. $H^q(\mathrm{Br}\, C_m) = \bigoplus_{i=0}^{\infty} H^{q-i}(\mathrm{Br}(m-i))$.

Theorem D. *Let $K(a, b)$ denote the kernel of the canonical homomorphism $H^a(\mathrm{Br}(b); \mathbb{Z}) \to H^a(\mathrm{Br}(b-1); \mathbb{Z})$. Then*

$$H^q(\mathrm{Br}\, D_m; \mathbb{Z}) = H^q(\mathrm{Br}(m); \mathbb{Z}) \oplus \left[\bigoplus_{i=0}^{\infty} K(q-2i, m-2i)\right] \oplus \left[\bigoplus_{i=0}^{\infty} H^{q-2i-3}(\mathrm{Br}(m-2i-3); \mathbf{Z}_2)\right].$$

Corollary. $H^q(\mathrm{Br}\, C_\infty; \mathbb{Z}) = \bigoplus_{i=0}^{\infty} H^{q-i}(\mathrm{Br}; \mathbb{Z})$,

$$H^q(\mathrm{Br}\, D_\infty; \mathbb{Z}) = H^q(\mathrm{Br}; \mathbb{Z}) \oplus \left[\bigoplus_{i=1}^{\infty} H^{q-2i-3}(\mathrm{Br}; \mathbb{Z}_2)\right].$$

According to [59], [106], the complement of the discriminant of a D_m-singularity is a $K(\pi, 1)$-space, where π is the Brieskorn braid group D_m. In conjunction with Theorem D, this yields the cohomology of such a complement. Similarly, Theorem C yields the cohomology of the complement of the discriminant of a boundary C_m-singularity (see [21]).

6.8. The Stable Cohomology of Complements of Level Bifurcation Sets. The cohomology ring of the complement of the discriminant variety Σ in the space of a versal deformation is defined for any finite-multiplicity singularity of a function; this ring does not depend on the choice of a versal deformation. Any adjacency of singularities defines a ring homomorphism: the complement of the discriminant of the simpler singularity embeds in the complement of the discriminant of the more complex singularity. (For example, Fig. 39 shows the embedding of the complement of the discriminant of the real singularity A_2 in the analogous space for A_3.) The hierarchy of singularities allows one to pass to stable objects, namely, the stable cohomology rings of complements of discriminants. In doing so one can stabilize in the class of functions of a fixed number of variables as well as with respect to all functions of arbitrarily many variables. Let us define and describe these stable cohomology rings.

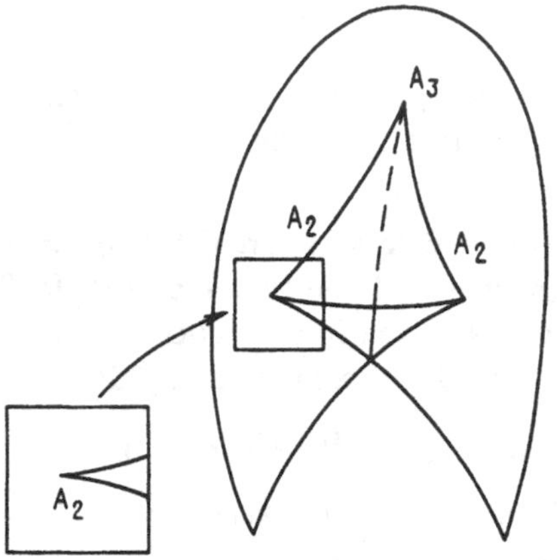

Fig. 39

Let $f: (\mathbb{C}^n, 0) \to (\mathbb{C}, 0)$ be a germ of smooth function with a singularity at the point 0, $F: \mathbb{C}^n \times \mathbb{C}^l \to \mathbb{C}$ a versal deformation of f (in particular, $F(\cdot, 0) \equiv f$), and Σ its discriminant.

The *local* (near 0) *cohomology ring* of the complement of Σ is defined as the ring $H^*(U \setminus \Sigma)$, where U is a sufficiently small ball centered at $0 \in \mathbb{C}^l$. We denote this ring by $J^*(F)$. (Thanks to the analyticity of Σ, this definition does not depend on the choice of a ball U, provided U is small enough.)

Let g be another singularity of finite multiplicity and $G: \mathbb{C}^n \times \mathbb{C}^k \to \mathbb{C}$ a versal deformation of g. Suppose that g is adjacent to f; this means that in any neighborhood of $0 \in \mathbb{C}^k$ the set $\{f\}$, consisting of the values of the parameter $\varkappa \in \mathbb{C}^k$ for which one of the critical points of the function $G(\cdot, \varkappa)$ close to 0 is biholomorphically equivalent to the singularity of f, is nonempty. For any nonsingular point $\varkappa_0$ of the set $\{f\}$ that is sufficiently close to 0, and any transversal L to $\{f\}$ at $\varkappa_0$, the family of functions $G(\cdot, \varkappa)$, $\varkappa \in L$, is a versal deformation of the singular point of the function $G(\cdot, \varkappa_0)$. Consequently, the deformation F is equivalent to that induced from this family by some local holomorphic map $\varphi: (\mathbb{C}^l, 0) \to (L, \varkappa_0)$. Moreover, $\varphi^{-1}(\Sigma(G)) \subset \Sigma(F)$ and hence there arises a ring homomorphism $\varphi^*: J^*(G) \to J^*(G)$. The latter depends on the choice of the point $\varkappa_0$ and of the inducing map φ.

Example. Let $n = 1$, $f = x^m$, $g = x^{m+1}$. Then the resulting homomorphism $H^*(\mathrm{Br}(m+1)) \to H^*(\mathrm{Br}(m))$ coincides with the homomorphism induced by the obvious injection $\mathrm{Br}(m) \to \mathrm{Br}(m+1)$ (see subsection 6.2).

Definition. A *stable cohomology class of complements of discriminants of singularities* of functions of n variables is a rule that assigns to each deformation F of an arbitrary singularity of finite multiplicity an element of the ring $J^*(F)$, in such a manner that for each pair f, g of adjacent singularities any of the homomorphisms $J^*(G) \to J^*(F)$ constructed according to the scheme described above takes the class corresponding to the deformation G into the class corresponding to the deformation F.

The set of all such rules forms a graded ring, denoted by ${}_n\mathscr{H}^*$.

In particular, ${}_1\mathscr{H}^*$ is the stable cohomology ring $H^*(\mathrm{Br})$ of braid groups; see subsection 6.2.

Further, to each singularity f in n variables $x_1, \ldots, x_n$ and each deformation $F(x_1, \ldots, x_n, \lambda)$ of f one can assign the function $f + x_{n+1}^2$ and its deformation $F(x_1, \ldots, x_n, \lambda) + x_{n+1}^2$; under this assignment the discriminants of the two deformations correspond to the same values of the parameter λ. These assignments define homomorphisms

$$\cdots \leftarrow {}_n\mathscr{H}^* \leftarrow {}_{n+1}\mathscr{H}^* \leftarrow \cdots.$$

Definition. The *"big" stable cohomology ring $\mathscr{H}^*$ of complements of discriminants of singularities* is $\varprojlim {}_n\mathscr{H}^*$, taken with respect to the homomorphisms given above.

The analogous stable cohomology groups with coefficients in $\mathbb{Z}_q$ are denoted by ${}^q_n\mathscr{H}^*$ and ${}^q\mathscr{H}^*$.

Theorem ([359], [362]). *For each i,*

$$\mathscr{H}^i = \bigoplus_{k=0}^{i} H^{i-k}(S(k), T_k),$$

where T_k is the unique nontrivial representation $S(k) \to \mathrm{Aut}(\mathbb{Z})$ for $k \geqslant 2$, and $T_0 = T_1 = \mathbb{Z}$.

Corollary. *All groups $\mathscr{H}^i$ are finite, except for $\mathscr{H}^0 = \mathscr{H}^1 = \mathbb{Z}$.*

Corollary. $\mathscr{H}^2 = 0$, $\mathscr{H}^3 = \mathscr{H}^4 = \mathbb{Z}_2$, $\mathscr{H}^5 = \mathbb{Z}_6 \oplus \mathbb{Z}_2$.

Since $T_k \otimes \mathbb{Z}_2 = \mathbb{Z}_2$, the results of [182] provide full information on the groups ${}^2\mathscr{H}^i$; in particular, for $i = 1, \ldots, 8$, $\dim {}^2\mathscr{H}^i = 1, 1, 2, 3, 4, 6, 9, 12$.

For the groups ${}_n\mathscr{H}^*$ with fixed n, the following statements hold.

Theorem ([364a], [364c]). ${}_n\mathscr{H}^* = H^*(\Omega^{2n}S^{2n+1})$, *where $\Omega^i(\cdot)$ denotes the i-th loop space.*

(For $n = 1$, one obtains Segal's theorem; see subsection 6.4.)

Corollary 1. *For each i,*

$$H^i(\Omega^\infty S^{\infty+1}) = \bigoplus_{k=0}^{i} H^{i-k}(S(k), T_k).$$

In fact, a more general relation holds:

$$H^i(\Omega^\infty S^{\infty+t}) = \bigoplus_{k\geqslant 0} H^{i-kt}(S(k), T_k^{\otimes t})$$

($T_k^{\otimes t} = T_k$ for t odd, and $= \mathbb{Z}$ for t even).

Corollary 2 (finiteness theorem, see [9]). *All groups ${}_n\mathscr{H}^i$ are finite, except for ${}_n\mathscr{H}^0 = {}_n\mathscr{H}^1 = \mathbb{Z}$.*

Corollary 3 (stability theorem, see [9]). *For $i \leqslant 2n - 2$, ${}_n\mathscr{H}^i = \mathscr{H}^i$.*

Theorem (second stability theorem). *For each pair m, n there exists a singularity in n variables such that for any of its versal deformations F the obvious homomorphism from ${}_n\mathscr{H}^i$ into $J^i(F)$ is an isomorphism for all $i \leqslant m$.*

A similar formula holds for the stable cohomology ring ${}_n\tilde{\mathscr{H}}^*$ of the complements of caustics (see subsection 1.1.11) of holomorphic function-germs of n complex variables:

$${}_n\tilde{\mathscr{H}}^* = H^*(\Omega^{2n}\Sigma^{2n}\Lambda(n)),$$

where Σ^i denotes the ith suspension and $\Lambda(n) = U(n)/O(n)$ is the Lagrangian Grassmann manifold; see [364a], [364c].

6.9. Characteristic Classes of Milnor Cohomology Bundles. We remind the reader that a remarkable vector bundle is defined over the space $U \setminus \Sigma \subset \mathbb{C}^l$: the Milnor cohomology bundle, whose fibre over a point λ is the space $H^{n-1}(V_\lambda; \mathbb{R})$, where $V_\lambda = \{x: F(x, \lambda) = 0\} \cap B_\varepsilon$; see subsection 2.3.3. Moreover, stabilization of singularities allows us to regard $U \setminus \Sigma$ as the common base of a whole series of Milnor cohomology bundles corresponding to the singularities $f, f + x_{n+1}^2, \dots, f + x_{n+1}^2 + \cdots + x_{n+s}^2, \dots$. As follows readily from the Picard-Lefschetz formulas, all such bundles corresponding to numbers s of the same parity are isomorphic to one another; accordingly, we shall term *even* [resp. *odd*] the bundles corresponding to an even [resp. odd] value of $n + s$.

Theorem ([359], [362]). *All Stiefel-Whitney classes of the Milnor cohomology bundles are stable classes (i.e., define elements of the groups ${}^2\mathscr{H}^*$, ${}^2_n\mathscr{H}^*$).*

Corollary. *All canonical homomorphisms ${}^2_n\mathscr{H}^* \to {}^2_1\mathscr{H}^*$* are surjective.

This is an immediate consequence of the first theorem of subsection 6.2.

Theorem. *All Stiefel-Whitney classes of even bundles, the squares of all Stiefel-Whitney classes of odd bundles, and the Pontryagin classes of arbitrary Milnor cohomology bundles are trivial.*

The last two assertions follow from the trivializability of the complexified Milnor bundles (see subsection 2.3.6).

The Stiefel-Whitney classes of the odd bundles are all nontrivial and admit the following realization. Consider the set of points of transverse k-fold self-intersection of the discriminant Σ (this set is nonempty for any sufficiently complicated singularity, for example for a singularity of type A_{2k-1}). Near any point of this set the pair $(\mathbb{C}^\mu, \Sigma)$ is diffeomorphic to the product of a linear space and the pair ($\mathbb{C}^k$, the union of the coordinate planes in $\mathbb{C}^k$). Consequently, for a small ball W centered at any point of this set the group $H_k(W \setminus \Sigma, \mathbb{Z}_2)$ is one-dimensional and is generated by the class of an embedded k-dimensional torus.

Theorem ([362]). *The k-dimensional Stiefel-Whitney class of an odd Milnor cohomology bundle takes a nonzero value on this torus.*

6.10. Stable Irreducibility of Strata of Discriminants. Let f and g be singular function-germs $(\mathbb{C}^n, 0) \to (\mathbb{C}, 0)$ such that g is adjacent to f, i.e., in the base of the versal deformation of g the origin lies in the closure of the set $\{f\}$ (see subsection 6.8). It may happen that $\{f\}$ has several components: for example, in the discriminant D_4 the singularity A_3 is represented by three curves. As it turns out, this cannot happen if g is sufficiently complicated in comparison with f; moreover, an analogous result is valid for strata of multisingularities.

Definition. The isolated singularity g is said to *envelop* the singularity f if for any singularity h adjacent to g, the set $\{f\}$ in the versal deformation of h is nonempty and irreducible.

Theorem ([362]). *For any isolated singularity f there exists an enveloping singularity g. Specifically, it suffices to take* $g = x_1^t + \cdots + x_n^t$, *with* $t \geqslant \mu(f) + 2$, *where* $\mu(f)$ *is the Milnor number of f.*

To state an analogous theorem for strata of multisingularities we need to relax the notion of adjacency.

Definition. Two singularities $\varphi, \psi\colon (\mathbb{C}^n, 0) \to (\mathbb{C}, 0)$ are said to be *almost equivalent* if there is a constant $c \neq 0$ such that φ is equivalent to $c\psi$.

In the case of singularities that are equivalent to quasihomogeneous ones the notions of equivalence and almost equivalence coincide.

Theorem ([362]). *For any set* $f_1, \ldots, f_k, g$ *of isolated singularities in n variables there is an isolated singularity h with the properties that h is adjacent to g and the set of points of any versal deformation of h that correspond to functions having near 0 exactly k distinct critical points with critical value 0 which are almost equivalent to* $f_1, \ldots, f_k$, *respectively, is nonempty. Specifically, it suffices to take* $h = x_1^t + \cdots + x_n^t$, *where* $t > n[k(1 + \max \mu(f_i)) - 1]$ *and is also sufficiently large to ensure that h is adjacent to g.*

Chapter 3
Basic Properties of Maps

One of the tasks that must be tackled repetedly in singularity theory is the investigation of how certain properties of map-germs behave with respect to various equivalence relations. Most often the equivalence of two germs means that they belong to the same orbit of the action of some group of diffeomorphisms on a function space.

In this chapter we treat the general theory of equivalence of differentiable maps. We give the classical results of Mather on stability of maps. We state the standard theorems on versality and finite determinacy. We also indicate sufficient conditions on the equivalence relations for these standard theorems to hold. Finally, in various situations in which the relation of differentiable equivalence becomes burdensome for smooth maps we examine the consequences of extending this equivalence to a topological one.

§ 1. Stable Maps and Maps of Finite Multiplicity

1.1. The Left-Right Equivalence. Let $f: N \to P$ be a map between real manifolds. We say that f is *smooth* if it is continuously differentiable the needed number of times (for example, it belongs to the class C^∞).

Definition. A point $x \in N$ is called a *critical point* of f if the rank of the derivative $df_x: T_xN \to T_{f(x)}P$ of f at x is smaller than the maximal one (i.e., is smaller than min(dim N, dim P)).

Maps between manifolds can be regarded as specified up to various equivalence relations (see § 2). If neither the source nor the target manifold is endowed with additional structures that the equivalence relation should preserve, then the most natural (though not the most convenient to work with) is the left-right (or $\mathscr{A}$-) equivalence.

Definition. Two maps $f_1: N_1 \to P_1$ and $f_2: N_2 \to P_2$ are said to be *$\mathscr{A}$-equivalent* if there exists a commutative diagram

$$\begin{array}{ccc} N_1 & \xrightarrow{f_1} & P_1 \\ \downarrow{\scriptstyle h} & & \downarrow{\scriptstyle k} \\ N_2 & \xrightarrow{f_2} & P_2 \end{array}$$

in which the vertical maps are diffeomorphisms.

An analogous definition is given for germs. The equivalence class of a map-germ at a critical point is called a *singularity*.

Let $\Omega(N, P)$ denote the space of all smooth maps from N to P, and let $\mathscr{D}(N)$ denote the group of diffeomorphisms of the manifold N. The $\mathscr{A}$-equivalence is precisely the equivalence of elements $f \in \Omega(N, P)$ under the action of the group $\mathscr{A} = \mathscr{D}(N) \times \mathscr{D}(P)$ of pairs (h, k) of diffeomorphisms given by the rule

$$(h, k)f = k \circ f \circ h^{-1}.$$

The groups of diffeomorphisms of the source, $\mathscr{D}(N)$, and of the target, $\mathscr{D}(P)$, are called the *groups of right* and *left equivalence* and are denoted by $\mathscr{R}$ and $\mathscr{L}$. In the local case, when one considers map-germs at a specified point in the source and with a specified image of that point, $\mathscr{R}$ and $\mathscr{L}$ are the groups of germs of diffeomorphisms of the source and the target manifold, respectively, that keep those points fixed.

1.2. Stability. Consider a smooth map $f: N \to P$ of a closed (i.e., compact boundaryless) manifold N into a manifold P.

Definition. The *map* f is said to be *stable* if any map that is close enough to f is $\mathscr{A}$-equivalent to f.

The closeness of two maps is understood as the closeness of the maps themselves and of a sufficiently large number of their derivatives.

The stability condition for f means that the $\mathscr{A}$-orbit of f in the space $\Omega(N, P)$ is open.

Definition. The *germ* of a smooth map $f: N \to P$ at a point x of the source manifold N is said to be *stable* if for any map $N \to P$ that is sufficiently close to f there is a point in the vicinity of x at which its germ is $\mathscr{A}$-equivalent to the germ of f at x.

In other words, for any sufficiently small neighborhood U of x there exists a neighborhood E of f in $\Omega(N, P)$ such that for any $\tilde{f} \in E$ there is a point $\tilde{x} \in U$ with the property that the germ of $\tilde{f}$ at $\tilde{x}$ is $\mathscr{A}$-equivalent to the germ of f at x. In the local situation at hand one can assume that N and P are domains in Euclidean spaces and E is defined by inequalities of the form $\|\tilde{f} - f\|_k \leqslant \varepsilon$, where $\|g\|_k = \sup_{|\alpha| \leqslant k} |\partial^\alpha g/\partial x^\alpha|$.

Example. Whitney's Theorem ([378]). A map of two-dimensional manifolds is stable at a point if and only if, in suitable local coordinates (x_1, x_2) in the source and (y_1, y_2) in the target, it can be written in one of the following three forms:

1. $y_1 = x_1, y_2 = x_2$ (regular point);
2. $y_1 = x_1^2, y_2 = x_2$ (*fold* or *pleat*);
3. $y_1 = x_1^3 + x_1 x_2, y_2 = x_2$ (*cusp*).

In all three cases the point in question is the origin.

Figure 40 shows the realization of Whitney's cusp as the singularity of the projection map of the surface $y_1 = x_1^3 + x_1 x_2$ in three-dimensional space on the plane. The sets of critical points and critical values of this map are distinguished. At all critical points, except for the origin, the singularity is of fold type.

Example (Whitney [380]). The image of a stable map $\mathbb{R}^2 \to \mathbb{R}^3$ is a surface that may have lines along which two of its sheets intersect transversally, as well as isolated points of transverse intersection of three sheets (of the type of the intersection of the coordinate planes in $\mathbb{R}^3$). Besides these singularities, a stable map may have only one more singularity, which in suitable coordinates takes the form

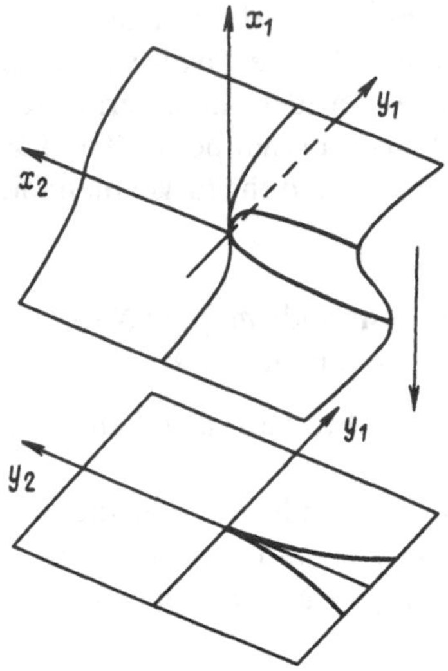

Fig. 40

$$y_1 = x_1 x_2, \qquad y_2 = x_2, \qquad y_3 = x_1^2.$$

The image of this map, i.e., the surface $y_1^2 = y_3 y_2^2$, $y_3 \geqslant 0$, shown in Fig. 41, is called *Whitney's umbrella* (or *Cayley's umbrella*).

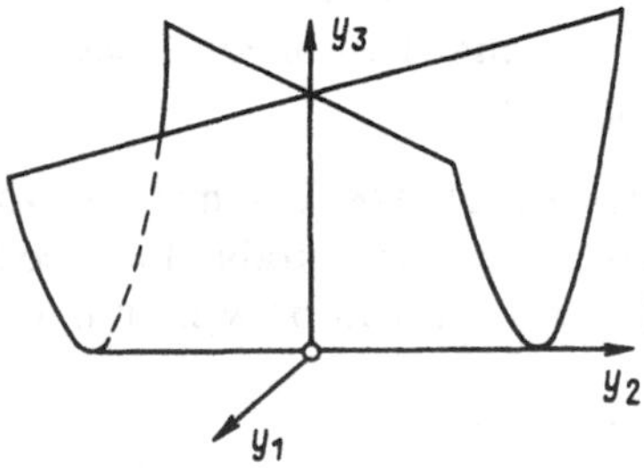

Fig. 41

In subsection 1.1.5 we defined the k-jet $j_a^k f$ of a function f at a point a. This notion carries over without modification to the case of maps. Once we fix a coordinate system, the k-jet is simply the set of Taylor polynomials (of degree k) of the coordinate functions of f.

Copying the definition given for functions, we say that the *k-jet of a map-germ* $f: (\mathbb{R}^n, 0) \to (\mathbb{R}^p, 0)$ *is $\mathscr{A}$-sufficient* if any other map-germ $\tilde{f}: (\mathbb{R}^n, 0) \to (\mathbb{R}^p, 0)$ with the same k-jet as f is $\mathscr{A}$-equivalent to f. In particular, if this is the case, then f is $\mathscr{A}$-equivalent to its k-jet. A *germ* that has an $\mathscr{A}$-sufficient jet of finite order is said to be *finitely determined* (or, briefly, *finite*).

We have the following rather crude estimate.

Theorem ([236, Part IV]). *The* $(p+1)$*-jet of any stable smooth germ* $f:(\mathbb{R}^n,0)\to(\mathbb{R}^p,0)$ *is* $\mathcal{A}$*-sufficient.*

Example. The germ of x^3 at zero is not stable.

1.3. Transversality. In the example of maps of two-dimensional manifolds discussed in the preceding subsection the stable singularities of distinct types were observed on subsets of distinct dimensions: the folds on curves and the cusps at isolated points. This phenomenon, and also the absence in the stable case, in the setting of the example at hand, of other types of degeneracies, are explained by Thom's transversality theorem.

Definition. Let $f: A\to B$ be a smooth map of manifolds and let C be a smooth submanifold of B. The map f is said to be *transverse* to C at the point $a\in A$ if either $f(a)\notin C$ or the image of the tangent space to A at a under the differential df_a is transverse to the tangent space to C:

$$df_a(T_aA)+T_{f(a)}C=T_{f(a)}B.$$

The map f is said to be *transverse to* C if it is transverse to C at any point $a\in A$. If this is the case, then $f^{-1}(C)$ is a smooth submanifold of A and its codimension in A is equal to the codimension of C in B.

Remark. Transversality of the image $f(A)$ to C does not guarantee the transversality of the map f itself to C.

Definition. The set $J^k(N,P)$ of all k-jets of smooth maps of a fixed manifold N into a fixed manifold P (at all points of N) is called the *space of k-jets* of maps of N into P.

Definition. The *k-jet extension* j^kf of a smooth map $f: N\to P$ is the map of N into the k-jet space $J^k(N,P)$ which sends each point of N into the k-jet of f at that point.

(Thom's) Strong Transversality Theorem [37]. *Let N be a closed manifold and C a closed submanifold of the k-jet space* $J^k(N,P)$. *Then the set of all maps* $f: N\to P$ *whose k-jet extensions are transverse to C is open and dense in the space of all smooth maps of N into P.*

The $k=0$ version of this statement is known as the *weak transversality theorem.*

The proof of the transversality theorem is carried out by reduction to the well-known *theorem of Bertini and Sard,* which asserts that the set of critical values of a sufficiently smooth map has measure zero.

Remark. 1°. If C is not closed, then in Thom's theorem one must replace "open" by "countable intersection of open sets".

2°. When the source manifold is not compact, the transversality theorem remains valid provided that one endows the space of maps with Whitney's fine topology (to guarantee that the set of transverse maps is open C must be closed).

We recall that a basis of that topology is obtained by declaring as open any of the sets consisting of all C^∞-maps $f: N \to P$ with the property that the k-jet of f at any point of N belongs to V, where V is an arbitrary open subset of the jet space $J^k(N, P)$, and k is arbitrary.

3°. One frequently encounters situations in which C is a manifold with singularities rather than a smooth submanifold.

Definition. A *stratified submanifold* of a smooth manifold is a finite union of pairwise-disjoint smooth path-connected manifolds (strata) with the property that the closure of any stratum consists of that stratum and a finite union of strata of lower dimensions.

A map is said to be *transverse to a stratified submanifold* if it is transverse to each of its strata.

The transversality theorem extends also to the case where C is a stratified submanifold. However, in this case it guarantees only that the set of transverse maps is the intersection of a countable number of open sets and dense (instead of being open and dense).

In order that the maps transverse to a stratified analytic submanifold form an open dense set it suffices that the stratification satisfy the following additional condition: any embedding that is transverse to a stratum of lower dimension is transverse to all adjacent strata of higher dimension in some neighborhood of that lower-dimensional stratum[1].

An example of such a stratification is provided by the partition of the set of all matrices of a given size according to rank. Specifically, in the space of all $n \times p$ matrices the manifold of the matrices of rank r has codimension $(n - r)(p - r)$ (product-of-coranks formula). This implies that any smooth map $g: N^n \to P^p$ can be approximated by a map f with the property that the set $\Sigma^i(f)$ of critical points of f at which the differential has rank $r = n - i$ is a smooth submanifold of N^n of codimension $(n - r)(p - r) = i(p - n + i)$. The same conclusion follows also from the general theorems on transverse maps proved by V.I. Bakhtin, which generalize Thom's transversality theorem.

Definition. Two smooth *maps* $f: A \to B$ and $g: C \to B$ of manifolds A and C into the same manifold B are said to be *transverse* if for any pair of points $a \in A$, $c \in C$ such that $f(a) = f(c) = b \in B$ the images of the corresponding tangent spaces under the differentials of f and g are transverse, i.e.,

$$df_a(T_aA) + df_c(T_cC) = T_bB.$$

Theorem ([44]). *Let the space $C^1(M, P)$ of continuously differentiable maps $M \to P$ be endowed with Whitney's fine topology. Then for any given proper C^1-map $g: N \to P$, the set of the C^1-maps $M \to P$ that are transverse to g is open and dense in $C^1(M, P)$.*

---

[1] This is also true for stratified manifolds obtained by means of analytic diffeomorphisms, but is not always true for C^∞-stratified manifolds [333].

Theorem ([44]). *Let $g: N \to J^k(M, P)$ be a proper continuously differentiable map of N into the space of k-jets of maps from M to P. Then in the space $C^{k+1}(M, P)$ (endowed with the fine topology) the set of maps whose k-jet extensions are transverse to g is open and dense.*

In [206] these two theorems were stated without the requirement that g be proper; thus one could conclude only that the set of all maps transverse to g is a countable intersection of open dense sets (rather than open).

Theorem ([44]). *Let the k-times ($k \geqslant 1$) continuously differentiable maps $f: M \to P$ and $g: N \to P$ be transverse. Then the set Q of all pairs $(x, y) \in M \times N$ such that $f(x) = g(y)$ is a C^k-submanifold of $M \times N$ of dimension* $\dim M + \dim N - \dim P$. *If, in addition, g is proper, then the projection $\pi_M: Q \to M$ is proper.*

It is natural to call the map π_M the preimage of g under the action of f.

Theorem ([44]). *In the space of $n \times p$ matrices consider the set of all matrices whose rank is less than or equal to some fixed number k. This set is the image of a smooth proper map, and the preimage manifold has the same dimension as the set itself.*

The map whose existence is asserted in the theorem can be constructed explicitly as follows. Identify the set of $n \times p$ matrices with the space $\mathrm{Hom}(\mathbb{R}^p, \mathbb{R}^n)$ of linear operators $\mathbb{R}^p \to \mathbb{R}^n$. To the matrices of rank at most r there correspond the operators whose kernels contain at least one $(p - r)$-dimensional subspace. The $(p - r)$-dimensional subspaces of $\mathbb{R}^p$ form the Grassmann manifold G. Now take as source the set N of all pairs $(A, L) \in \mathrm{Hom}(\mathbb{R}^p, \mathbb{R}^n) \times G$ such that L is contained in the kernel of A, and define the desired map by the rule $(A, L) \mapsto A$.

1.4. The Thom-Boardman Classes. Let $I = (i_1, i_2, \ldots, i_k)$ be a finite nonincreasing sequence of nonnegative integers. Given a smooth map $f: N^n \to P^p$, with $n \geqslant i_1$, define the set $\Sigma^I(f)$ inductively as follows.

Definition (Thom). Suppose $\Sigma^I(f) = \Sigma^{i_1, \ldots, i_k}(f) \subset N$ is a smooth submanifold. Then

$$\Sigma^{i_1, \ldots, i_k, i_{k+1}}(f) = \Sigma^{i_{k+1}}(f | \Sigma^I(f))$$

is the set of all points at which the kernel of the differential of the restriction of f to $\Sigma^I(f)$ has dimension i_{k+1}.

One has the chain of inclusions $N \supset \Sigma^{i_1} \supset \Sigma^{i_1, i_2} \supset \cdots \supset \Sigma^{i_1, i_2, i_3} \supset \cdots$.

Boardman [54] proposed a different definition of $\Sigma^I(f)$ in terms of jet spaces, removing the requirement that $\Sigma^I(f)$ be smooth. Specifically, for any nonincreasing set of nonnegative integers $I = (i_1, \ldots, i_k)$ he defined a smooth submanifold Σ^I (not necessarily closed) in the k-jet space $J^k(N^n, P^p)$; his definition involves no particular map f.

To construct Σ^I according to Boardman we need the following

Definition. Let B be an ideal in the ring $\mathscr{E}_n$ of germs at zero of C^∞-functions on $\mathbb{R}^n$. The *Jacobian extension* $\Delta_k(B)$ of B is the ideal generated by B and all Jacobians $\det(\partial\varphi_i/\partial x_j)$ of order k built from partial derivatives of functions belonging to B relative to some fixed coordinates in $(\mathbb{R}^n, 0)$.

It is readily verified that the ideal $\Delta_k(B)$ does not depend on the choice of the coordinates.

Clearly, $\Delta_{k+1}(B) \subseteq \Delta_k(B)$ for all k. A Jacobian extension $\Delta_k(B)$ is said to be *critical* if the order k of the Jacobians is the smallest for which the extension does not coincide with the whole of $\mathscr{E}_n$: $\Delta_k(B) \neq \mathscr{E}_n = \Delta_{k-1}(B)$.

Example. Take $n = 4$ and let B be the ideal generated by x_1, x_2, x_3^2, x_4^2. The successive critical extensions of B are $\Delta_3(B) = \mathfrak{m}_4$ and $\Delta_5(\Delta_3(B)) = \Delta_3(B)$.

For our purposes it is convenient to relabel the extensions, setting $\Delta^k = \Delta_{m-k+1}$.

With this notation, in the example given above the critical extensions are Δ^2 and $\Delta^0\Delta^2$.

Definition. Let $I = (i_1, \ldots, i_k)$ be a nonincreasing set of nonnegative integers. The *ideal B is said to have Boardman symbol I* if its successive critical extensions are $\Delta^{i_1}B, \ldots, \Delta^{i_k}\Delta^{i_{k-1}} \ldots \Delta^{i_1}B$.

Definition. Let the map-germ $f\colon N^n \to P^p$ be given, in some local coordinates, by the formulas $y_i = f_i(x)$, $f(0) = 0$. We say that *f has a singularity of class Σ^I at zero* if the ideal of $\mathscr{E}_n$ generated by the p coordinate functions f_i of f has Boardman symbol I.

One can readily verify that the condition imposed in this definition is actually a condition on the k-jet of f at zero and as such does not depend on the choice of coordinates. The set of all k-jets satisfying this condition is precisely the intersection of the sought-for set $\Sigma^I \subset J^k(N, P)$ and the fibre of the bundle $J^k(N, P) \to N \times P$.

The codimension of the submanifold Σ^I in $J^k(N, P)$ is given by the formula

$$v_I(n, p) = (p - n + i_1)\mu(i_1, i_2, \ldots, i_k) - (i_1 - i_2)\mu(i_2, i_3, \ldots, i_k) - \cdots$$
$$- (i_{k-1} - i_k)\mu(i_k),$$

where $\mu(i_1, \ldots, i_k)$ denotes the number of nonincreasing sequences of integers $j_1, j_2, \ldots, j_k$ such that $0 \leqslant j_r \leqslant i_r$ for $r = 1, 2, \ldots, k$ and $j_1 > 0$.

Concerning the Boardman manifolds see also [239], [251].

A map f is said to be *nice* if its k-jet extension is transverse to the manifolds Σ^I.

Boardman proved the following result.

Theorem ([54]). 1) *If f is a nice map, then $\Sigma^I(f) = (j^kf)^{-1}(\Sigma^I)$. In other words, $\Sigma^I(f)$ is a submanifold of codimension $v_I(n, p)$ in N and $x \in \Sigma^I(f)$ if and only if the k-jet of f at x belongs to Σ^I.*

2) *Any smooth map $f\colon N^n \to P^p$ can be arbitrarily well approximated, together with any number of its derivatives, by a nice map.*

Example. For $I = \underbrace{(1, 1, \ldots, 1)}_{k}$ we have $\mu(1, 1, \ldots, 1) = k$ and $\nu_I(n, p) = (p - n + 1)k$. In particular, the Whitney cusp (a singularity of class $\Sigma^{1,1}$) has codimension 2 and consequently for maps of the plane into itself it cannot be removed (by a small perturbation of the map) at isolated points.

1.5. Infinitesimal Stability. We indicated earlier the conditions under which a Thom-Boardman class may occur in a stable fashion for smooth maps of manifolds. However, the classification of germs according to Thom-Boardman types is coarser than the classification up to $\mathscr{A}$-equivalence.

Let us now describe a linearization method for settling the stability question for smooth map-germs.

A smooth map $f: N^n \to P^p$ induces a vector bundle f^*TP with base N and p-dimensional fibre. The space of sections of this bundle can be canonically identified with the tangent space at the point f to the set $\Omega(N, P)$ of all smooth maps from N to P, i.e., with the space of variations (infinitesimal deformations) of the map f:

$$\Gamma(f^*TP) = T_f\Omega(N, P).$$

A part of these infinitesimal deformations of f are obtained from the action of the infinitesimal diffeomorphisms of the source and target on f, i.e., from the action of smooth vector fields on N and P (of elements of $\Theta(N) = \Gamma(TN)$ and $\Theta(P)$). The linear operator of this action is given by the rule

$$(h, k) \mapsto \alpha[h] + \omega[k],$$

where $\alpha: \Theta(N) \to \Gamma(f^*TP)$ and $\omega: \Theta(P) \to \Gamma(f^*TP)$ are defined by the relations $\alpha[h](x) = -df_x(h(x))$ and $\omega[k](x) = k(f(x))$, respectively.

Definition. A map f is said to be *infinitesimally stable* if

$$\alpha(\Theta(N)) + \omega(\Theta(P)) = \Gamma(f^*TP).$$

In other words, one obtains any variation of f by letting some element of the Lie algebra of the group $\mathscr{A}$ act on f.

In coordinates the infinitesimal stability condition means that the "homological equation"

$$u(x) = -\frac{\partial f}{\partial x} h(x) + k(f(x))$$

can be solved with respect to h and k for any deformation field u, i.e., for any element in $\Gamma(f^*TP)$.

Definition. A *germ* f of a smooth, formal, real-analytic, or holomorphic map is said to be *infinitesimally stable* if the homological equation for the germs h and k of the same class as f can be solved for every germ u (again of the same class as f).

Stability Theorem (Mather [236, Part II]). *An infinitesimally stable map is stable.*

This theorem is valid for germs, as well as for maps between compact manifolds N and P; in the noncompact case it is valid provided that stability is understood in the sense of Whitney's fine topology. The converse also holds: A stable map is infinitesimally stable.

1.6. The Groups $\mathscr{C}$ and $\mathscr{K}$. Let us introduce two more notions of equivalence.

Consider first the group $\mathscr{C}$ of germs at zero of smooth n-parameter families $\{h_x\}$ of diffeomorphisms of $(\mathbb{R}^p, 0)$, or, equivalently, the group of germs of diffeomorphisms H of the direct product $(\mathbb{R}^n \times \mathbb{R}^p, 0)$ that act as the identity on the space $\mathbb{R}^n \times 0$ and preserve the projection onto the first factor $\mathbb{R}^n$. An element $\{h_x\} \in \mathscr{C}$ acts on a map-germ $f\colon (\mathbb{R}^n, 0) \to (\mathbb{R}^p, 0)$ by sending it into the germ of the map $x \mapsto h_x(f(x))$ (that is to say, two maps are $\mathscr{C}$-equivalent if their graphs, which lie in $\mathbb{R}^n \times \mathbb{R}^p$, are mapped into one another by the corresponding diffeomorphism $H \in \mathscr{C}$). Note that by requiring that each member of the family $\{h_x\}$ be linear (i.e., $h_x \in GL(\mathbb{R}^p)$) we obtain the same notion of equivalence for maps.

Next, let $\mathscr{K}$ denote the semidirect product of the group $\mathscr{R}$ of germs of diffeomorphisms of the space $(\mathbb{R}^n, 0)$ and the group $\mathscr{C}$. The elements of $\mathscr{K}$ are germs of diffeomorphisms of $(\mathbb{R}^n \times \mathbb{R}^p, 0)$ that preserve the projection onto $\mathbb{R}^n$ (thereby inducing a diffeomorphism of the base $\mathbb{R}^n$) and map the plane $\mathbb{R}^n \times 0$ into itself. Thus, two germs $f, \tilde{f}\colon (\mathbb{R}^n, 0) \to (\mathbb{R}^p, 0)$ are $\mathscr{K}$-equivalent if there exist a diffeomorphism-germ g of the source and a map-germ $M\colon (\mathbb{R}^n, 0) \to GL(\mathbb{R}^p)$ of the source into the manifold of nonsingular matrices of order p, such that $\tilde{f}(x) \equiv M(x)f(g(x))$.

$\mathscr{K}$-equivalence is also called *contact equivalence*, for the graphs of $\mathscr{K}$-equivalent maps in $\mathbb{R}^n \times \mathbb{R}^p$ have diffeomorphic intersections (contacts) with the plane $\mathbb{R}^n \times 0$. Martinet [233] called $\mathscr{K}$-equivalence *V-equivalence*, for it takes the germ of the *variety* $\tilde{f}^{-1}(0)$ into the germ $f^{-1}(0)$ and the term "contact group" is traditionally used for a different object (see [34]).

Let $I_f = \mathscr{E}_n\langle f_1, \ldots, f_p\rangle$ be the ideal generated by the coordinate functions of the map f.

Theorem ([236]). *Two map-germs f and $\tilde{f}$ are $\mathscr{K}$-equivalent if and only if the corresponding ideals in $\mathscr{E}_n$, I_f and $I_{\tilde{f}}$, are taken into one another by [the map induced by] a suitable local diffeomorphism of the source.*

Let us extend the notion of $\mathscr{K}$-equivalence by allowing translations in the source (see subsection 2.1). To proceed further, we need to describe the tangent space $T_e\mathscr{K}(f)$ to the extended $\mathscr{K}$-equivalence class of the germ f.

To that end let us choose local coordinates $x_1, \ldots, x_n$ and $y_1, \ldots, y_p$ in the source $(\mathbb{R}^n, 0)$ and the target $(\mathbb{R}^p, 0)$, respectively. The set of variations of map-germs (identified with the set of all germs of smooth maps from $(\mathbb{R}^n, 0)$ to $\mathbb{R}^p$) is the free $\mathscr{E}_n$-module $\mathscr{E}_n^p = (\mathscr{E}_n)^p$ with p generators $e_i = \partial_{y_i}$.

In the notations introduced above

$$T_e\mathcal{K}(f) = \mathcal{E}_n\langle \partial f/\partial x_1, \ldots, \partial f/\partial x_n\rangle + I_f\mathcal{E}_n^p \subset \mathcal{E}_n^p.$$

The first component is the tangent space $T_e\mathcal{R}(f)$ to the extended right-equivalence class of the germ f, while the second is the space $T\mathcal{C}(f)$.

1.7. Normal Forms of Stable Germs. Let $f: (\mathbb{R}^n, 0) \to (\mathbb{R}^p, 0)$ be a stable germ of rank r at zero. Then f is obviously $\mathcal{K}$-equivalent to a germ of the form

$$\begin{cases} y = \varphi(x), & (x, \lambda) \in \mathbb{R}^s \times \mathbb{R}^r = \mathbb{R}^n, \\ z = \lambda, & (y, z) \in \mathbb{R}^t \times \mathbb{R}^r = \mathbb{R}^p, \end{cases}$$

where the differential of the map $\varphi: (\mathbb{R}^s, 0) \to (\mathbb{R}^t, 0)$ at zero is equal to zero.

Definition. The germ φ is called a *genotype* of the germ f.

A transversal to the tangent space $T_e\mathcal{K}(\varphi)$ in the space of maps $\mathcal{E}_s^t$ is spanned over the ground field by the t basis vectors $e_1, \ldots, e_t$ and u elements (maps) $\alpha_1, \ldots, \alpha_u$ that vanish at zero. The stability of the original map f implies that $u \leqslant r$ ([236]).

Theorem ([236]). *The map-germ f is $\mathcal{A}$-equivalent to the germ $y = \varphi(x) + \sum_{i=1}^{u} \lambda_i\alpha_i(x)$, $z = \lambda$.*

Corollary ([236]). *$\mathcal{K}$-equivalent stable germs are $\mathcal{A}$-equivalent.*

Definition. The *local algebra* of the germ $f: (\mathbb{R}^n, 0) \to (\mathbb{R}^p, 0)$ is the quotient algebra $Q_f = \mathcal{E}_n/I_f$.

It is readily seen that the local algebra of a germ and the local algebra of its genotype are isomorphic. Moreover, one has the following:

Theorem ([236]). *The $\mathcal{A}$-equivalence class of a stable germ $f: (\mathbb{R}^n, 0) \to (\mathbb{R}^p, 0)$ of rank r at zero is completely determined by the numbers n, p and the finite-dimensional $\mathbb{R}$-algebra $Q_{f,r+2} = \mathcal{E}_n/(I_f + \mathfrak{m}_n^{r+3})$, the latter being regarded as unique up to an isomorphism of $\mathbb{R}$-algebras.*

Any finite-dimensional local $\mathbb{R}$-algebra Q can be realized as the algebra $Q_{f,r+2}$ for a suitable stable germ f.

1.8. Examples

Definition. A *trivial extension* of the germ $f: (\mathbb{R}^n, 0) \to (\mathbb{R}^p, 0)$ is defined to be a map-germ

$$F: (\mathbb{R}^n \times \mathbb{R}^m, 0) \to (\mathbb{R}^p \times \mathbb{R}^m, 0) \quad \text{such that} \quad F(x, u) \equiv (f(x), u).$$

Theorem ([250]). *Any stable germ $f: (\mathbb{R}^n, 0) \to (\mathbb{R}^p, 0)$ for which both coranks are equal to 1 is $\mathcal{A}$-equivalent to a trivial extension of a generalized Whitney cusp*

$$\begin{cases} y = x^k + \lambda_1 x^{k-2} + \cdots + \lambda_{k-2} x, \\ z = \lambda \end{cases}$$

where $k \leqslant n + 1$.

Theorem ([250]). *Any stable germ* $f: (\mathbb{R}^n, 0) \to (\mathbb{R}^p, 0)$, $n \leqslant p$, *whose corank in the source is equal to* 1 *is* $\mathscr{A}$*-equivalent to a trivial extension of a generalized Whitney umbrella*

$$\begin{cases} y_1 = x^k + 0 \quad + \lambda_{1,2} x^{k-2} + \cdots + \lambda_{1,k-1} x \\ y_2 = \qquad \lambda_{2,1} x^{k-1} + \lambda_{2,2} x^{k-2} + \cdots + \lambda_{2,k-1} x, \\ \cdots\cdots\cdots\cdots\cdots\cdots\cdots\cdots\cdots\cdots \\ y_t = \qquad \lambda_{t,1} x^{k-1} + \lambda_{t,2} x^{k-2} + \cdots + \lambda_{t,k-1} x, \\ z = \lambda, \end{cases}$$

where $t(k-1) \leqslant n$.

Remark. Under the assumptions of the theorem $r = n - 1$, $t = p - r = p - n + 1$. For $n = 2$ and $p = 3$ we have $r = 1$, $t = 2$, $k = 2$ and hence $y_1 = x^2$, $y_2 = \lambda x$, $z = \lambda$, i.e., a map whose image is the ordinary umbrella (see subsection 1.2) $y_2^2 = y_1 z^2$.

Theorem ([250]). *Suppose that a given genotype is a function of* $s = n - p + 1$ *variables whose second differential at zero has corank* 1. *Then any stable germ* $(\mathbb{R}^n, 0) \to (\mathbb{R}^p, 0)$ *with that genotype is* $\mathscr{A}$*-equivalent to a trivial extension of the germ of the following combination of Morse and Whitney singularities*:

$$\begin{cases} y = x_1^k + \lambda_1 x_1^{k-2} + \cdots + \lambda_{k-2} x_1 \pm x_2^2 \pm x_3^2 \pm \cdots \pm x_s^2, \\ z = \lambda, \end{cases}$$

where $k \leqslant p + 1$.

A byproduct of the theorems formulated above is the classification of all generic map-germs between manifolds N^n, P^p with $n, p \leqslant 3$. As it turns out, the $\mathscr{A}$-equivalence class of such a germ is completely determined by its Thom-Boardman class and the signature of the quadratic form appearing in the last theorem. However, starting with maps between four-dimensional manifolds, when the class Σ^2 comes into play, new invariants arise. These invariants are connected with the behavior of the quadratic part of the map on the kernel of its differential.

Thus, let us consider a germ $f: (\mathbb{R}^n, 0) \to (\mathbb{R}^n, 0)$ of type Σ^2 at zero. Let l be an arbitary linear form on the target space that vanishes on the $(n-2)$-dimensional subspace Im df. Restrict the quadratic part of the function $f^*(l) = l \circ f$ to Ker df (regarded as a subspace of $\mathbb{R}^n$). The resulting quadratic form on Ker df is invariantly defined by the map f and the form l and depends linearly on l. We thus get a linear mapping of the space $\mathbb{R}^2$ of all such forms l into the three-dimensional space of all quadratic forms on Ker df. The latter space has a canonical stratification: the set of degenerate quadratic forms is a cone. The mutual disposition of this cone and the image L of $\mathbb{R}^2$ under the map described above is an invariant of the singularity, and the following situations are possible: L is

1) a plane that intersects the cone only at the point 0;
2) a plane that intersects the cone along two lines;
3) a plane tangent to the cone;
4) a line lying inside the cone;
5) a line lying outside the cone;
6) a line lying on the cone;
 and
7) the point 0.

In the last two cases the singularity belongs not only to the class Σ^2, but also to $\Sigma^{2,1}$ or $\Sigma^{2,2}$. However, we confine ourselves to the classification of $\Sigma^{2,0}$ and consider the first five possibilities, i.e., we are concerned with the classification of the generic singularities that lie outside the manifold $\Sigma^{2,1}$ of codimension 7 in the source (see subsection 1.4). The codimensions of the sets of singular points of these five types are 4, 4, 5, 6 and 6, respectively. The singularities of the first type are called elliptic, those of the second type, hyperbolic, and those of the remaining three types, degenerate.

Theorem ([277]). *For a generic map $f: \mathbb{R}^n \to \mathbb{R}^n$ in a neighborhood of a generic point of type $\Sigma^{2,1}$ the closure of the set of degenerate points of type Σ^2 is diffeomorphic to the product of a linear space (of the same dimension as $\Sigma^{2,1}$) and a Whitney umbrella. Moreover, in any transversal to $\Sigma^{2,1}$ the partition according to types 1 to 5 is structured as shown in Fig. 42.*

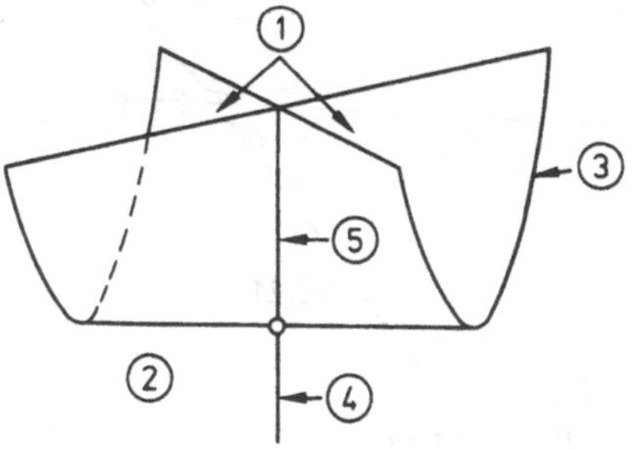

Fig. 42

The disposition of the types 4 and 5 is in agreement with fact that, under a deformation of the singularity, the line L that appears in the definition of these classes unfolds into a plane that can take different positions with respect to the cone. More precisely, there are three such possible positions when the line lies outside the cone, and only one (that of intersection) when the line lies inside the cone.

Theorem ([277]). *A generic map $f: N^n \to P^n$ can be written, in suitable coordinates in a neighborhood of an elliptic [resp. hyperbolic] singular point 0 as*

$$y_1 = x_1, \ldots, y_{n-2} = x_{n-2}, \qquad y_{n-1} = x_{n-1}x_n,$$
$$y_n = x_{n-1}^2 - x_n^2 + x_{n-3}x_{n-1} + x_{n-2}x_n$$

[resp. $y_n = x_{n-1}^2 + x_n^2 + \cdots$].

Together with the theorems about singularities of corank one this assertion provides the classification of singularities of stable maps $N^4 \to P^4$.

1.9. Nice and Semi-Nice Dimensions

Definition. A *pair of dimensions* (n, p) is called *nice* (see [236, Part VI]) if the set of stable maps is dense in the set of all smooth maps $N^n \to P^p$. A pair (n, p) called *semi-nice* (see [371]) if only the (larger) set of the maps whose germs at all points of the source are finitely $\mathscr{A}$-determined is dense.

The dotted line and the continuous line in Fig. 43 represent the boundary of the region of nice dimensions [236, Part VI] (the integer points on the boundary

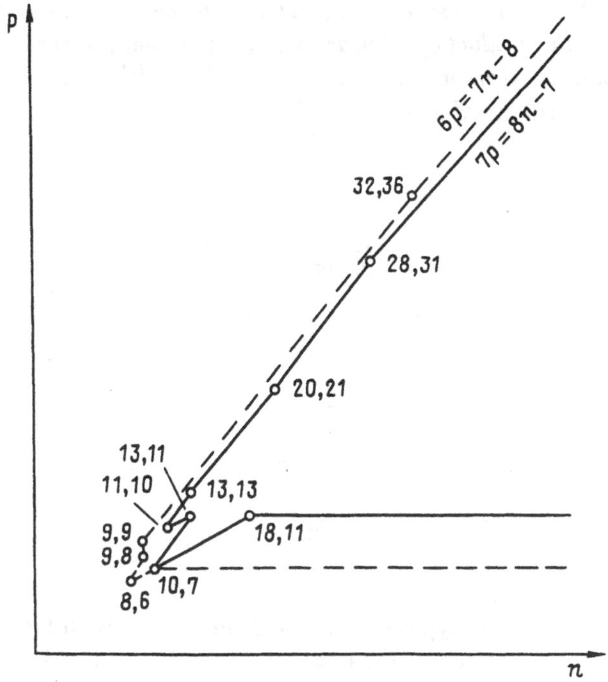

Fig. 43

do not belong to that region) and respectively the boundary of the region of semi-nice dimensions [273], [371] (in the second case the boundary belongs to the region). The partition of the set of pairs (n, p) is based on the classification of simple and unimodal genotypes [236], [237], [108], [371] (the notions of simplicity and modality, introduced earlier for critical points of functions, admit a natural generalization to the case of $\mathscr{K}$-equivalence of maps – see § 1.2 in [39]).

1.10. Maps of Finite Multiplicity. Consider the local algebra $Q_f = \mathcal{O}_n/I_f$ of a holomorphic map-germ $f: (\mathbb{C}^n, 0) \to (\mathbb{C}^n, 0)$ of spaces of the same dimension.

Definition. The *algebraic multiplicity* of the germ f is the dimension of its local algebra, $\mu = \dim_{\mathbb{C}} Q_f$.

This number is infinite if and only if zero is a nonisolated solution of the system of holomorphic equations $f = 0$.

A germ f is said to be of *finite multiplicity* if its algebraic multiplicity is finite.

In the smooth, formal, real-analytic and holomorphic settings one has the following frequently used result.

The Weierstrass Preparation Theorem. *Let $y = f(x)$ be a map of finite multiplicity μ and let $e_1, \ldots, e_\mu$ be a system ofgenerators of its local algebra Q_f. Then any function α admits a decomposition*

$$\alpha(x) = c_1(f(x))e_1(x) + \cdots + c_\mu(f(x))e_\mu(x),$$

where the c_k are appropriate functions of y.

Definition. The *geometric multiplicity* of a holomorphic germ $f: (\mathbb{C}^n, 0) \to (\mathbb{C}^n, 0)$ is the number of points near zero in the preimage of a generic point in the vicinity of zero in the target.

One obtains the same number if one defines the geometric multiplicity of a map of finite multiplicity as the index of the vector field f, i.e., as the degree of the map $f/\|f\|: S_\varepsilon^{2n-1} \to S_1^{2n-1}$ from a sufficiently small sphere $\|x\| = \varepsilon$ around zero in the source to the unit sphere in the target.

Theorem ([255], [204], [37]). *The algebraic multiplicity of a holomorphic germ of finite multiplicity is equal to its geometric multiplicity.*

The real-case analogue of this assertion is as follows.

Theorem ([256]). *Let $f: (\mathbb{R}^n, 0) \to (\mathbb{R}^n, 0)$ be a smooth map-germ of finite multiplicity. Then the number of points in the preimage of any point under f does not exceed the multiplicity μ of f, and is congruent* mod 2 *to μ for almost all points.*

Of course, here again we have in mind points that are close to zero in the preimage of a point that is close to zero in the target.

Example ([256]). The germ at zero of the map

$$\begin{cases} y_1 = x_1, \\ y_2 = x_2^3 + a(x_1)x_2, \end{cases}$$

where $a(x_1) = \sin(\pi/x_1)\exp(-1/x_1^2)$, has finite multiplicity ($\mu = 3$) and in a punctured neighborhood of zero it has only stable singularities. Nevertheless, this germ is not finitely $\mathscr{A}$-determined. Moreover, even if one allows homeomorphic changes of coordinates in the source and in the target, this germ does not become equivalent to an analytic germ.

In the holomorphic setting the above phenomenon cannot occur.

Definition. Fix a finite collection of points $x_1, \ldots, x_r$ in the source N. Two maps defined on open sets that contain these points are said to be *equivalent* if they coincide in some neighborhood of the collection. Such an equivalence class of maps is called a *multigerm* at the points $x_1, \ldots, x_r$.

Two multigerms, $f\colon N \to P$ at points $x_1, \ldots, x_r$ and $f'\colon N \to P$ at points $x'_1, \ldots, x'_r$, are said to be *$\mathscr{A}$-equivalent* if there exist multigerms (at the corresponding points) of diffeomorphisms h of N and k of P such that $f' = k \circ f \circ h^{-1}$. The definition of a stable multigerm is obvious (cf. subsection 1.2).

Theorem ([237], [137]). *Let $f\colon (\mathbb{C}^n, 0) \to (\mathbb{C}^p, 0)$ be a finitely $\mathscr{K}$-determined holomorphic germ. The germ f is finitely $\mathscr{A}$-determined if and only if for a representative of f one can find neighborhoods U of zero in the source and V of zero in the target such that, for any $y \in V \setminus \{0\}$, if $f^{-1}(y) \cap \Sigma \cap U = \{x_1, \ldots, x_r\}$, then the multigerm of f at the points $x_1, \ldots, x_r$ is stable.*

Here Σ denotes the subset of all points in the source at which the differential of f is not surjective (in particular, $\Sigma = \mathbb{C}^n$ whenever $n < p$).

1.11. The Number of Roots of a System of Equations. Consider a system of equations $f_1 = \cdots = f_n = 0$, where f_i are Laurent polynomials in n variables (i.e., elements of the ring $\mathbb{C}[x_1, x_1^{-1}, \ldots, x_n, x_n^{-1}]$). In this subsection we give a formula for the number of solutions of such a system in the torus $(\mathbb{C} \setminus 0)^n$.

Consider the support of the function $f_i = \sum c_\alpha x^\alpha$, i.e., the (finite) subset of all points α of the lattice $\mathbb{Z}^n$ for which $c_\alpha \neq 0$. The convex hull Γ_i of the support is called the *Newton polyhedron* of the polynomial f_i.

Definition. Take as unit of volume V the volume of the unit cube of the lattice $\mathbb{Z}^n$. The *mixed Minkowski volume* v of the sets $\Gamma_1, \ldots, \Gamma_n$ is the number given by the formula

$$n!v(\Gamma_1, \ldots, \Gamma_n) = (-1)^{n-1} \sum_i V(\Gamma_i) + (-1)^{n-2} \sum_{i<j} V(\Gamma_i + \Gamma_j) + \cdots$$
$$+ V(\Gamma_1 + \cdots + \Gamma_n).$$

Here the sum of two sets Γ_1 and Γ_2 is defined to be $\{\gamma_1 + \gamma_2 \colon \gamma_1 \in \Gamma_1, \gamma_2 \in \Gamma_2\}$. v is a symmetric multilinear form that coincides with the volume V on the diagonal $\Gamma_1 = \cdots = \Gamma_n$. The formula for v in the definition given above is the standard polarization formula that recovers the value of such a form from its diagonal part.

Theorem ([50]). *For almost any system of equations with given polyhedra $\Gamma_1, \ldots, \Gamma_n$ the number of solutions in the torus equals $n!v(\Gamma_1, \ldots, \Gamma_n)$.*

Let us formulate the nondegeneracy condition for the system $f_1 = \cdots = f_n = 0$.

Let $q = (q_1, \dots, q_n)$ be a nonzero rational linear form on $\mathbb{Z}^n$ and let S be a finite subset of $\mathbb{Z}^n$. Let m be the minimum of q on S, and let $S_q = \{\alpha \in S: q(\alpha) = m\}$ denote the intersection of S with the supporting plane in the direction q. For a Laurent polynomial $f = \sum_{\alpha \in S} c_\alpha x^\alpha$ we put $f_q = \sum_{\alpha \in S_q} c_\alpha x^\alpha$; f_q is the *principal part* of f with respect to q.

Given a system $F = (f_1, \dots, f_n)$ with corresponding polyhedra $\Gamma_1, \dots, \Gamma_n$, consider the system $F_q = (f_{1q}, \dots, f_{nq})$ of respective principal parts. In the generic case the system $F_q = 0$ has no roots in $(\mathbb{C} \setminus 0)^n$.

Theorem ([50]). a) *If the system F_q has no roots in $(\mathbb{C} \setminus 0)^n$ for all $q \neq 0$, then all the roots of the system F are isolated, and there are $n!v(\Gamma_1, \dots, \Gamma_n)$ such roots.*

b) *If F_q has a root for some $q \neq 0$, then the number of isolated roots of the system F, counting multiplicities, is strictly smaller than $n!v(\Gamma_1, \dots, \Gamma_n)$.*

Note that in order to verify condition a) it suffices to take one value of q for each face of the sum $\Gamma_1 + \cdots + \Gamma_n$.

A generalization of the theorem on the solutions of a system of equations to the case where the number of equations is smaller than the number of variables is provided by the following:

Theorem ([51]). *For almost any set $f_1, \dots, f_p$ of Laurent polynomials with given Newton polyhedra $\Gamma_1, \dots, \Gamma_p$, $p \leqslant n$, the Euler characteristic of the set $f_1 = \cdots = f_p = 0$ in the torus $(\mathbb{C} \setminus 0)^n$ is equal to*

$$(-1)^{n-p} n! \sum v(\Gamma_1, \Gamma_2, \dots, \Gamma_p, \Gamma_{i_1}, \dots, \Gamma_{i_{n-p}}),$$

where the sum runs over all $i_1, \dots, i_{n-p}$ such that $1 \leqslant i_1 \leqslant \cdots \leqslant i_{n-p} \leqslant p$.

For other results in this direction (computation of the arithmetic and geometric genera of complete intersections, of Hodge numbers of the mixed Hodge structure in the cohomology of complete intersections, and so on) refer to [103] and [195].

1.12. The Index of a Singular Point of a Real Germ, and Polynomial Vector Fields. Let $f: (\mathbb{R}^n, 0) \to (\mathbb{R}^n, 0)$ be a smooth map-germ of finite multiplicity. The index of f is defined in much the same way as the index of a holomorphic map, with the only difference that the map $f/\|f\|$ is considered between $(n-1)$-dimensional spheres.

Let $J = \det(\partial f/\partial x)$, the Jacobian of f, computed in some orientation-compatible coordinates. The following two assertions are well known (see, e.g., [37]).

Theorem. *The Jacobian J does not belong to the ideal I_f.*

Let $\alpha: Q_f \to \mathbb{R}$ be some linear form on the local algebra of f. We associate with α a bilinear form q_α on Q_f, defined as $q_\alpha(g, h) = \alpha(gh)$ (cf. subsection 2.5.6).

Theorem. *The bilinear form q_α is nondegenerate if and only if the value of α on J is different from zero.*

Theorem (signature formula [193], [121]). *The signature of the bilinear form q_α is equal to the index of the singular point zero of the germ f, provided that $\alpha(J) > 0$.*

In subsection 2.4.12 we gave an estimate of the index of a singular point of a gradient vector field. Let us state an analogous result for arbitrary polynomial fields.

Thus, let $V = (P_1, \ldots, P_n)$ be a vector field on $\mathbb{R}^n$ with polynomial components P_j, and let P_0 be one more polynomial. We are interested in three numbers: ind, ind$^+$, and ind$^-$, defined as the sum of the indices of all singular points of V in $\mathbb{R}^n$ and in the domains $\{P_0 > 0\}$ and $\{P_0 < 0\}$, respectively. We say that the pair V, P_0 has *degree at most* [resp. *equal to*] m, m_0, where $m = (m_1, \ldots, m_n)$, if the degree of the polynomial P_i is at most [resp. equal to] m_i for $i = 0, 1, \ldots, n$. The pair V, P_0 is said to be *nondegenerate* if the level set $\{P_0 = 0\}$ contains no singular points of V and if the real singular points of V have multiplicity ± 1 and "lie in the finite part of $\mathbb{R}^n$". The meaning of the last condition is as follows. Let $\tilde{P}_i$ be a homogeneous polynomial of degree m_i in the variables $x_0, x_1, \ldots, x_n$ such that $\tilde{P}_i(1, x_1, \ldots, x_n) \equiv P_i(x_1, \ldots, x_n)$; then the system $\tilde{P}_1 = \cdots = \tilde{P}_n = x_0 = 0$ is required to have only the trivial solution $x_0 = x_1 = \cdots = x_n = 0$.

We introduce the following notations:

$\Delta(m)$ is the parallelipiped $0 \leqslant y_1 \leqslant m_1 - 1, \ldots, 0 \leqslant y_n \leqslant m_n - 1$ in $\mathbb{R}^n$;

$\mu = m_1 \cdot \ldots \cdot m_n$ is the number of integer points in $\Delta(m)$;

$\Pi(m)$ is the number of integer points in the central section $y_1 + \cdots + y_n = \frac{1}{2}(m_1 + \cdots + m_n - n)$ of $\Delta(m)$;

$\Pi(m, m_0)$ is the number of integer points of $\Delta(m)$ lying in the layer $\frac{1}{2}(m_1 + \cdots + m_n - n - m_0) \leqslant y_1 + \cdots + y_n \leqslant \frac{1}{2}(m_1 + \cdots + m_n - n + m_0)$;

$O(m, m_0)$ is the number of integer points of $\Delta(m)$ lying in the layer $\frac{1}{2}(m_1 + \cdots + m_n - n - m_0) \leqslant y_1 + \cdots + y_n \leqslant \frac{1}{2}(m_1 + \cdots + m_n - n)$.

Note that $O(m, m_0) = \frac{1}{2}(\Pi(m, m_0) + \Pi(m))$ and $\Pi(m) \equiv \Pi(m, m_0) \equiv \mu \pmod 2$.

Theorem ([196]). *For a nondegenerate pair V, P_0 of degree m, m_0 the numbers $a =$ ind, $b =$ ind$^+$ $-$ ind$^-$, and $c =$ ind$^+$ satisfy the inequalities $|a| \leqslant \Pi(m)$, $|b| \leqslant \Pi(m, m_0)$, and $|c| \leqslant O(m, m_0)$, and the congruences $a \equiv b \equiv \mu \pmod 2$. Conversely, for any number a [resp. b, c] subject to these constraints there exists a nondegenerate pair V, P_0 of degree m, m_0 for which* ind $= a$ [*resp.* ind$^+$ $-$ ind$^-$ $= b$, ind$^+$ $= c$].

Corollary ([196]). *The index* ind *of an isolated singular point of a vector field $V = (P_1, \ldots, P_n)$ with homogeneous components of degree $m = (m_1, \ldots, m_n)$ satisfies the inequality* $|\text{ind}| \leqslant \Pi(m)$ *and the congruence* ind $\equiv \mu \pmod 2$; *moreover, these are the only constraints on the number* ind.

One can also estimate the index for vector fields V with "singular points at infinity". The number ind$^+$ is meaningful whenever the domain $\{P_0 > 0\}$ contains only isolated singular points of V. The number ind is defined provided all singular points of V are isolated.

Theorem ([196]). *Suppose that* V, P_0 *is a pair of degree at most* m, m_0 *for which* ind^+ *is defined. If* $m_0 + \cdots + m_n \equiv n(\text{mod } 2)$, *then* $|\text{ind}^+| \leqslant O(m, m_0)$; *in this case the number* ind^+ *is subject to no other constraints. If* $m_0 + \cdots + m_n \not\equiv n(\text{mod } 2)$ *and* m_0 *is even, then* $|\text{ind}^+| \leqslant O(m, m_0 + 1)$; *moreover, in this case there exist pairs* V, P_0 *for which* $\text{ind}^+ = \pm O(m, m_0 + 1)$.

Corollary ([196]). *Let* V *be a vector field of degree at most* $m = (m_1, \ldots, m_n)$ *with isolated singular points. Then for* $m_0 + \cdots + m_n \equiv n(\text{mod } 2)$ *one has the estimate* $|\text{ind}| \leqslant \Pi(m)$, *whereas for* $m_0 + \cdots + m_n \not\equiv n(\text{mod } 2)$ *one has the estimate* $|\text{ind}| \leqslant O(m, 1)$. *Both estimates are sharp.*

For examples realizing all the numbers allowed by these theorems and their corollaries refer to [196].

§ 2. Finite Determinacy of Map-Germs, and Their Versal Deformations

In analyzing a concrete equivalence relation specified by the action of some group of diffeomorphisms $\mathscr{G}$ on the space of map-germs, one is inevitably faced with two standard questions, along with the classification problem:

1) Under which conditions is a germ finitely determined, i.e., equivalent to any other germ with the same jet of sufficiently high order (and, in particular, when does it have a polynomial normal form)?

2) Does a germ have a versal (with respect to the equivalence relation under consideration) deformation, i.e., a family of germs, depending on a finite number of parameters, that contains up to equivalence all small perturbations of the germ?

As we already saw in Chapter 1, in the case of right equivalence of functions these two questions have affirmative answers if and only if the tangent space to the equivalence class of the germ has finite codimension in the tangent space to the whole space of germs. We begin the present section with a study of the actions of the groups $\mathscr{R}$, $\mathscr{L}$, $\mathscr{A}$, $\mathscr{C}$, and $\mathscr{K}$ on map-germs (in subsections 2.1–2.3, $\mathscr{G}$ will denote one of these groups). The situation turns out to be completely analogous to that of the action of $\mathscr{R}$ on functions. In subsection 2.5 we will consider sufficient conditions for the situation to remain equally pleasant in a more general case ("nice geometric" groups). According to these conditions the bad cases include, for example, the natural equivalence of diagrams of maps containing cyclic or divergent subdiagrams (such diagrams arise in e.g., the problem of classifying self-maps of a manifold or the problem on the envelope of a family of hypersurfaces – see § 1.6 in [39]).

The treatment given here is for smooth maps, but it carries over word for word to the analytic and holomorphic settings.

2.1. Tangent Spaces and Codimensions. Thus, suppose that the group $\mathscr{G}$ acts on the set $\mathfrak{m}_n\mathscr{E}_n^p$ of all germs of maps from $(\mathbb{R}^n, 0)$ to $(\mathbb{R}^p, 0)$. The tangent space $V(f)$ to this set at any one of its points f is canonically identified with the set $\mathfrak{m}_n\mathscr{E}_n^p$ itself. $V(f)$ contains the tangent space $T\mathscr{G}(f)$ to the $\mathscr{G}$-orbit of the germ f. According to the preceding section,

$$T\mathscr{R}(f) = \mathfrak{m}_n\langle \partial f/\partial x_1, \ldots, \partial f/\partial x_n \rangle,$$

$$T\mathscr{L}(f) = f^*(\mathfrak{m}_p)\langle \partial_{y_1}, \ldots, \partial_{y_p} \rangle,$$

$$T\mathscr{C}(f) = I_f\mathscr{E}_n^p, \qquad T\mathscr{A}(f) = T\mathscr{R}(f) + T\mathscr{L}(f), \qquad T\mathscr{K}(f) = T\mathscr{R}(f) + T\mathscr{C}(f).$$

Let us extend the tangent spaces by including the initial velocity vectors of all one-parameter deformations of maps and of the identity diffeomorphisms of the source and target.

Definition. We call *l-parameter deformation* of a map-germ $f: (\mathbb{R}^n, 0) \to (\mathbb{R}^p, 0)$ a germ of smooth map $F: (\mathbb{R}^n \times \mathbb{R}^l, 0) \to (\mathbb{R}^p, 0)$ whose restriction to the plane $\mathbb{R}^n \times 0$ coincides with f. The space $\mathbb{R}^l$ is called the *base* of the deformation.

Note that the deformation of f is not carried out inside $\mathfrak{m}_n\mathscr{E}_n^p$: for $\lambda \neq 0$ the map $F(\cdot, \lambda)$ is not required to take $0 \in \mathbb{R}^n$ into $0 \in \mathbb{R}^p$. Correspondingly, the tangent space to the set of maps is extended to $V_e(f) = \mathscr{E}_n^p$, and for the tangent spaces to the orbits we have (cf. subsection 1.5)

$$T_e\mathscr{R}(f) = \mathscr{E}_n\langle \partial f/\partial x_1, \ldots, \partial f/\partial x_n \rangle,$$

$$T_e\mathscr{L}(f) = f^*(\mathscr{E}_p)\langle \partial_{y_1}, \ldots, \partial_{y_p} \rangle, \qquad T_e\mathscr{C}(f) = T\mathscr{C}(f),$$

$$T_e\mathscr{A}(f) = T_e\mathscr{R}(f) + T_e\mathscr{L}(f), \qquad T_e\mathscr{K}(f) = T_e\mathscr{R}(f) + T_e\mathscr{C}(f).$$

The same extended spaces are obtained if instead of the actions of the groups $\mathscr{R}$ and $\mathscr{L}$ on $\mathfrak{m}_n\mathscr{E}_n^p$ one considers the actions on $\mathscr{E}_n^p$ of the pseudogroups $\mathscr{R}_e$ and $\mathscr{L}_e$ of local diffeomorphisms of $\mathbb{R}^n$ and $\mathbb{R}^p$ and one takes the tangent spaces to the corresponding orbits; $\mathscr{C}_e = \mathscr{C}$, $\mathscr{A}_e = \mathscr{R}_e \times \mathscr{L}_e$, $\mathscr{K}_e = \mathscr{R}_e\mathscr{C}_e$.

Definition. The numbers $d(f, \mathscr{G}) = \dim_{\mathbb{R}}(V(f)/T\mathscr{G}(f))$ and $d_e(f, \mathscr{G}) = \dim_{\mathbb{R}}(V_e(f)/T_e\mathscr{G}(f))$ are called the *$\mathscr{G}$-codimension* and the *extended $\mathscr{G}$-codimension* of the germ f, respectively.

Remark. Together with a deformation $F(x, \lambda)$ of a germ $f(x)$ one often considers its unfolding $(x, \lambda) \mapsto (F(x, \lambda), \lambda)$

2.2. Finite Determinacy. All definitions necessary for the ensuing discussion were given in the end of subsection 1.2: one has only to replace the group $\mathscr{A}$ by $\mathscr{G}$.

Theorem (on finite determinacy). *For any map-germ f and any group $\mathscr{G}$ the following assertions are equivalent*:

i) *f is finitely $\mathscr{G}$-determined*;

ii) *for some k, $T\mathscr{G}(f) \supset \mathfrak{m}_n^k V_e(f)$*;

iii) $d(f, \mathscr{G}) < \infty$;

iv) $d_e(f, \mathscr{G}) < \infty$;

The conditions of finite determinacy with respect to the groups $\mathcal{R}$, $\mathcal{L}$, or $\mathcal{C}$ can be reduced to the conditions of $\mathcal{A}$- and $\mathcal{K}$-finiteness. For example, if the rank of the map f is less than the maximal rank possible, they by [137], [369]:

1°. f is $\mathcal{R}$-finite $\Leftrightarrow p = 1$ and f is $\mathcal{K}$-finite;

2°. f is $\mathcal{C}$-finite $\Leftrightarrow p \geqslant n$ and f is $\mathcal{K}$-finite;

3°. f is $\mathcal{L}$-finite $\Leftrightarrow p \geqslant 2n$ and f is $\mathcal{A}$-finite.

In the first two cases the k-$\mathcal{K}$-determinacy implies the $n(k+1)$-$\mathcal{R}$-determinacy and the $p(k+1)$-$\mathcal{C}$-determinacy, respectively.

Definition. A map-germ is said to be *infinitesimally $\mathcal{G}$-stable* if its extended $\mathcal{G}$-codimension is equal to zero.

For example, a map f is infinitesimally $\mathcal{C}$-stable only outside the set $f^{-1}(0)$.

An analogous definition is given for multigerms.

In the complex case one has the following geometric criterion (cf. subsection 1.10).

Theorem (see [369]). *A holomorphic germ $f\colon (\mathbb{C}^n, 0) \to (\mathbb{C}^p, 0)$ is $\mathcal{G}$-finite if and only if the point $0 \in \mathbb{C}^n$ has a neighborhood U such that the multigerm of f at any finite subset of $U \setminus \{0\}$ is infinitesimally $\mathcal{G}$-stable.*

2.3. Versal Deformations. We repeat the definitions from subsection 1.1.6 in a more general setting.

Consider an l-parameter *deformation* F of a map $f\colon (\mathbb{R}^n, 0) \to (\mathbb{R}^p, 0)$, i.e., a family of maps $\{F_\lambda\colon \mathbb{R}^n \to \mathbb{R}^p, \lambda \in \mathbb{R}^l\}$. A *deformation $\mathcal{G}$-equivalent to F* is, by definition, a family $\{g_\lambda F_\lambda\}$, where $\{g_\lambda\}$ is some deformation of the identity element of the group $\mathcal{G}$ (g_λ is an element of the corresponding pseudogroup $\mathcal{G}_e$).

Let $\varphi\colon (\mathbb{R}^{l'}, 0) \to (\mathbb{R}^l, 0)$ be a smooth germ. By definition, the *deformation induced from F by the map φ* is the l'-parameter deformation $\{F_{\varphi(\lambda')}\}$.

Definition. A deformation F of a germ f is said to be *$\mathcal{G}$-versal* if any deformation of f is $\mathcal{G}$-equivalent to a deformation induced from F.

In the case of $\mathcal{A}$-equivalence this means that every deformation F' of f can be represented as

$$F'(x, \lambda') \equiv k(F(h(x, \lambda'), \varphi(\lambda')), \lambda'),$$

where $h(x, \lambda')$ and $k(y, \lambda')$ are deformations of the identity diffeomorphisms of the source $\mathbb{R}^n$ and target $\mathbb{R}^p$, respectively (it is not required that $h(0, \lambda') = 0$ and $k(0, \lambda') =$ for $\lambda' \neq 0$) and φ is as above.

For the group $\mathcal{K}$ the definition means that

$$F'(x, \lambda') \equiv M(x, \lambda')F(h(x, \lambda'), \varphi(\lambda')),$$

where $M(x, 0)$ is the identity matrix.

The explicit meaning of the notion of $\mathcal{G}$-versality for the groups $\mathcal{R}$, $\mathcal{L}$, and $\mathcal{C}$ is obvious.

Let F be a deformation of a germ f and let $\lambda_1, \ldots, \lambda_l$ be coordinates in its base, $\lambda(0) = 0$. Consider the initial velocities of F:

$$\dot{F}_i = \partial F(x, \lambda_1, \ldots, \lambda_l)/\partial \lambda_i|_{\lambda=0}, \qquad i = 1, \ldots, l.$$

Definition. The deformation F of the germ f is said to be *infinitesimally $\mathscr{G}$-versal* if its initial velocity vectors together with the extended tangent space $T_e\mathscr{G}(f)$ to the orbit of f span the full extended space $V_e(f)$ of variations of f.

In other words, the map $\lambda \mapsto F_\lambda$ of the base $(\mathbb{R}^l, 0)$ of the deformation into $(\mathscr{E}_n^p, f)$ is required to be transverse to the $\mathscr{G}_e$-orbit of f.

The conditions of infinitesimal $\mathscr{G}$-versality for each of the groups considered here amount to the following relations being fulfilled:

$$\mathscr{R}\colon \mathscr{E}_n^p = \mathscr{E}_n\langle \partial f/\partial x_1, \ldots, \partial f/\partial x_n\rangle + \mathbb{R}\langle \dot{F}_1, \ldots, \dot{F}_l\rangle;$$

$$\mathscr{L}\colon \mathscr{E}_n^p = f^*(\mathscr{E}_p)\langle \partial_{y_1}, \ldots, \partial_{y_p}\rangle + \mathbb{R}\langle \dot{F}_1, \ldots, \dot{F}_l\rangle;$$

$$\mathscr{C}\colon \mathscr{E}_n^p = I_f\mathscr{E}_n^p + \mathbb{R}\langle \dot{F}_1, \ldots, \dot{F}_l\rangle;$$

$$\mathscr{A}\colon \mathscr{E}_n^p = \mathscr{E}_n\langle \partial f/\partial x_1, \ldots, \partial f/\partial x_n\rangle + f^*(\mathscr{E}_p)\langle \partial_{y_1}, \ldots, \partial_{y_p}\rangle + \mathbb{R}\langle \dot{F}_1, \ldots, \dot{F}_l\rangle;$$

$$\mathscr{K}\colon \mathscr{E}_n^p = \mathscr{E}_n\langle \partial f/\partial x_1, \ldots, \partial f/\partial x_n\rangle + I_f\mathscr{E}_n^p + \mathbb{R}\langle \dot{F}_1, \ldots, \dot{F}_l\rangle.$$

Theorem (versality). *A deformation is $\mathscr{G}$-versal if and only if it is infinitesimally $\mathscr{G}$-versal.*

Theorem (uniqueness of the versal deformation). *Any l-parameter $\mathscr{G}$-versal deformation of a germ f is $\mathscr{G}$-equivalent to the deformation induced from any other, l-parameter $\mathscr{G}$-versal deformation of f by a diffeomorphic mapping of the bases.*

2.4. Examples. The theorems on finite determinacy, versality, and uniqueness of the versal deformation have been carried over word-for-word to many other cases of actions of various groups $\mathscr{G}$ on linear function spaces $\mathscr{F} \simeq V(f)$ (only assertion ii) of the finite-determinacy theorem has to be slightly modified to suit the concrete situation at hand). We pause to discuss some of these cases.

1°. Equivariant germs ([19], [52], [53], [275], [280], [283]). Consider two representations of a compact or reductive Lie group G on n- and p-dimensional linear spaces. Take as function space the set $(\mathfrak{m}_n\mathscr{E}_n^p)^G$ of G-equivariant germs of maps from the first representation space into the second that send 0 into 0 (for a reductive group we consider only the analytic case). Here $V_e(f) = (\mathscr{E}_n^p)^G$. The equivalence group is any of the G-equivariant groups $\mathscr{R}^G$, $\mathscr{L}^G$, $\mathscr{C}^G$, $\mathscr{A}^G$, or $\mathscr{K}^G$. For example, in the case of the group $\mathscr{R}^G$ the extended tangent space to the orbit of a germ f is $\Theta_n^G(f)$, where Θ_n^G denotes the algebra of G-equivariant vector fields on the source space.

2°. Distinguished parameters ([384], [19], [157], [151], [373], [372]; see also § 1.3 in [39]). To suit various applications in natural sciences, say, physics, one has often to endow the source space $\mathbb{R}^n$ with a fibre bundle structure $\mathbb{R}^n = \mathbb{R}^{n-r} \times \mathbb{R}^r \to \mathbb{R}^r$, $(x, \mu) \mapsto \mu$. Then the equivalence group is required to preserve the bundle structure. For instance, if one considers the corresponding right equivalence on $\mathfrak{m}_n\mathscr{E}_n^p$, then it will coincide with the natural right equivalence of

r-parameter deformations of germs in $\mathfrak{m}_{n-r}\mathscr{E}^p_{n-r}$ (which gives an arbitrary diffeomorphism on the base of the deformation). In this case

$$T_e\mathscr{G}(f) = \mathscr{E}_{x,\mu}\langle \partial f/\partial x_1, \ldots, \partial f/\partial x_{n-r}\rangle + \mathscr{E}_\mu\langle \partial f/\partial \mu_1, \ldots, \partial f/\partial \mu_r\rangle.$$

Note that one may require that an entire sequence of projections $\mathbb{R}^n \to \mathbb{R}^{n_1} \to \cdots \to \mathbb{R}^{n_s}$, $n > n_1 > \cdots > n_s$, rather than a single one, be preserved.

3°. Composition of multigerms ([112], [113]). Consider a graph that is the disjoint union of a finite number of trees, whose edges are oriented towards the corresponding roots (Fig. 44). Assign to each vertex α the space-germ $(\mathbb{R}^{n_\alpha}, 0)$, and assign to each edge $\alpha \to \beta$ a germ $f_{\alpha\beta}$ of a map of the corresponding spaces. An arbitrary element $\{g_\alpha\}$ in the direct product of the groups of diffeomorphisms of the space-germs $(\mathbb{R}^{n_\alpha}, 0)$ acts on the collection of maps $\{f_{\alpha\beta}\}$ according to the rule

$$\{g_\alpha\}\cdot\{f_{\alpha\beta}\} = \{g_\beta \circ f_{\alpha\beta} \circ g_\alpha^{-1}\}.$$

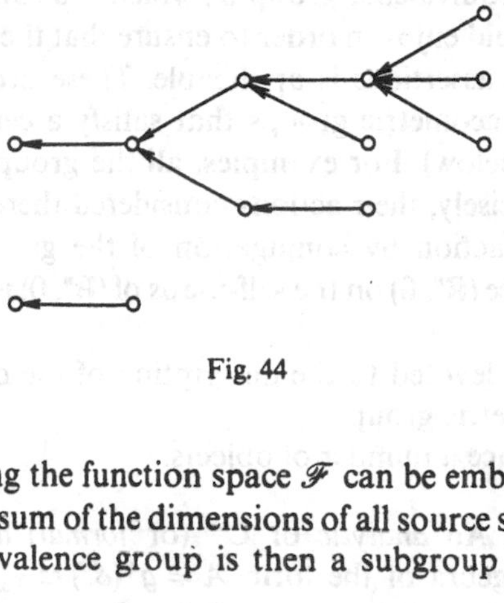

Fig. 44

In the present setting the function space $\mathscr{F}$ can be embedded in $\mathfrak{m}_n\mathscr{E}^p_n$, where n [resp. p] denotes the sum of the dimensions of all source spaces $\mathbb{R}^{n_\alpha}$ [resp. target spaces $\mathbb{R}^{n_\beta}$]. The equivalence group is then a subgroup of the corresponding left-right group.

Remark. The theorems in which we are interested here are also valid for various combinations of the equivalence relations considered in examples 1°–3° (of the type "composition of G-equivariant multigerms" or "G-equivariant germs with distinguished parameters" [152]).

4°. Nonisolated singularities ([305]–[308], [257], [258]; see also §1.4 in [39]). Let $\mathscr{F}$ be the space of all germs of analytic functions $f: (\mathbb{C}^n, 0) \to (\mathbb{C}, 0)$ that have a fixed analytic set $(\Sigma, 0)$ as critical set. Let $I \subset \mathcal{O}_x$ denote the ideal of all functions that vanish on Σ.

Definition ([257]). The *primitive ideal* of the ideal I is

$$\int I = \{f \in \mathcal{O}_x : f, \partial f/\partial x_1, \ldots, \partial f/\partial x_n \in I\}.$$

For instance, if Σ is a complete intersection, then $\int I = I^2$.

The group $\mathscr{D}_I = \{h \in \mathscr{R}: h^*(I) = I\}$, which maps $(\Sigma, 0)$ into itself, acts on $\mathscr{F} = \int I$. By [257], the standard theorems hold true for this action. To construct a versal deformation in the class of functions with a fixed critical set one has to set $V_e(f) = \mathscr{F}$ and then extend the group $\mathscr{D}_I$ to the pseudogroup of all germs of biholomorphisms $h: (\mathbb{C}^n, 0) \to \mathbb{C}^n$ that preserve Σ. The Lie algebra of that pseudogroup is the set of all germs of vector fields v on $(\mathbb{C}^n, 0)$ that preserve the ideal $I: vI \subset I$.

2.5. Geometric Subgroups. A systematic approach to the versality and finite-determinacy problems was developed by Mather in a series of works [236, Parts II, III, IV, VI]. An even nicer proof of the existence of a versal deformation for the groups $\mathscr{A}$ and $\mathscr{K}$ was proposed by Martinet in [233]. For all cases considered in the preceding subsection the versality and finite-determinacy theorems can be proved by the Mather-Martinet scheme, suitably adapted to fit the situation at hand. In [94], [96] Damon formalized those sufficient properties that the action of an equivalence group $\mathscr{G}$, which is a subgroup of the left-right or contact group, should enjoy in order to ensure that the traditional scheme for proving the standard assertions is applicable. These properties single out the class of the so-called geometric groups that satisfy a certain additional "niceness" condition (see below). For examples, all the groups encountered in subsection 2.4 (more precisely, their actions considered therein) are nice geometric groups, whereas the action by conjugation of the group of germs of diffeomorphisms of the space $(\mathbb{R}^n, 0)$ on the self-maps of $(\mathbb{R}^n, 0)$ is not nice (see also § 1.6 in [39]).

This subsection is devoted to the description of the conditions imposed by Damon on nice geometric groups.

First, let us introduce a number of objects.

Definition ([228]). An *analytic* or C^∞ (or *formal*) *algebra* [for brevity, a *DA-algebra*] is an algebra of the form $A = g^*(\mathscr{E}_y) \subset \mathscr{E}_x$, where $g: x \mapsto y$ is a map-germ and $\mathscr{E}_x$, $\mathscr{E}_y$ denote the spaces of function germs of the corresponding class. A *homomorphism of DA-algebras* $\alpha: A \to B$ is a homomorphism of $\mathbb{R}$- or $\mathbb{C}$-algebras that can be lifted to form a commutative diagram

$$\begin{array}{ccc} \mathscr{E}_y & \xrightarrow{\tilde{\alpha}} & \mathscr{E}_{y'} \\ \Big\downarrow{\scriptstyle g^*} & & \Big\downarrow{\scriptstyle g'^*} \\ A & \xrightarrow{\alpha} & B \end{array},$$

where the homomorphism $\tilde{\alpha}$ is induced by some map-germ $y' \mapsto y$ (again of the corresponding class).

Any DA-algebra A contains a unique maximal ideal, denoted $\mathfrak{m}_A$, which consists of the images of functions that vanish at zero.

The most important property of the DA-algebras and their homomorphisms is that for them the preparation theorem holds (cf. subsection 1.10):

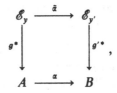

Theorem ([228], [236], Part [III]). *Let $\alpha: A \to B$ be a homomorphism of DA-algebras and let M be a finitely generated B-module. If the linear space $M/\mathfrak{m}_A M$ is finite-dimensional, then M is a finitely generated A-module.*

Besides the algebras $\mathscr{E}_x$ themselves, important examples of DA-algebras are algebras of invariants. Consider two linear representations of a compact or reductive Lie group G. Also, let $f: x \mapsto y$ be a G-equivariant map of one representation space into the other.

Theorem ([52], [223], [241], [275], [294], [94]). *The algebra $\mathscr{E}_x^G$ of G-invariant function-germs is a DA-algebra (for a reductive group one excludes the C^∞ case); the induced map $f^*: \mathscr{E}_y^G \to \mathscr{E}_x^G$ is a homomorphism of DA-algebras.*

We also need the space of deformations of elements of a DA-algebra A inside A itself and the space of deformations of elements of a finitely generated A-submodule $M \subset \mathscr{E}_x^r$ inside M itself. If λ denotes the deformation parameters, these spaces are $A_\lambda = A \otimes E_\lambda$ and $M_\lambda = M \otimes E_\lambda$, respectively, where the tensor products are taken over the ground number field.

Definition. Let $(\mathfrak{A}, <)$ be a partially ordered finite set. A *system of rings* associated with $(\mathfrak{A}, <)$ is family $\{R_\alpha, \alpha \in \mathfrak{A}\}$ of rings, together with connecting homomorphisms $\varphi_{\alpha\beta}: R_\alpha \to R_\beta$, defined whenever $\alpha \leqslant \beta$, such that $\varphi_{\beta\gamma} \circ \varphi_{\alpha\beta} = \varphi_{\alpha\gamma}$ and $\varphi_{\alpha\alpha} = \mathrm{id}$. A *system* $\{I_\alpha\} \subset \{R_\alpha\}$ *of ideals* is a family $I_\alpha \subset R_\alpha$ of ideals such that $\varphi_{\alpha\beta}(I_\alpha) \subseteq I_\beta$ whenever $\alpha \leqslant \beta$. A finitely generated *module M over the system* $\{R_\alpha\}$ is a direct sum $\oplus M_\alpha$, where M_α is a finitely generated R_α-module for any $\alpha \in \mathfrak{A}$.

The definition of a submodule of a module over a system of rings is obvious. For example, given a system $\{I_\alpha\}$ of ideals, the set $\{I_\alpha\}M = \oplus I_\alpha M_\alpha$ is a submodule of M.

Definition. A *homomorphism $M \to N$ of modules over the same system of rings* is defined to be the sum of any family $M_\alpha \to N_\beta$, $\alpha \leqslant \beta$, of maps which are homomorphism over $\varphi_{\alpha\beta}$.

Definition. A *system of rings* associated with a partially ordered set $(\mathfrak{A}, <)$ is said to be *adequately ordered* if every element in A has at most one direct predecesor.

To a given system of rings one can associate a directed graph in which a vertex α corresponds to the ring R_α, and an edge $\alpha \to \beta$, where α is a direct predecessor of β, corresponds to the homomorphism $\varphi_{\alpha\beta}$. For an adequately ordered system of rings this graph is the disjoint union of a finite number of trees in each of which the edges are oriented from the roots to the other vertices (the orientation is opposite the one of the graph shown in Fig. 44). See also example 3° in subsection 2.4, where $R_\alpha = \mathscr{E}_{n_\alpha}$ and $\varphi_{\alpha\beta} = f_{\beta\alpha}^*$.

We now turn to the description of the notion of a geometric group.

Thus, suppose that a subgroup $\mathscr{G}$ of the left-right or the contact group acts on a linear subspace of map-germs $\mathscr{F} \subset \mathfrak{m}_n \mathscr{E}_n^p$. Suppose also that for any number of parameters $l > 0$ there is given a deformation group $\mathcal{G}_d(l)$ (a subgroup in the group of all possible l-parameter deformations of elements in $\mathcal{A}$ or $\mathcal{K}$) which acts on a given linear subspace of deformations of maps, $\mathcal{F}_d(l) \subset \mathfrak{m}_{n+l}\mathcal{E}^p_{n+l}$ (cf. subsection 2.3). The reason for introducing this assumption is that there is no canonical way of specifying deformations of elements of a group or function space. In what follows, whenever possible, we shall omit the number of parameters in notations.

Definition ([94], [96]). The group $\mathcal{G}$ acting on the space $\mathcal{F}$ (together with the deformation groups $\mathcal{G}_d$ acting on $\mathcal{F}_d$) is said to be a *geometric subgroup* of the group $\mathcal{A}$ or $\mathcal{K}$ if it possesses the following four properties:

1°. Naturalness with respect to pull-back;
2°. Algebraic structure of the tangent spaces;
3°. Exponential map.
4°. Filtration property.

Each of these properties, which describe relationships between groups and function spaces, requires some comments.

1°. Naturalness with respect to pull-back. The meaning of this property is that the l'-parameter deformation induced from an l-parameter deformation $F \in \mathcal{F}_d(l)$ by a map $\varphi: (\mathbb{R}^{l'}, 0) \to (\mathbb{R}^l, 0)$ is an element of the space $\mathcal{F}_d(l')$, and similarly for groups.

2°. Algebraic structure of the tangent spaces. The embeddings $\mathbb{R}^l \to \mathbb{R}^{l+1}$ of parameter spaces allow us to regard the elements of the group $\mathcal{G}_d(l+1)$ and of the space $\mathcal{F}_d(l+1)$ as one-parameter deformations of elements of $\mathcal{G}_d(l)$ and $\mathcal{F}_d(l)$, respectively (and as deformations of elements of the group $\mathcal{G}$ and the space $\mathcal{F}$ for $l = 0$). Further, we proceed as in subsection 2.1 and declare the spaces of all possible initial velocity vectors of such one-parameter deformations of the identity element of $\mathcal{G}_d(l)$ and of an arbitrary element $F \in \mathcal{F}_d(l)$ to be the extended Lie algebra $T_e\mathcal{G}_d$ and the extended space of variations $V_e(F)$, respectively (the latter does not depend on F, thanks to linearity).

The condition in question is that these extended spaces must satisfy the following requirements. First of all, in the ring $\mathcal{E}_{n+p}$ there must exist a family $\{R_\alpha\}$ of DA-algebras such that, for any number of parameters λ, the corresponding spaces $T_e\mathcal{G}_d$ and $V_e(F)$ are finitely generated $\{R_{\alpha,\lambda}\}$-modules. Moreover, these modules must contain as finitely generated $\{R_{\alpha,\lambda}\}$-submodules the ordinary tangent spaces $T\mathcal{G}_d$ and $V(F) \cong \mathcal{F}_d$ and (in view of property 1°) the $\{R_\alpha\}$-submodules $T_e\mathcal{G}$ and $V_e(f)$, $f \in \mathcal{F}$.

Next, any l-parameter family of maps $F \in \mathcal{F}_d$ must transform the family $\{R_{\alpha,\lambda}\}$ into a system of DA-algebras in which the connecting homomorphisms are homomorphisms of $\mathcal{E}_\lambda$-algebras. Moreover, the tangent map $T_e\mathcal{G}_d \to V_e(f)$, which sends the initial velocity vector $(dg_t/dt)_{t=0}$ of the one-parameter deformation g_t of the identity element of $\mathcal{G}_d$ into the initial velocity vector $(d(g_tF)/dt)_{t=0}$ of the evolution of F under the action of g_t, must be a homomorphism of $\{R_{\alpha,\lambda}\}$-modules. The image $T_e\mathcal{G}_d(F)$ of the tangent map is the tangent space to the extended $\mathcal{G}_d$-orbit of F.

Also, the canonical maps of restriction to $\lambda = 0$, $T_e\mathcal{G}_d/\mathfrak{m}_\lambda T_e\mathcal{G}_d \to T_e\mathcal{G}$ and $V_e(F)/\mathfrak{m}_\lambda V_e(F) \to V_e(f)$, where $f = F|_{\lambda=0}$, are required to be homomorphisms of $\{R_\alpha\}$-modules.

Finally, it is required that the germs appearing in deformations differ from germs from $\mathcal{G}$ or $\mathcal{F}$ by at most translations of the origin in the target and source:

$$\{\mathfrak{m}_\alpha\} V_e(f) \subset V(f) \qquad \text{and} \qquad \{\mathfrak{m}_\alpha\} T_e\mathcal{G} \subset T\mathcal{G},$$

where $\mathfrak{m}_\alpha \subset R_\alpha$ is the maximal ideal.

All definitions given in subsection 2.3 extend naturally to the equivalence group $\mathcal{G}$ considered here. For example, an infinitesimally $\mathcal{G}$-versal deformation of a germ f is defined as one for which

$$V_e(f) = T_e\mathcal{G}(f) + \mathbb{R}\langle \dot{F}_1, \ldots, \dot{F}_l \rangle.$$

3°. Exponential map. In proving the versality theorem following the scheme of Mather and Martinet one needs to establish the following fact. Let F be an infinitesimally $\mathcal{G}$-versal deformation of a germ f and let F' be a one-parameter deformation of F. Then F' is equivalent to a trivial one-parameter deformation of F (i.e., to a deformation that does not depend on the additional parameter). Here the equivalence is understood up to an element of the deformation group and a diffeomorphic change of the deformation parameters.

The main step in proving this assertion is establishing the existence of an exponential map, i.e., the possibility of integrating any element v of the extended Lie algebra $T_e\mathcal{G}_d(l)$ and any germ v' of vector field on the parameter space $(\mathbb{R}^l, 0)$ so as to obtain a one-parameter deformation $\{(g_t, g_t')\}$ of the pair (identity element of $\mathcal{G}_d(l)$, identity diffeomorphism of $\mathbb{R}^l$) such that

$$\partial g_t/\partial t = v \circ (g_t, g_t'), \qquad \partial g_t'/\partial t = v' \circ g_t'.$$

Moreover, $\{g_t\}$ is an element of the group $\mathcal{G}_d(l+1)$.

In the general case, for a geometric group the existence of an analogous map – the restriction of the exponential map for the group $\mathcal{A}$ or $\mathcal{K}$ – is required.

4°. Filtration property. In view of the linearity, $\mathcal{F}_d \cong V(F)$ and the space of deformations is also an $\{R_{\alpha,\lambda}\}$-module. The last condition is that the deformation group $\mathcal{G}_d$ preserve the filtration $\{\{\mathfrak{m}_\alpha^k\}\mathcal{F}_d\}$ on $\mathcal{F}_d$ and induce an action on the quotient space of "$(k-1)$-jets" $\mathcal{F}_d/\{\mathfrak{m}_\alpha^k\}\mathcal{F}_d$ for all $k \geqslant 0$.

This completes Damon's definition of the notion of geometric group.

A *geometric group* is said to be *nice* if the corresponding system $\{R_\alpha\}$ of DA-algebras is adequately ordered (all the groups considered in subsection 2.4 are nice).

According to [94], [96], the finite-determinacy theorem (subsection 2.2) and the theorems on versality and uniqueness of the versal deformation (subsection 2.3) are valid for nice geometric subgroups of $\mathcal{A}$ and $\mathcal{K}$.

Remarks. 1. Condition ii) of the finite-determinacy theorem must be replaced by "$T\mathcal{G}(f) \supset \{\mathfrak{m}_\alpha^k\} V_e(f)$". Here a germ f is considered to be k-$\mathcal{G}$-determined whenever it is $\mathcal{G}$-equivalent to any other germ $g \in \mathcal{F}$ such that

$g = f \bmod \{\mathfrak{m}_\alpha^{k+1}\}\mathscr{F}$ (this is usually compatible with the condition on the jets of the germs).

2. The standard proof of the finite-determinacy [resp. versality] theorem uses all properties [resp. all properties except the filtration property] of the geometric group.

3. The adequate ordering of a system $\{R_\alpha\}$ allows one to apply in the proof Nakayama's lemma, which extends to modules over such systems (see [96]): Let M be a finitely generated module over a commutative ring with unit 1, let I be an ideal such that $1 + r$ is invertible for all elements $r \in I$, and let A be a submodule of M. Then the relation $A + I \cdot M = M$ implies $A = M$.

Concerning groups with nonadequately ordered systems of DA-algebras see § 1.6 in [39].

2.6. The Order of a Sufficient Jet

Definition. The *order of $\mathscr{G}$-determinacy* of a map-germ f is the minimal order of a $\mathscr{G}$-sufficient jet of f.

Here we are again dealing with ordinary germs rather than those considered at the end of the preceding subsection.

We have already given one of the first results of Tougeron and Mather on estimates of the order of determinacy (subsections 1.1.5 and 1.2). The investigations in this direction were subsequently continued in a more general setting (see Wall's survey [369], which contains an extensive list of relevant references, and also [48], [69], [137]–[139], [141], [142], [271], [370], [381], [382]). Let us state here one of the results on the order of determinacy.

Theorem ([139], [369]). *Let $\varepsilon = 1$ for $\mathscr{G} = \mathscr{R}, \mathscr{C}, \mathscr{K}$ and $\varepsilon = 2$ for $\mathscr{G} = \mathscr{L}, \mathscr{A}$. Then:*

i) *if the germ f is r-$\mathscr{G}$-determined, then $T\mathscr{G}(f) \supset \mathfrak{m}_n^{r+1}\mathscr{E}_n^p$;*

ii) *if $T\mathscr{G}(f) \supset \mathfrak{m}_n^{r+1}\mathscr{E}_n^p$, then f is $(\varepsilon r + 1)$-$\mathscr{G}$-determined;*

iii) *if $d(f, \mathscr{G}) = d < \infty$, then $T\mathscr{G}(f) \supset \mathfrak{m}_n^{(d+1)\varepsilon}\mathscr{E}_n^p$;*

iv) *f is r-$\mathscr{G}$-determined if and only if*

$$\mathfrak{m}_n^{r+1}\mathscr{E}_n^p \subset T\mathscr{G}(g) + \mathfrak{m}_n^{\varepsilon r+2}\mathscr{E}_n^p$$

for any germ g with the same r-jet as f.

We see that, whereas for the groups $\mathscr{R}$, $\mathscr{C}$, and $\mathscr{K}$ the first two assertions of the theorem allow us to specify the order of determinacy to within one unit, for the left and left-right equivalences the situation is worse. A new approach to the determination of the order of a sufficient jet was proposed in [72]; it is based on the theory of unipotent groups and allows to derive good estimates and frequently also exact results for various equivalence relations (including the groups $\mathscr{L}$ and $\mathscr{A}$). This is precisely the approach to which we devote the present subsection. Whenever an assertion is cited below without a reference this means that its proof is given in [72].

We continue to take a subgroup $\mathscr{G}$ of the contact group $\mathscr{K}$ as the group of equivalence of map germs.

Two series of groups are connected with $\mathscr{K}$, namely, the finite-dimensional groups $J^r\mathscr{K}$, $r \geqslant 1$, of r-jets of elements of $\mathscr{K}$, and the infinite-dimensional groups $\mathscr{K}_r$ consisting of the elements of $\mathscr{K}$ whose r-jets are the identity: $J^r\mathscr{K} = \mathscr{K}/\mathscr{K}_r$. The action of the contact group on the space of germs of maps $(\mathbb{R}^n, 0) \to (\mathbb{R}^p, 0)$ induces a well-defined action of the group $J^r\mathscr{K}$ on the finite-dimensional space of r-jets of such germs.

Correspondingly, for the group $\mathscr{G}$ we introduce the groups $\mathscr{G}_r = \mathscr{G} \cap \mathscr{K}_r$ and $J^r\mathscr{G}$, defined as the image of $\mathscr{G}$ under the projection $\mathscr{K} \to J^r\mathscr{K}$.

The situation in which we are interested is as follows. Suppose we already know that the germ of a given map is finitely $\mathscr{G}$-determined. We wish to estimate its order of $\mathscr{G}$-determinacy. To this end we may consider an affine space of jets of sufficiently high order r (what really matters is that the space must be finite dimensional) and the action of the group $J^r\mathscr{G}$ on this space. We would like the $J^r\mathscr{G}$-orbit of the point j^rf to contain the affine subspace that passes through j^rf and consists of the r-jets of the maps that have the same jet as f, of the lowest possible order.

As it turns out, in this case the action of $J^r\mathscr{G}$ can be replaced by the action of one of its unipotent subgroups. Recall that a group is said to be *unipotent* if it admits a faithful (i.e., with zero kernel) representation whose image lies, up to conjugation, in a group of upper triangular matrices with all diagonal entries equal to 1 (if this is the case, then all its representations have that property). The Lie algebra of a unipotent group is nilpotent (i.e., for any element one of its powers is equal to zero). Any unipotent affine algebraic group is isomorphic, as a variety, to an affine space. The orbits of an algebraic action of such a group on an affine space are closed in the Zariski topology.

If the ground field is $\mathbb{C}$ one has the following:

Proposition. *Suppose that an affine algebraic group acts algebraically and transitively on an affine space A. Then its unipotent radical (i.e., the unique maximal unipotent subgroup) already acts transitively on A.*

In the real case this assertion remains valid when the group and the affine space are replaced by their sets of rational points (for the definition see [177]).

The unipotency criterion for subgroups of jets of transformations in $\mathscr{K}$ is simple enough.

Proposition. *Let H be a closed connected subgroup of the group $J^s\mathscr{K}$ for some $s \geqslant 1$. Then H is a unipotent algebraic group if and only if the group J^1H is unipotent.*

J^1H is the image of the projection of H on $J^1\mathscr{K} \cong GL(n) \times GL(p)$.

The constructions carried out in [72] rely heavily on the following result.

Lemma (Mather [236, Part III]). *Let G be a Lie group that acts smoothly on a finite-dimensional manifold M. Let W be a connected submanifold of M. Then*

W is contained in a single orbit if and only if the following two conditions are satisfied:

i) $T_x W \subset T_x(Gx)$ *for all* $x \in W$;

ii) $\dim T_x(Gx)$ *does not depend on* $x \in W$.

Here is an example showing the necessity of the second condition. Consider the action of the group of linear-fractional transformations $z \mapsto (az + b)(cz + d)^{-1}$ on the complex z-plane. The real line and the (open) upper half-plane are distinct orbits of this action. Now in the upper half-plane consider a curve that is tangent to the real line ($z = t + it^2$ will do; see Fig. 45). The tangent to this curve at any of its points lies in the tangent space of the corresponding orbit, yet the curve passes from one orbit to the other.

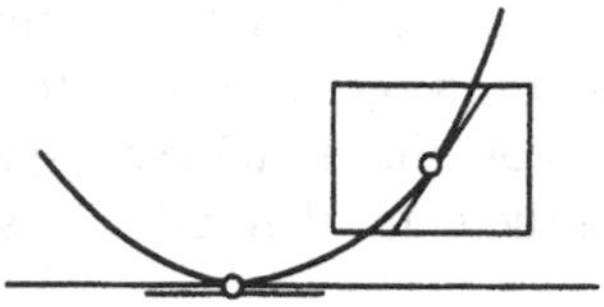

Fig. 45

For unipotent groups one derives from Mather's lemma the following:

Corollary. *Let U be a unipotent affine algebraic group that acts on an affine space A. Let B be a vector subspace of the vector space V_A associated with A. Let TU denote the Lie algebra of U. If $x \in A$, then the affine subspace $x + B$ is contained in a single U-orbit whenever*

i) $B \subset TU(x)$, *and*

ii) $TU(y) \subset TU(x)$ *for all points* $y \in x + B$.

Let us formulate the main result of [72].

Suppose that the subgroup $\mathcal{G}$ of the contact group acts on an affine space $\mathcal{F}$ of map-germs. Let B be a vector subspace of $V_{\mathcal{F}}$.

Definition (generalization of the notion of r-determinacy). A germ f is said to be *B-$\mathcal{G}$-determined* if the affine subspace $f + B$ is contained in a single $\mathcal{G}$-orbit.

Theorem. *Suppose that all jet subgroups $J^r\mathcal{G} \subset J^r\mathcal{K}$, $r \geqslant 1$, are closed. Let $f \in F$ be an s-$\mathcal{G}$-determined germ. Further, assume that the Lie algebra of the group $J^s\mathcal{G}$ contains a subspace L with the following properties:*

0) $J^1 L$ *acts nilpotently on* $\mathbb{R}^n \oplus \mathbb{R}^p$,

i) $J^s B \subset L(j^s f)$,

and either

ii′) L *is a Lie algebra and* $L(j^s(f + b)) \subset L(j^s f)$ *for all* $b \in B$

or

ii″) $l(j^s(f + b)) - l(j^s f) \in J^s B$ *for all* $l \in L$ *and all* $b \in B$.

Then f is B-$\mathcal{G}$-determined.

Remarks. 1°. Under more stringent assumptions (specifically, that all subgroups of r-jets be closed in the Zariski topology and there should exist a well-defined action of G on the quotient space $\mathscr{F}/B$) conditions 0) and i) are necessary and sufficient for the B-$\mathscr{G}$-determinacy of the germ f.

2°. Suppose that $J^s(T\mathscr{G}) = T(J^s\mathscr{G})$ for all $s \geqslant 1$, the action of $\mathscr{G}$ on $\mathscr{F}$ satisfies the hypotheses of the finite-determinacy theorem of subsection 2.2, and for some r the space B contains all elements of $V_{\mathscr{F}}$ with r-jet equal to zero. Then the theorem remains valid if f is not required beforehand to be s-$\mathscr{G}$-determined and one replaces the s-jets by the germs themselves throughout the statement.

Let us give the variant of the assertion for the traditional groups $\mathscr{R}$, $\mathscr{L}$, $\mathscr{A}$, $\mathscr{C}$ and $\mathscr{K}$.

Definition. A subgroup $\mathscr{H}$ of the group $\mathscr{G}$ is said to be *strongly closed* in $\mathscr{G}$ if all elements of $\mathscr{G}$ sufficiently close to the identity lie in $\mathscr{H}$, i.e., there is an s such that $\mathscr{G}_s \subset \mathscr{H}$, and if, for the same s, the jet group $J^s\mathscr{H}$ is a closed subgroup of $J^s\mathscr{G}$. If, in addition, $J^s\mathscr{H}$ is algebraic, then $\mathscr{H}$ is said to be *strongly Z-closed.*

Theorem. *Let $\mathscr{G}$ be one of the groups $\mathscr{R}$, $\mathscr{L}$, $\mathscr{A}$, $\mathscr{C}$ or $\mathscr{K}$. Let $\mathscr{H}$ be a strongly Z-closed subgroup of $\mathscr{G}$ and f a map-germ $(\mathbb{R}^n, 0) \to (\mathbb{R}^p, 0)$. Then for any given $r < \infty$, the germ f is r-$\mathscr{H}$-determined if and only if there exists a strongly closed subgroup $U \subset \mathscr{H}$ in $\mathscr{G}$ such that the group J^1U is unipotent and $\mathfrak{m}_n^{r+1}\mathscr{E}_n^p \subset TU(f)$.*

Here the most interesting cases are $\mathscr{G} = \mathscr{L}$ and $\mathscr{G} = \mathscr{A}$.

Example. For the germs $f_1(x, y) = (x, y^3 + xy)$ and $f_2(x, y) = (x, y^4 + xy)$ one has that

$$T\mathscr{A}_1(f_1) + \mathbb{R}(x^3 + xy, 0) \supset \mathfrak{m}_2^4\mathscr{E}_2^2$$

and

$$T\mathscr{A}_1(f_2) + \mathbb{R}(y^4 + xy, 0) + \mathbb{R}(x\partial f_2/\partial y) \supset \mathfrak{m}_2^5\mathscr{E}_2^2.$$

Consequently, f_1 and f_2 are 3- and respectively 4-$\mathscr{A}$-determined. The 3-$\mathscr{A}$-determinacy of f_1 follows also from its $\mathscr{A}$-stability. At the same time, assertion ii) of the theorem given at the beginning of the present subsection merely yields the 5-$\mathscr{A}$-determinacy of f_1 and the 7-$\mathscr{A}$-determinacy of f_2.

We remark that if the group $\mathscr{H}$ is the same as in the last theorem, then a real-analytic germ is r-$\mathscr{H}$-determined if and only if its complexification is r-$\mathscr{H}_{\mathbb{C}}$-determined.

Weighted determinacy. Upon assigning positive integral weights to the coordinates $x_1, \ldots, x_n$ in $\mathbb{R}^n$ one defines a decreasing filtration on the space of function-germs (see subsection 1.3.2). Now giving one more set of natural numbers $(d_1, \ldots, d_p)$ one can define a decreasing filtration $\{F_s\mathscr{E}_n^p\}$ for map-germs as follows: the filtration of $f = (f_1, \ldots, f_p)$ is $\min_i((\text{order } f_i) - d_i)$. The group $\mathscr{G}$, which acts on map-germs, is itself endowed with a decreasing filtration $\{F_r\mathscr{G}\}$: $g \in F_r\mathscr{G}$ if the operator $g - \text{id}$ increases the order of any map-germ by at least r.

Theorem. *Let $\mathcal{G} = \mathcal{R}, \mathcal{L}, \mathcal{A}, \mathcal{C}$ or $\mathcal{K}$. A germ $f \in F_0\mathcal{E}_n^p$ is $F_s\mathcal{E}_n^p$-determined with respect to the action of the group $F_r\mathcal{G}$ ($s \geqslant 0, r \geqslant 1$) if and only if $F_s\mathcal{E}_n^p \subset T(F_r\mathcal{G})(f)$.*

Example. The map $f(x, y) = (x, y^5 + xy)$ has order zero for the normalization wt $x = 4$, wt $y = 1$ and $d_1 = 4$, $d_2 = 5$. One readily verifies that $T(F_1\mathcal{A})(f)$ contains $F_3\mathcal{E}_2^2$. Consequently, the germ f is $F_3\mathcal{E}_2^2$-$\mathcal{A}$-determined (and hence 7-$\mathcal{A}$-determined; cf. [271]).

2.7. Determinacy with Respect to Transformations of Finite Smoothness. Let $\mathcal{G} = \mathcal{R}, \mathcal{L}, \mathcal{A}, \mathcal{C}$ or $\mathcal{K}$. Along with the action of the group $\mathcal{G}$ of infinitely differentiable transformations on the space of C^∞-map-germs we may consider the equivalence of the same germs with respect to analogous transformations, but of class C^k for some $k \geqslant 0$. Let us formulate a number of conditions (where $0 \leqslant k \leqslant \infty$).

a_k) The germ f is ∞-C^k-$\mathcal{G}$-determined, meaning that any flat additive perturbation of f (i.e., perturbation by an element of the space $\mathfrak{m}_n^\infty\mathcal{E}_n^p$) can be killed by a C^k-element, added in the group $\mathcal{G}$.

b_k) The germ f is finitely C^k-$\mathcal{G}$-determined.

t) $T\mathcal{G}(f) \supset \mathfrak{m}_n^\infty\mathcal{E}_n^p$.

g) there exists a neighborhood U of the origin in the source such that the multigerm of f at the points of any finite subset of $U \setminus \{0\}$ is $\mathcal{G}$-stable (see subsection 2.2).

To formulate the last of our conditions we introduce, for each group, its own ideal in the ring $\mathcal{E}_n$: $I_\mathcal{R}(f)$ and $I_\mathcal{L}(f)$ are generated by all $p \times p$ and respectively $n \times n$ minors of the Jacobian matrix of f, $I_\mathcal{C}(f) = f^*(\mathfrak{m}_p)\mathcal{E}_n$, and $I_\mathcal{K}(f) = I_\mathcal{R}(f) + I_\mathcal{C}(f)$.

e) $I_\mathcal{G}(f) \supset \mathfrak{m}_n^\infty$ (the so-called ellipticity condition for the ideal $I_\mathcal{G}(f)$).

For the group $\mathcal{L}$ in condition *e*) one has to require, in addition, that the ideal in $\mathcal{E}_{2n}$ generated by the p elements $f_i(x) - f_i(y)$ be elliptic on the diagonal $\{x = y\}$ of $\mathbb{R}^{2n}$, i.e., that it contain all functions $g(x, y)$ that are $O(\|x - y\|^k)$ for any k. For the statement of the corresponding condition for the group $\mathcal{A}$ refer to [382].

Theorem (see [369]). i) *Let f be a C^∞-map-germ $(\mathbb{R}^n, 0) \to (\mathbb{R}^p, 0)$ and let $\mathcal{G} = \mathcal{R}, \mathcal{C}, \mathcal{K}$ or $\mathcal{L}$. Then the following conditions are equivalent:*

$$a_k\ (0 \leqslant k \leqslant \infty),\ b_k\ (0 \leqslant k < \infty),\ t, \text{ and } e.$$

ii) *For a $\mathcal{K}$-finite germ f and the group $\mathcal{A}$ the following conditions are equivalent:*

$$a_k\ (\min(n + 1, p + 1) \leqslant k \leqslant \infty),\ t, \text{ and } e.$$

If the germ f is analytic then in i) *and* ii) *one must add condition g.*

In the holomorphic case condition t is replaced by

$$t')\ T\mathcal{G}(f) \supset \mathfrak{m}_n^k\mathcal{O}_n^p \text{ for some finite } n.$$

Theorem ([369]). *Let $f: (\mathbb{C}^n, 0) \to (\mathbb{C}^p, 0)$ be a holomorphic germ and $\mathscr{G} = \mathscr{R}$, $\mathscr{C}$, $\mathscr{K}$, or $\mathscr{L}$. Then conditions b_k $(0 \leqslant k \leqslant \infty)$, t', and g are equivalent, except possibly condition b_0 in the case $\mathscr{G} = \mathscr{L}$ and $p \geqslant 2n - 1$.*

Concerning the C^k-$\mathscr{G}$-determinacy see also [48].

§ 3. The Topological Equivalence

3.1. The Topologically Stable Maps are Dense. In a number of instances the relation of smooth equivalence of smooth maps is too delicate, for it yields many continuous invariants – moduli. For example, not every typical map is $\mathscr{A}$-stable (see subsection 1.9). It is therefore reasonable to enlarge the equivalence group, including in it not only the diffeomorphic transformations, but also the respective homeomorphisms. This is exactly the way in which R. Thom proposed that one should proceed in the case of the left-right equivalence. He hoped that in the generic situation the topological equivalence does not admit continuous moduli and that the topologically stable maps form a dense set in the space of all smooth maps between manifolds of arbitrary dimensions. As it turned out, his expectations were justified.

Theorem (Mather; see [238], [240], [144]). *Let N and P be smooth manifolds, with N compact. Then the topologically stable maps constitute a dense set in the space of all smooth maps of N into P, endowed with the Whitney topology.*

3.2. Whitney Stratifications. The proof of Mather's theorem uses the theory of stratified sets and stratified maps developed by Thom, in particular, Thom's isotopy lemmas.

First of all we note that the relationships between strata may satisfy various regularity conditions. Here are two of them.

Let U and V be two strata of a stratified subset of $\mathbb{R}^n$, with $V \subset \overline{U}$, and let $x \in V$.

Definition ([379]). The adjacency of V to U is said to satisfy *Whitney's condition a* at the point x if for every sequence $\{y_i\}$ of points of the submanifold U that converges to x and for which the sequence of tangent spaces $T_{y_i}U$ has a limit (in the Grassmann manifold of (dim U)-planes in $\mathbb{R}^n$), that limit contains the space $T_x V$.

Definition ([376]). Let $\{x_i\}$ and $\{y_i\}$ be sequences of points of the submanifolds V and U, respectively, which both converge to x. Suppose that the sequence of secants $\overline{x_i y_i}$ converges (in the $(n-1)$-dimensional projective space) and that the sequence of tangent spaces $T_{y_i}U$ also converges. The adjacency of V to U is said to satisfy *Whitney's condition b* at the point x if for any choice of sequences of points with the indicated properties the limiting secant lies in the limiting tangent space.

Condition b implies condition a (see [334]). The standard counterexample to the converse implication, described next, is given in [376]. Consider the surface $u^2 = w^2v^2 + v^3$ in $\mathbb{R}^3$, with the partition into two strata: the axis $0w$ and its complement (Fig. 46). Now take for $\{y_i\}$ any sequence of points on the curve $u = v + w^2 = 0$ that converges to zero, and take for $\{x_i\}$ the w-coordinates of these points.

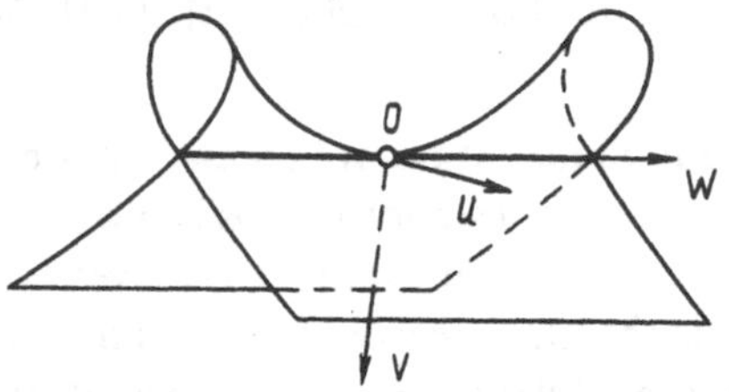

Fig. 46

For other regularity conditions for stratifications and comparisons between those conditions refer to [335], which also provides a list of references on this subject.

In what follows we will be interested only in Whitney's condition b. This condition is invariant under local diffeomorphisms of $\mathbb{R}^n$ (incidentally, so is condition a). This allows us to give the following

Definition. A stratification of a subset W of a manifold M is called a *Whitney stratification* if for any pair U, V of its strata, with $V \subset \overline{U}$, Whitney's condition b is satisfied at any point of V.

Thom's first isotopy lemma ([325], [326], [238]). *Let N, P be manifolds, $f\colon N \to P$ a smooth map, and W a closed subset of N endowed with a Whitney stratification. Suppose $f|_W\colon W \to P$ is a proper map and $f|_U\colon U \to P$ is a submersion for every stratum U of the stratification of W. Then $f|_W\colon W \to P$ is a topologically locally trivial fibration.*

Definition. A *map* $f\colon A \to B$ of two stratified subsets is said to be *stratified* if it extends to a smooth map of neighborhoods of A and B in the ambient manifolds and if it maps every stratum U of the source into some stratum of the target, the restriction $f|_U$ having the maximal possible rank. A stratified map is said to be a *Thom map* if for any pair U, V, $V \subset \overline{U}$, of strata of the source the following holds: for any point $x \in V$ and any sequence of points $\{y_i\} \subset U$ such that $y_i \to x$ and the sequence of kernels $\operatorname{Ker} d(f|_U)_{y_i}$ has a limit, the limit contains the space $\operatorname{Ker} d(f|_V)_x$.

Let P be a connected manifold with the trivial stratification consisting of the single stratum P.

Thom's second isotopy lemma (see [238]). *Let $f\colon A \to B$ and $g\colon B \to P$ be proper stratified maps, and let f be a Thom map. For each $p \in P$, set*

$A_p = (g \circ f)^{-1}(p)$ and $B_p = g^{-1}(p)$. Then, for any pair of points $p, q \in P$, the maps $A_p \to B_p$ and $A_q \to B_q$ obtained by restricting f to A_p and A_q, respectively, are topologically left-right equivalent.

3.3. The Topological Classification of Smooth Map-Germs. As early as 1962 Thom constructed the following example of a family of polynomial maps $f_s\colon \mathbb{R}^3 \to \mathbb{R}^3$, depending on a real parameter s, that are pairwise topologically nonequivalent [326]:

$$f_s\colon (x, y, z) \mapsto (X, Y, Z),$$

$$\begin{cases} X = [x(x^2 + y^2 - a^2) - 2ayz]^2[(sy + x)(x^2 + y^2 - a^2) - 2az(y - sx)^2], \\ Y = x^2 + y^2 - a^2, \\ Z = z. \end{cases}$$

This phenomenon is, however, highly atypical. To give this claim a rigorous meaning, consider the space $J^r(n, p)$ of r-jets of germs at zero of smooth maps of $\mathbb{R}^n$ into $\mathbb{R}^p$.

Theorem ([341], [342]). *For any natural number r there exists a partition of $J^r(n, p)$ into disjoint semialgebraic subsets $V_0, V_1, V_2, \ldots$, with the following properties:*

1°. *If the r-jets of the maps f_1 and f_2 belong to the same subset V_i, $i > 0$, then the germs of f_1 and f_2 are topologically $\mathscr{A}$-equivalent.*

2°. *Any germ whose r-jet belongs to V_i, $i > 0$, is a simplicial map with respect to suitable triangulations of $\mathbb{R}^n$ and $\mathbb{R}^p$.*

3°. *The codimension of V_0 in $J^r(n, p)$ tends to infinity as $r \to \infty$.*

4°. *Any germ whose r-jet belongs to V_i, $i > 0$, admits a finite-dimensional topologically $\mathscr{A}$-versal deformation.*

5°. *If the r-jets of the germs f_1 and f_2 belong to the same set V_i, $i > 0$, then f_1 and f_2 admit topologically equivalent topologically $\mathscr{A}$-versal deformations.*

The notion of a topologically versal deformation is introduced in much the same manner as that of differentiably versal deformation (see also subsection 3.5). Note that there exist pairs (n, p) such that the germs which admit no finite-dimensional differentiably $\mathscr{A}$-versal deformations form a set of finite codimension in the space of germs $(\mathbb{R}^n, 0) \to \mathbb{R}^p$ (this is the case, for example, when the pair (n, p) lies outside the regions of "nice" and "semi-nice" dimensions – see subsection 1.9).

The papers [74], [75], [242] give the topological classification of germs in the following setting. Consider a family of smooth functions $f(\cdot, y)$ on a closed manifold that depends on an n-dimensional parameter y, and take the maximum function $F(y) = \max_x f(x, y)$.

Theorem. *For generic families f the function F is topologically equivalent to a Morse function.*

For generic families $f(\cdot, y, z)$ the function $\min_y \max_x f(x, y, z)$ is also topologically equivalent to a Morse function [242].

In problems connected with topologically-Morse functions, and also in game theory, there arise so-called positive strictly minimal finite families of continuous, convex, homogeneous functions on $\mathbb{R}^n$ (a family is said to be *positive* if for any point different from zero it contains a function that is positive at that point, and is said to be *strictly minimal* if, in addition to this, for any function in the family there is a point at which all the other functions in the family are negative; a function φ is said to be *homogeneous* if $\varphi(tx) = t\varphi(x)$ for all $t \geqslant 0$).

Theorem ([243]). *Let $\{\varphi_1, \ldots, \varphi_{m+1}\}$ be a positive strictly minimal family of continuous, convex, homogeneous functions. Then, by means of a homeomorphism of $\mathbb{R}^n$, the function $\min\{\varphi_1, \ldots, \varphi_{m+1}\}$ can be reduced to the form $|x_1| + \cdots + |x_{n-m}| - |x_{n-m+1}| - \cdots - |x_n|$ (here $x_1, \ldots, x_n$ are coordinates in $\mathbb{R}^n$).*

Corollary ([243]). *For $1 \leqslant m \leqslant n$, the union of any family of $m + 1$ convex open subsets of the $(n - 1)$-dimensional sphere with the property that the intersection of any m of its members is not empty but the intersection of all $m + 1$ sets is empty, is homeomorphic to the direct product of an $(m - 1)$-dimensional sphere and an $(n - m)$-dimensional disk.*

For instance, when $n = m$ any such family covers the entire $(n - 1)$-dimensional sphere.

3.4. Topological Invariants. The natural setting for considering the topological invariants of the left-right equivalence is that of stable germs (to avoid unpleasant phenomena such as, say, the topological equivalence of the germs x and x^3).

A considerable amount of information on invariants is provided by the local algebra of the map.

Theorem ([131]). *If the local algebras of the topologically stable C^∞-germs f, $g\colon (\mathbb{R}^n, 0) \to (\mathbb{R}^p, 0)$ are isomorphic, then f and g are topologically $\mathscr{A}$-equivalent.*

Example. Consider the map $(x, \lambda) \mapsto (F(x, \lambda), \lambda)$, where F is a truncated $\mathscr{R}$-versal deformation of a parabolic function f (see Chapter 1). According to subsection 1.7, this map is differentiably $\mathscr{A}$-stable. For close values of the modulus of f one obtains germs that are topologically, but not smoothly $\mathscr{A}$-equivalent.

For C^∞-stable germs $f\colon (\mathbb{R}^n, 0) \to (\mathbb{R}^p, 0)$ the following result holds.

Theorem ([89], [91]). *The Thom-Boardman type Σ^i is a topological invariant. For $p \geqslant n + \binom{i}{2}$ the type $\Sigma^{i,j}$ is also an invariant. For $n > p$ the types $\Sigma^{n-p+1,j}$ and $\Sigma^{n-p+1,2,j}$ are also invariants.*

Note that according to [221], the type $\Sigma^{3,2,1,1,j}$ is not a topological invariant for C^∞- or holomorphically stable germs.

Any C^∞-stable germ $\mathbb{R}^n \to \mathbb{R}^p$, $n > p$, of class Σ^{n-p+1} is contact-equivalent to a function-germ. By [91], the minimum of the positive and negative indices of inertia of the quadratic part of that function is a topological invariant.

We turn now to C^∞-stable contact-simple germs. For $n \leqslant p$ a topological invariant is the Hilbert-Samuel function h of the real local algebra Q_f of such a germ f ($h(k) = \dim_{\mathbb{R}} Q_f/\mathfrak{m}^{k+1}$) [89], [90]. More precisely, one has the following

Theorem ([90], [91]). *For C^∞-stable contact-simple germs $f\colon \mathbb{R}^n \to \mathbb{R}^p$:*

a) *if $n \leqslant p$, then the complexification $Q_f \otimes_{\mathbb{R}} \mathbb{C}$ of the local algebra is a topological invariant;*

b) *in the region of nice dimensions $n \leqslant p$, the $\mathbb{R}$-algebra Q_f is itself a topological invariant;*

c) *for $n > p$, the C^∞- and the topological $\mathscr{A}$-classifications coincide.*

Thus, in the entire region of nice dimensions the aforementioned classifications coincide.

3.5. Topological Triviality and Topological Versality of Deformations of Semi-Quasihomogeneous Maps. Let $F(x, \lambda)$ be a deformation of a smooth germ $f\colon (\mathbb{R}^n, 0) \to (\mathbb{R}^p, 0)$. We say that F is *topologically $\mathscr{A}$-trivial* if it is topologically $\mathscr{A}$-equivalent to a trivial deformation of f, i.e., if it is representable in the form

$$F(x, \lambda) = K(f(H(x, \lambda)), \lambda),$$

where K and H are continuous deformations of germs of the identity homeomorphisms of $\mathbb{R}^p$ and $\mathbb{R}^n$, respectively.

Analogous definitions can be given for any other equivalence group of maps (for example, for a geometric group or the corresponding deformation group; see subsection 2.5).

In subsection 2.5.11 we gave Looijenga's theorem asserting that a versal deformation of a germ of parabolic function is topologically trivial along the parameter that corresponds to the modulus of the singularity. This result has been generalized by Wirthmüller as follows.

Theorem ([383]). *Let f be a quasihomogeneous function with a nonsimple isolated singularity. Then any $\mathscr{R}$-versal deformation of f is topologically trivial along the parameter that corresponds to the Hessian* $\det(\partial^2 f/\partial x_i \partial x_j)$.

Ronga [284] showed that any contact-versal deformation of the intersection of three quadrics in three-dimensional space is topologically trivial along the parameter that corresponds to the modulus of the singularity. Also, a series of topological triviality results are known for functions, quasihomogeneous complete intersections (see subsection 1.2.9 in [39]) and some other singularities (see [92], [93], [97], [98], [101], [42]).

The proofs of all the aforementioned results (except for Ronga's theorem) rely on the idea, going back to the papers [202], [322], of the solvability of the so-called localized homological equation [92], [99], [101]. In what follows, $\mathscr{G}$ is a nice geometric group, for which in addition this solvability holds, for example, any of the groups considered in subsection 2.4 (for the general definition see [99], [100]).

Suppose the coordinates $x_1, \dots, x_n$ in $\mathbb{R}^n$ are assigned positive integer weights $w_1, \dots, w_n$. A map $g: (\mathbb{R}^n, 0) \to (\mathbb{R}^p, 0)$ is said to be *quasihomogeneous* if each coordinate function g_i of g is quasihomogeneous, say, of weight d_i. Let f be a map that is not quasihomogeneous with respect to the given weights $w_1, \dots, w_n$. Consider its principal part f_0, consisting of the principal parts of the coordinate functions (see subsection 1.3.2).

Definition. The map f is said to be *semi-quasihomogeneous* with respect to the given set of weights and the equivalence group $\mathscr{G}$, which acts on the function space $\mathscr{F}$, if the $\mathscr{G}$-codimension of its principal part, $\dim_{\mathbb{R}} V(f_0)/T\mathscr{G}(f_0)$, is finite.

A (semi-)quasihomogeneous map f defines a quasihomogeneous filtration on the space $\mathscr{E}_n^p = \mathscr{E}_n\langle \partial_{y_1}, \dots, \partial_{y_p}\rangle$ of all germs of maps from $(\mathbb{R}^n, 0)$ to $\mathbb{R}^p$: wt $\partial_{y_i} = -d_i$. In this filtration (the principal part of) f has weight zero.

Definition. A deformation $f(x) + \varphi(x, \lambda)$ of the semi-quasihomogeneous map f is said to be *a deformation of nonnegative weight* if the map φ has nonnegative filtration (all λ_j are assigned the weight zero).

Theorem (topological triviality [99], [100]). *Any deformation of nonnegative weight of a semi-quasihomogeneous map is topologically $\mathscr{G}$-trivial.*

Corollary (topological determinacy [100]). *Any two semi-quasihomogeneous maps with equal principal parts are topologically $\mathscr{G}$-equivalent.*

In particular, any semi-quasihomogeneous map is topologically $\mathscr{G}$-equivalent to its principal part.

Let M be the set of the maps in $\mathscr{F}$ that have a fixed quasihomogeneity type $w_1, \dots, w_n, d_1, \dots, d_p$ and are of finite $\mathscr{G}$-codimension.

Corollary (topological equivalence [100]). *Any two elements of the function space $\mathscr{F}$ whose principal parts lie in the same connected component of the set M are topologically $\mathscr{G}$-equivalent.*

Now let f be a quasihomogeneous map of finite $\mathscr{G}$-codimension. The space

$$N = V_e(f)/T_e\mathscr{G}(f)$$

inherits the quasihomogeneous grading from $\mathscr{E}_n^p$. Let $\varphi_1, \dots, \varphi_r \in \mathscr{E}_n^p$ be a quasihomogeneous basis of the linear subspace of N spanned by all elements of weight less than k.

Definition. The deformation

$$F(x, \lambda) = f(x) + \lambda_1\varphi_1(x) + \cdots + \lambda_r\varphi_r(x)$$

is said to be *versal in weight* $<k$.

A deformation F is said to have *finite $\mathscr{G}$-codimension* if in the space $\mathscr{F}_d(r)$ of all possible r-parameter deformations the subspace of variations of F under the action of the natural equivalence group of such deformations has finite codimension (see property 2° in subsection 2.5). This is equivalent to the requirement that the number

$$\dim_{\mathbf{R}} V_e(F)/(T_e\mathscr{G}_d(F) + \mathscr{E}_\lambda\langle\varphi_1, \ldots, \varphi_r\rangle)$$

be finite.

Let us assign to the parameters λ_i of the deformation F the weights $-\mathrm{wt}\,\varphi_i$ (so that in the end the map F itself will have weight zero). If $k \geqslant 0$, then among these weights there may be nonpositive ones. If that is the case, then we say that the deformation F has *finite $\mathscr{G}$-codimension in positive weights* if

$$\dim_{\mathbf{R}} V_e(F_+)/(T_e\mathscr{G}_d(F_+) + \mathscr{E}_{\lambda_+}\langle\varphi_1, \ldots, \varphi_{r_+}\rangle) < \infty,$$

where r_+ denotes the dimension of the space of parameters λ_+ of positive weight and F_+ is the restriction of the deformation F to that space.

Theorem (topological versality [99], [100]). *Suppose that the deformation F is versal in weight $<k$ for some $k \geqslant 0$. Suppose further that F has either finite $\mathscr{G}$-codimension or finite $\mathscr{G}$-codimension in positive weights. Then the deformation F is topologically $\mathscr{G}$-versal.*

In particular, a $\mathscr{G}$-versal deformation is topologically $\mathscr{G}$-trivial along the parameters that correspond to the basis elements in N of weight $\geqslant k$.

Corollary ([100]). *Let $f \in \mathscr{F}$ be a (semi-) quasihomogeneous map such that all elements in N have nonnegative weights. Then f is deformation-topologically $\mathscr{G}$-stable.*

Papers [99], [100] also give variants of the assertions made above for the case of a broken filtration, in which the principal part of the map relative to the filtration is required to have finite $\mathscr{G}$-codimension (this is an analogue of the Γ-nondegeneracy property of a function; see subsection 2.3.12). As a byproduct one obtains, for example, a topological proof of A.G. Kushnirenko's theorem asserting that the Milnor number of a Γ-nondegenerate function depends only on its Newton polyhedron (subsection 2.3.12).

In [101] there are considered deformations of critical points of finite multiplicity of functions for which the Γ-nondegeneracy requirement is removed.

Chapter 4
The Global Theory of Singularities

In this chapter we describe topological and numerical characteristics of singular sets of smooth maps, such as cohomology classes dual to sets of critical points and critical values; the invariants of maps defined by these classes; their connections with standard topological characteristics of the source and target manifolds; the structure of spaces of smooth maps without singularities of some given type; restrictions on the number and coexistence of singular points.

§ 1. Thom Polynomials for Maps of Smooth Manifolds

Let M, N be smooth manifolds of dimensions m, n, and let Σ^I be an arbitrary Thom-Boardman class of maps $\mathbb{R}^m \to \mathbb{R}^n$ (see § 3.1). Let $v = v_I(m, n)$ be the codimension of the corresponding set Σ^I in the jet space $J^k(M, N)$. Then, for almost any map $f\colon M \to N$, the element $[\Sigma^I(f)]$ dual to the closure of the set $\Sigma^I(f)$ is defined in the group $H^v(M; \mathbf{Z}_2)$. This element does not change under smooth deformations of the map f. Moreover, such elements give invariants of the pair $[M, f]$ in the m-dimensional nonoriented bordism group of the manifold N. For any I the element $[\Sigma^I(f)]$ can be expressed through the characteristic classes of the manifolds M, N; this expression is called the *Thom polynomial* for the class Σ^I.

1.1. Cycles of Singularities and Topological Invariants of Maps. Let S be an arbitrary regular analytic[1] subset in the space $J^k(M, N)$ of k-jets of maps $M \to N$, i.e., in a neighborhood of any of its points S is given by analytic equations. To the set S there corresponds, in a unique manner, its *fundamental homology class*, i.e., a class α in the homology group $H^{\mathrm{f}}_{\dim S}(S; \mathbb{Z}_2)$ of S with closed supports, such that, for every regular point $x \in S$, the image of α in the group $H_{\dim S}(S, S \setminus x; \mathbb{Z}_2)$ generates the latter (see [175], [55]). The inclusion $S \to J^k(M, N)$ maps α into an element of the group $H^{\mathrm{f}}_{\dim S}(J^k(M, N); \mathbb{Z}_2)$; then, using the Poincaré duality, we obtain an element of the group $H^v(J^k(M, N); \mathbb{Z}_2)$, $v = \operatorname{codim} S$, which is called the *class dual to the set* S and is denoted by $[S]$. Recall (see § 3.1) that the k-jet extension f_k of f, $f_k\colon M \to J^k(M, N)$ is defined for any smooth map $f\colon M \to N$ and any positive integer k.

Theorem (see [175]). *Suppose that the map $f\colon M \to N$ is such that its k-jet extension $f_k\colon M \to J^k(M, N)$ is transverse to the stratified set S. Then the set $S(f) = f_k^{-1}(S)$ has a fundamental cycle, and consequently it defines a dual class $[S(f)] \in H^v(M; \mathbb{Z}_2)$; moreover, $[S(f)]$ coincides with the class $f_k^*([S])$.*

Corollary. *The class $[S(f)]$ does not change under smooth homotopies of the map f.*

Indeed, a homotopy of maps defines a homotopy of their jet extensions.

Note that, in addition, we now have a recipe for defining the class dual to the set $S(f)$ even if the map f is not generic: by definition, this class is always equal to $f_k^*([S])$.

Now let us describe the bordism invariants defined by the classes $[\Sigma^I]$.

---

[1] In view of smoothing theorems, M and N, and together with them $J^k(M, N)$, can be assumed to be analytic manifolds.

Definition. Let N be a smooth manifold. The m-dimensional *nonoriented bordism group* $\mathfrak{N}_m(N)$ of N is the abelian group generated by the pairs of the form [closed (i.e., compact boundaryless) m-dimensional manifold M, smooth map $f: M \to N$], subject to the relations generated by the pairs $[M, f]$ with the property that M is the boundary of some compact manifold M_1 and there is a smooth map $f_1: M_1 \to N$ such that $f = f_1|_{\partial M_1}$.

In particular, if f, $g: M \to N$ are homotopic maps, then the sum $[M, f] + [M, g]$ is bordant to zero.

Let $J^k(\mathbb{R}^n_0, \mathbb{R}^p_0)$ denote the space of k-jets of maps $(\mathbb{R}^n, 0) \to (\mathbb{R}^p, 0)$. Fix a number r and assume that an $\mathscr{A}$-invariant analytic subset $S = S(n, n-r)$ is defined in the space $J^k(\mathbb{R}^n_0, \mathbb{R}^{n-r}_0)$ for any n; see subsection 3.1.1. Let $\mathbb{1}$ denote the identity map $\mathbb{R} \to \mathbb{R}$. The system of sets S is said to be *stable* if, for any n, the k-jet of a map $f: (\mathbb{R}^n, 0) \to (\mathbb{R}^{n-r}, 0)$ belongs to $S(n, n-r)$ if and only if the jet of the map $f \times \mathbb{1}: (\mathbb{R}^n \times \mathbb{R}, 0) \to (\mathbb{R}^{n-r} \times \mathbb{R}, 0)$ belongs to $S(n+1, n-r+1)$.

The sets $\overline{\Sigma^I}$ form a stable system.

Proposition. *Suppose that for each $i = 1, \ldots, t$ a stable system S_i of $\mathscr{A}$-invariant analytic subsets in the space $J^k(\mathbb{R}^n_0, \mathbb{R}^{n-r}_0)$ is given such that the codimension of the set $S_i(n, n-r)$ is equal to a_i for all n, and $m = \sum a_i$. Then for any map f of an m-dimensional closed manifold M into an $(m-r)$-dimensional manifold N the value of the element $[S_1(f)] \cup \cdots \cup [S_t(f)] \in H^m(M; \mathbb{Z}_2)$ on the fundamental cycle of M is an invariant of the element $[M, f]$ in the group $\mathfrak{N}_m(M)$.*

1.2. Thom's Theorem on the Existence of Thom Polynomials. Let $f: M^m \to N^n$ be a map of smooth manifolds. The spaces of all tangent vectors to these manifolds, TM and TN, are vector bundles over M and N, and so (see subsection 2.6.2) they define Stiefel-Whitney classes[2] $w_i(M) \in H^i(M; \mathbb{Z}_2)$, $i = 0, 1, \ldots, m$ and $w_j(N) \in H^j(N; \mathbb{Z}_2)$, $j = 0, 1, \ldots, n$, respectively. The map f allows us to pull back the classes $w_j(N)$ to M. The resultant classes $f^*w_j(N) = w_j(f^*TM)$ will be denoted by $w_j'(N)$.

Theorem (see [175]). *Let S be an arbitrary analytic $\mathscr{A}$-invariant subset of $J^k(\mathbb{R}^m_0, \mathbb{R}^n_0)$. Then there exists a universal polynomial T_S over $\mathbb{Z}_2$ in $m+n$ variables, which depends only on S, m and n, such that for any smooth manifolds M^m, N^n and almost any map $f: M \to N$ the class in $H^*(M; \mathbb{Z}_2)$ dual to the singular set $S(f) = f_k^{-1}(S)$ is equal to the value of the polynomial $T_S(w_1(M), \ldots, w_m(M), w_1'(N), \ldots, w_n'(N))$.*

This assertion is usually applied in the case where S is the closure of one of the Thom-Boardman classes. The polynomial T_S is called the *Thom polynomial* (T.P.) for the set S.

---

[2] Henceforth, all the results discussed in this section admit a natural "complexification": the assertions remain valid if real manifolds and bundles are replaced by complex ones, the homology with coefficients in $\mathbb{Z}_2$ by that with coefficients in $\mathbb{Z}$, the Stiefel-Whitney classes by the Chern classes, and so on.

Example. Let $m = n$, $S = \overline{\Sigma^1}$ = the set of all singular jets of maps. Then, for any f, the class dual to $S(f)$ is the difference of the first Stiefel-Whitney classes of the bundles TM and f^*TN: $[S(f)] = w_1(TM) - w_1'(TN)$. If N is orientable, then this class "keeps up with" the orientation of the manifold M: M is orientable if and only if the set $S(f)$ is homologous to zero in M.

The present status of the computations of the T.P. for the classes Σ^I is as follows. For singularities of the first order ($\Sigma^I = \Sigma^i$) the T.P. are completely known, see [276]. For the case $I = (i, j)$ an algorithm for the computation of the T.P. was proposed in [276], [282]; in [282] this computation has been carried out for a number of examples. For more complex singularities there are only separate results (see subsection 1.5 below). In all these cases one succeeds in finding an expression for the class $[S(f)]$ that does not involve all classes $w_i(TM)$, $w_j'(TN)$, but only the ratio of the total Stiefel-Whitney classes of the bundles TM and f^*TN. Let us recall this notion.

The *total Stiefel-Whitney class* of an l-dimensional vector bundle $L \to M$ is the element $1 + w_1(L) + \cdots + w_l(L) \in H^*(M; \mathbb{Z}_2)$, denoted by $w(L)$. Let $A(M)$ be the subset of $H^*(M; \mathbb{Z}_2)$ consisting of all elements of the form $a = a_0 + a_1 + \cdots$, $a_i \in H^i(M; \mathbb{Z}_2)$, such that a_0 is the unit of the ring $H^*(M; \mathbb{Z}_2)$. $A(M)$ is an abelian group with respect to the cup product and contains the total Stiefel-Whitney classes of all vector bundles over M, the products and ratios of such classes, and so on.

Notation. Let $f: M \to N$ be a smooth map. Then $w(f)$ denotes the element in $A(M)$ equal to $w(TM)/f^*w(TN)$, i.e., the ratio of the total Stiefel-Whitney class of the bundle TM and the f-pullback of the total Stiefel-Whitney class of the bundle TN.

In most cases the computation of the T.P.s rests on the following construction, which is also interesting in its own right.

1.3. Resolution of the Singularities of the Closures of the Thom-Boardman Classes. As a rule, the closures of the sets $\Sigma^I \subset J^k(M, N)$, and hence those of the corresponding critical sets $\Sigma^I(f) \subset M$, are not smooth (an important exception is the case of the Morin maps – see subsection 1.5 below). However, many of the sets $\overline{\Sigma}^I$ admit canonical resolutions of singularities.

Definition. A *resolution of singularities* of the set $\overline{\Sigma^I} \subset J^k(M, N)$ is a pair consisting of a smooth manifold $\tilde{\Sigma}^I$ and a proper analytic mapping $p_I: \tilde{\Sigma}^I \to J^k(M, N)$, such that $\overline{\Sigma^I} = p_I(\tilde{\Sigma}^I)$ and p_I is a local diffeomorphism between $\tilde{\Sigma}^I$ and $\overline{\Sigma^I}$ almost everywhere over $\overline{\Sigma^I}$.

For example, in the case where the Thom-Boardman symbol I reduces to a single element i, the canonical resolution of singularities of $\overline{\Sigma^i}$ is constructed as follows (see [276]).

For each $i = 1, \ldots, m - 1$, *the i-th Grassmann bundle over M associated with the tangent bundle TM* of the manifold M is defined: it consists of all possible pairs of the form (point $x \in M$, i-dimensional subspace in the tangent space T_xM). Consider also the fibred product of this bundle and the bundle $J^k(M, N)$

$\to M$. (The total space of this last product consists of all triplets ($x \in M$, k-jet at x of a map $M \to N$, i-dimensional subspace in $T_x M$).) Now in that space consider the subset consisting of the triplets with the property that the k-jet belongs to $\overline{\Sigma^i}$ (and consequently the kernel of its differential has dimension at least i), and the i-dimensional subspace in $T_x M$ is contained in this kernel. This set together with the canonical projection onto its factor $J^k(M, N)$ (i.e., the map that forgets the third element of the triplet), is precisely the desired resolution of singularities of the set $\overline{\Sigma^i}$.

The resolution of singularities of some sets $\overline{\Sigma^I}$ of higher order is carried out in an analogous manner with the help of *flag bundles* over M: for instance, if $I = (i, j)$, then in the previous construction one can take, instead of the ith Grassmann bundle, the bundle whose total space consists of the triplets (point $x \in M$, i-dimensional subspace L in $T_x M$, j-dimensional subspace in L); see [276], [282]. However, this recipe does not work for all sets I: starting already with three-element indices it is, generally speaking, not true (or at least not proven) that the set $\tilde{\Sigma}^I$ produced in the described manner is smooth.

An even more general means of constructing a resolution of the sets $\overline{\Sigma^I}$ can be described as follows. For any point $x \in M$ and any map-germ $f: M \to N$ at x consider the algebra $\mathfrak{M}(N, f(x))$ of germs at the point $f(x)$ of functions on N that vanish at that point. The preimage of this algebra generates an ideal $i(f, x)$ in the algebra $\mathfrak{M}(M, x)$, and hence, for every positive integer k, the ideal $i_k(f, x) = (i(f, x) + \mathfrak{M}^{k+1}(M, x))/\mathfrak{M}^{k+1}(M, x)$ in the finite-dimensional algebra $\mathfrak{M}(M, x)/\mathfrak{M}^{k+1}(M, x)$; $i_k(f, x)$ depends only on the k-jet of the germ f. For almost any $(x, f) \in \Sigma^I \subset J^k(M, N)$, the ideal $i_k(f, x)$ has one and the same (maximal possible) dimension $d(I, k)$. Now consider the set of all triplets of the form (point $x \in M$, k-jet of a map $M \to N$ at x, $d(I, k)$-dimensional subspace in the algebra $\mathfrak{M}(M, x)/\mathfrak{M}^{k+1}(M, x)$). Forgetting the last element in such a triplet yields on this set a structure of fibre bundle over $J^k(M, N)$. Restrict this bundle to the subset $\Sigma^I \subset J^k(M, N)$. Over almost any point of Σ^I the ideal $i_k(f, x)$ gives a section of the bundle. We denote the closure of that section again by $\tilde{\Sigma}^I$. In the case $k = 1$ we recover the preceding construction, and as k grows its resolution power increases.

1.4. Thom Polynomials for Singularities of First Order.

Let $\overline{w}_l$ denote the lth homogeneous component of the class $w(f)$, so that $w(f) = 1 + \overline{w}_1 + \overline{w}_2 + \cdots$. Set $j = \min(m, n) - i$.

Theorem ([276]). *For any $i = 1, 2, \ldots$ the class $[\Sigma^I(f)] \in H^{i(n-m+i)}(M; \mathbb{Z}_2)$ is equal to the determinant of the matrix*

$$\begin{vmatrix} \overline{w}_{m-j} & \overline{w}_{m-j+1} & \cdots & \overline{w}_{m+n-2j-1} \\ \overline{w}_{m-j-1} & \overline{w}_{m-j} & \cdots & \overline{w}_{m+n-2j-2} \\ \cdots & \cdots & \cdots & \cdots \\ \overline{w}_{m-n+1} & w_{m-n+2} & \cdots & \overline{w}_{m-j} \end{vmatrix}. \qquad (1)$$

Example. Let $m < n$ and $i = m - 1$, i.e., $j = m - 1$. Replace $\bar{w}$ in (1) by the ordinary Stiefel-Whitney classes of the tangent bundle of M. Suppose that the resulting determinant (1) is not equal to zero as a class in $H^{n-m+1}(M; \mathbb{Z}_2)$. Then the manifold M cannot be immersed in $\mathbb{R}^n$: the set Σ^1 of singularities of any map $M \to \mathbb{R}^n$ is nonempty. (This is not surprising, since for $j = m - 1$ the determinant (1) constructed for an arbitrary element $a = 1 + a_1 + \cdots \in A(M)$ is equal to the $(n - m + 1)$-dimensional component of the element $a^{-1} \in A(M)$, i.e., in our case, to the $(n - m + 1)$-dimensional Stiefel-Whitney class of the normal bundle to M under any immersion in Euclidean space.)

The proof of formula (1) is based on the following operation.

The Gysin homomorphism in cohomology. Let X, Y be boundaryless manifolds (generally speaking, noncompact) and let $p: X \to Y$ be a proper map. Then the homomorphism $p_*: H^*(X; \mathbb{Z}_2) \to H^*(Y; \mathbb{Z}_2)$ is defined, which is the composition of the Poincaré isomorphism in X, the ordinary homomorphism $H^{\mathrm{f}}_*(X; \mathbb{Z}_2) \to H^{\mathrm{f}}_*(Y; \mathbb{Z}_2)$ in the homology with closed supports, and the Poincaré homomorphism in Y. If X and Y are oriented, then the homomorphism p_* is also defined in the integer cohomology. The operation p_* has the property that

$$p_*(p^*y \cup x) = y \cup p_*x$$

for any two elements $x \in H^*(X; \mathbb{Z}_2)$ and $y \in H^*(Y; \mathbb{Z}_2)$.

Now consider the canonical resolution $\tilde{\Sigma}^i$ of the set $\overline{\Sigma^i}$ (see subsection 1.3). $\tilde{\Sigma}^i$ lies in the total space $X^i = X^i(k, M, N)$ of the fibred product $G_i(M) \times J^k(M, N)$. The Gysin homomorphism associated to the obvious projection $X^i \to J^k(M, N)$ takes the class $[\tilde{\Sigma}^i] \in H^*(X^i, \mathbb{Z}_2)$ dual to $\tilde{\Sigma}^i$ precisely into the class $[\overline{\Sigma^i}] \in H^*(J^k(M, N); \mathbb{Z}_2)$, and so it remains to calculate the class $[\tilde{\Sigma}^i]$ and the action of the Gysin operator on the group $H^*(X^i; \mathbb{Z}_2)$. These calculations are carried out in [276] (see also [282]) and yield formula (1). An analogous method (applied to subtler resolutions of singularities of the sets $\overline{\Sigma^I}$) is used in almost all works dealing with the calculation of Thom polynomials.

1.5. Survey of Results on Thom Polynomials for Singularities of Higher Order. Let S be a stable system of subsets in $J^k(\mathbb{R}^n_0, \mathbb{R}^{n-r}_0)$. When the Thom polynomial T_s depends only on r, but not on n, we denote it by $T_{S,r}$ or, if $S = \Sigma^I$, by $T_{(I),r}$.

In [282] a method is indicated for computing the polynomials $T_{(I),r}$ for $I = (i, j)$, and the following results are obtained as examples.

Theorem[3]. For $r \leqslant 0$, $T_{(1,1),r} = \bar{w}_{1-r}^2 + \sum_{l=0}^{-r} 2^l \bar{w}_{l+2-r}\bar{w}_{-l-r}$

2. $T_{(2,1),0} = 2\begin{vmatrix} \bar{w}_3 & \bar{w}_2 \\ \bar{w}_5 & \bar{w}_4 \end{vmatrix} + 4\bar{w}_2\begin{vmatrix} \bar{w}_2 & \bar{w}_1 \\ \bar{w}_4 & \bar{w}_3 \end{vmatrix} + 2\begin{vmatrix} \bar{w}_1 & 1 \\ \bar{w}_3 & \bar{w}_2 \end{vmatrix}\begin{vmatrix} \bar{w}_2 & \bar{w}_1 \\ \bar{w}_3 & \bar{w}_2 \end{vmatrix}.$

3. $T_{(2,1),1} = 2\bar{w}_1(\bar{w}_1^2 - \bar{w}_2)$.

[3] Of course, here all the terms with even coefficients can be ignored, but the point is that these formulas remain valid after "complexification" (see the footnote on page 187).

Theorem ([140]). $T_{(1,1,1,1),0} = 6\overline{w}_4 + 9\overline{w}_3\overline{w}_1 + 2\overline{w}_2^2 + 6\overline{w}_2\overline{w}_1^2 + 4\overline{w}_1^4$.

Let α denote the subclass of Σ^2 that corresponds to the cloth of Whitney's umbrella (see subsection 3.1.8), and let β denote the union of the subclasses that correspond to the handle and the self-intersection of the umbrella.

Theorem. ([277]). $T_{\alpha,0} = 2(\overline{w}_1(\overline{w}_2^2 - \overline{w}_3\overline{w}_1) + \overline{w}_3\overline{w}_2 - \overline{w}_4\overline{w}_1)$.

$$T_{\beta,0} = (\overline{w}_1^2 - \overline{w}_2)(\overline{w}_2^2 - \overline{w}_3\overline{w}_1) + 2\overline{w}_1(\overline{w}_3\overline{w}_2 - \overline{w}_4\overline{w}_1) + 3(\overline{w}_3^2 - \overline{w}_4\overline{w}_2).$$

Definition. A map is said to be a *Morin map* if it has only Whitney-Morin singularities $S_k = \Sigma^{\overbrace{1,\dots,1}^{k}}$ and the corresponding jet extensions are transverse to all these classes.

Theorem (see [214]). *Let $f: M \to N$ be a Morin map, with* $\dim M = \dim N$. *Then the class $P_k \in H^k(M; \mathbb{Z}_2)$ dual to the set $\overline{S_k(f)}$ is given by the recursion formula*

$$P_k = \sum_{i=1}^{k} \frac{(k-1)!}{(k-i)!}\overline{w}_i P_{k-i}.$$

Theorem ([79]). *Suppose f is a Morin map, $k \geqslant 4$, $m > n$, and $m - n$ is odd. Then the submanifold $\overline{S_k(f)}$ is bordant to zero in $\overline{S_{k-2}(f)}$; in particular, the cohomology class dual to it is trivial.*

This theorem has led its author, D.S. Chess, to the following conjecture: under the assumptions of the theorem, f is homotopic to a map g for which the set $S_k(g)$ is not empty.

§ 2. Integer Characteristic Classes and Universal Complexes of Singularities

A Thom-Boardman class (or some other class of singularities) usually defines a dual element in the cohomology with coefficients in $\mathbb{Z}_2$, but not in that with coefficients in $\mathbb{Z}$. For example, the singular set of a typical map $\mathbb{R}P^2 \to \mathbb{R}^2$ does not define an element in $H^1(\mathbb{R}P^2; \mathbb{Z})$, the reason being that there is no invariant way of assigning the sign + or − to the points of transverse intersection of this set with oriented curves in $\mathbb{R}P^2$. In order for a class of singularities to define also an integer dual cohomology class it is necessary that it be a cocycle of the *universal complex* defined below (see [361], [357]); the generators of this complex are the cooriented classes of singularities, and the differentials are defined according to the scheme of adjacencies of classes of neighboring codimensions. To each "theory of singularities" – of smooth functions, of Lagrangian and Legendrian maps, of general maps of smooth manifolds, and so on – there corresponds its own universal complex, which gives a diverse topological information on the global properties of these singularities.

2.1. Examples: the Maslov Index and the First Pontryagin Class. Let N be an immersed Lagrangian submanifold in the total space of the contangent bundle T^*W of a manifold W. If the immersion $N \hookrightarrow T^*W$ is generic, then the singular set of the projection map $N \to W$ has codimension 1 in N. As it turns out, the index of intersection with that set already defines a class in the integer cohomology of N; see [235], [7]. (This implies, in particular, that $\mathbb{R}P^2$, as well as any other closed surface with odd Euler characteristic, does not admit a Lagrangian immersion in $T^*\mathbb{R}^2$.) This is connected with two remarkable properties of the simplest Lagrangian singularity (the fold A_2), namely, its *coorientability* and the *regular behavior of the coorientation* near more complex singularities. The meaning of these properties is as follows.

Since the immersion $N \to T^*W$ is generic, the singular set is the closure of the set of points of type A_2, it contains the set of points of type A_3 as a smooth subset of codimension 1, and the set of singularities of the remaining types has in it codimension $\geqslant 2$. It turns out that in the neighborhood of any point of type A_2 the two components into which the set $\{A_2\}$ locally divides N are not equivalent: there is an invariant way of calling one of them positive, and the other negative; see [235], [34]. A typical curve in N intersects the singular set only at points of the set $\{A_2\}$, and the intersections are transverse; the value of the Maslov index on a closed oriented curve is defined as the number of times the curve crosses from the negative to the positive side minus the number of times it crosses in the opoosite direction. It turns out that this number depends only on the class of the curve in $H_1(N; \mathbb{Z})$. This would be impossible if the transverse orientations of the components of the set $\{A_2\}$ were incompatible near the points of type A_3, for example, as shown in Fig. 47a. In fact, in that case the Maslov index of a small circle centered at the point A_3 would be equal to two, though such a circle is homologous to zero.

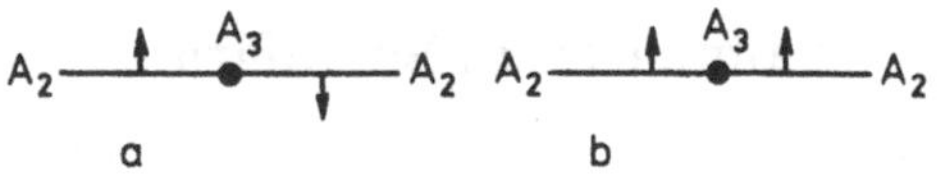

Fig. 47

However, the coorientations of A_2 are compatible, as in Fig. 47b. It is precisely this circumstance that allows one to define the Maslov index as an integer cohomology class. The compatibility condition of the orientations of the components of the singular set can be formalized and carried over to more complex singularities as the property of being a cocycle in the universal complex of singularities. For example, in that complex Fig. 47a would correspond to the formula $\delta(A_2) = 2A_3$, whereas the correct picture corresponds to the formula $\delta(A_2) = 0$.

A second important example concerns the theory of singularities of general (non-Lagrangian) maps. Let M^n be a compact manifold and $f: M^n \to \mathbb{R}^n$ a generic map.

Proposition. *The sets Σ^2_+ and Σ^2_- of elliptic and respectively hyperbolic singular points of type Σ^2 (see subsection 3.1.8) admit an invariant coorientation in M^n (i.e., an orientation for any germ of four-dimensional submanifold of M^n that is transverse to them).*

To prove that the union of these cooriented sets defines an integer class, one only needs to verify the compatibility of these coorientations in a neighborhood of the simplest degenerate points of Σ^2 (i.e., the points of the "cloth", see Fig. 41 in subsection 3.1.2). But near such points each of the classes Σ^2_+, Σ^2_- is represented by a single component, and so one is able to choose their standard coorientations in a compatible manner. Then the index of intersection with the cooriented set $\Sigma^2_+ \cup \Sigma^2_-$ gives the class $[\Sigma^2] \in H^4(M^n; \mathbb{Z})$.

Theorem (see [324]). *For an appropriate choice of compatible coorientations of the sets Σ^1_+, Σ^2_-, the class $[\Sigma^2]$ coincides, modulo torsion, with the first Pontryagin class of the manifold M.*

2.2. The Universal Complex of Singularities of Smooth Functions. The starting material in the construction of all our universal complexes consists of appropriate classifications of singularities, i.e., invariant partitions of the spaces of germs (or jets) of singular maps which satisfy certain regularity conditions. Let us examine in more detail the case of the theory of singularities of smooth functions.

Fix positive integers k, n. Let $J^k_0(n)$ denote the space of k-jets of smooth functions $f: (\mathbb{R}^n, 0) \to (\mathbb{R}, 0)$ such that $df(0) = 0$. The group R_0 of germs of diffeomorphisms $(\mathbb{R}^n, 0) \to (\mathbb{R}^n, 0)$ acts on $J^k_0(n)$.

Definition. An *R_0-classification* of the space $J^k_0(n)$ is any locally finite decomposition of this space into disjoint sets (classes) such that:

1) all classes are smooth manifolds and semialgebraic sets;

2) R_0-equivalent points lie in the same class;

3) for any two connected components of a given class there is a pair of R_0-equivalent points, each belonging to one of the component;

4) all adjacencies of classes satisfy Whitney's regularity conditions (see subsection 3.3.2); in particular, the bigger stratum has the same topological structure near all points of the smaller one.

Proposition. 1. *Any locally finite R_0-invariant semialgebraic decomposition of the space $J^k_0(n)$ is finite, and one can transform it into an R_0-classification by subdividing it in a suitable manner.*

2. *For any two R_0-classifications there exists a third one which is a subdecomposition of both.*

Example. According to § 1.2, the complete list of the real $\mu =$ const strata of codimension $\leqslant 7$ is as follows:

$$
\begin{gathered}
{}^rA_1,\ {}^rA_{2k},\ {}^rA_{2k+1}^{\pm}\ (k \geqslant 1),\ {}^rD_k^{\pm}\ (k \geqslant 4),\ {}^rE_6^{\pm},\ {}^rE_7,\ {}^rE_8, \\
{}^rP_8^{1,2},\ {}^rX_9^{\pm},\ {}^rX_9^{1,2},\ {}^rP_9^{\times\pm},\ {}^rP_9^{\cdot\pm},
\end{gathered}
\tag{2}
$$

where r is the positive index of inertia of the quadratic part of the singularity, so that, for example, ${}^2A_3^-$ is the stratum containing the function $-x_1^4 + x_2^2 + x_3^2 - x_4^2 - \cdots - x_n^2$. By the theorem on the sufficient jet (see subsection 1.1.5), these classes and their complement yield a uniquely determined decomposition of any of the spaces $J_0^k(n)$ with $k \geqslant 10$; here the classes P_8 and P_9 are empty if $n < 3$, and for $n = 1$ all classes (2) are empty, except for the classes A_i. By the preceding proposition, such a decomposition of $J_0^k(n)$ can be refined to an R_0-classification. It turns out that in order to achieve this it suffices to subdivide only the complement of the classes (2).

A *coorientation* of the class Σ is defined to be an orientation of its normal bundle in $J_0^k(n)$. The group R_0 acts on $J_0^k(n)$ preserving Σ, and hence it also acts on the normal bundle of Σ. A coorientation of a class in R_0 is called an R_0-*coorientation* if it is preserved by this action.

Theorem ([357]). *The only classes in the list* (2) *that are not R_0-coorientable are $A_{4k-1}^{\pm}$, $D_k^{\pm}$, and X_9.*

We now have all the ingredients needed to define the universal complex $\Omega = \Omega(n)$. Fix an R_0-classification Γ of the space $J_0^k(n)$. Set $\Omega(\Gamma) = \bigoplus_{l \geqslant 0} \Omega^l(\Gamma)$, where $\Omega^l(\Gamma)$ denotes the free abelian group generated by all R_0-cooriented classes of codimension l. Define the coboundary operator $\delta\colon \Omega^l(\Gamma) \to \Omega^{l+1}(\Gamma)$ as follows. Let Σ and Σ' be two cooriented classes of codimensions l and $l + 1$, respectively. Let L be an $(l + 1)$-dimensional transversal to Σ'. In L the class Σ is represented by a finite collection of curves. Each of these curves defines an orientation of L: this orientation is given by a tangent frame in which the first vector is directed along the curve toward a point of Σ', while the remaining l vectors give the chosen coorientation of the class Σ. Assign to the curve the number 1 or -1, depending on whether the indicated orientation coincides or not with the orientation corresponding to the chosen coorientation of Σ'. By definition, the incidence coefficient $[\Sigma, \Sigma']$ is equal to the sum of these numbers ± 1, taken over all branches of curves. (For example, in Fig. 48 the two lower

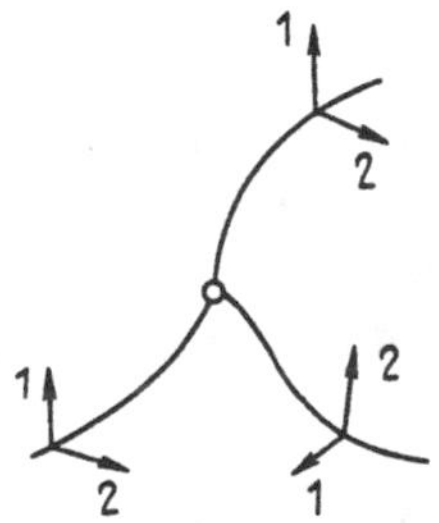

Fig. 48

branches of Σ give equal contributions in $[\Sigma, \Sigma']$, whereas the contribution of the upper branch is opposite.) Finally, set $\delta(\Sigma)$ equal to the sum of the terms $[\Sigma, \Sigma_i']\Sigma_i'$, taken over all $(l+1)$-dimensional generators Σ_i' of the group $\Omega^{l+1}(\Gamma)$.

Remarks. 1. The coefficient $[\Sigma, \Sigma']$ does not depend on the choice of the transversal L to the class Σ'.

2. If Σ' is not R_0-coorientable, then, upon giving an arbitrary orientation of the transversal L, one can define the coefficient $[\Sigma, \Sigma']$ in much the same manner as above, but it inevitably turns out to be zero.

Thus, we constructed a complex $\Omega(\Gamma)$ that depends on the choice of the number k and of the R_0-classification Γ of the space $J_0^k(n)$. Passing to the limit with respect to the natural maps $\Omega(\Gamma) \to \Omega(\Gamma')$, which arise whenever the classification Γ' is a subdecomposition of Γ, we obtain a complex $\Gamma_k(n)$ that depends only on k and n. The projections $J_0^{k+1}(n) \to J_0^k(n)$ induce homomorphisms of complexes $\to \Omega_k(n) \to \Omega_{k+1}(n) \to \cdots$; finally, passing to the limit with respect to these mappings, we obtain the desired universal complex $\Omega(n)$.

The complex N of noncooriented classes is constructed in amost the same manner as the complex Ω of cooriented classes, namely, it is the $\mathbb{Z}_2$-module generated by all classes (and not only the cooriented ones), and the incidence coefficient $[\Sigma, \Sigma']$ is defined as the number, reduced mod 2, of all components of Σ in the neighborhood of Σ'.

One of the applications of the complex $\Omega(n)$ (or similar complexes constructed by means of the singularities of maps in $\mathbb{R}^p$, $p > 1$) is that their cohomologies allows us to distinguish n-dimensional fibre bundles (and even foliations). Specifically, let $p: E \to M$ be a smooth locally trivial bundle, with both the base M and the fibre compact. Let $f: \mathrm{E} \to \mathbb{R}$ be a smooth function. For each class of singularities Σ let $\Sigma(f)$ denote the set of all points $x \in M$ such that the restriction of f to the fiber $p^{-1}(x)$ has a singular point of class Σ (i.e., it reduces to a function of class Σ when one chooses suitable local coordinates and adds a suitable constant).

Definition. A bundle $p: E \to M$ is said to be *cobordant to zero* if there exists a bundle $p_1: E_1 \to M_1$ (M_1 is compact) with the same fibre, such that $M = \partial M_1$, $E = \partial E_1$ and $p \equiv p_1$ on E.

Theorem. *Suppose the function $f: E \to \mathbb{R}$ is generic and Γ is an arbitrary R_0-classification of the space $J_0^k(n)$, where n is the dimension of the fibre of p. Then the following holds:*

1) *For any class Σ of the classification Γ the closure of the set $\Sigma(f)$ is a compact chain in M.*

2) *An R_0-coorientation of the class Σ induces a coorientation of the submanifold $\Sigma(f)$.*

3) *If $a_1\Sigma_1 + \cdots + a_t\Sigma_t$ is a cocycle of the complex $\Omega(n)$ [resp. $N(n)$], then intersection index with the chain $a_1\Sigma_1(f) + \cdots + a_t\Sigma_t(f)$ is well defined on M. The class of this intersection index in the cohomology of M does not depend on*

the choice of the function f and defines a homomorphism $H^*(\Omega(n)) \to H^*(M; \mathbb{Z})$ [*resp.* $H^*(N(n)) \to H^*(M; \mathbb{Z}_2)$].

4) *For any element* $c \in H^m(\Omega(n))$ [*resp.* $c \in H^m(N(n))$] *the value of the class* $c(f)$ *on the integer* [*resp.* $\mathbb{Z}_2$-] *fundamental cycle of the oriented* [*resp. non-oriented*] *m-dimensional manifold M is an invariant of oriented* [*resp. non-oriented*] *cobordism of bundles.*

5) *If the class* Σ *belongs to the group of m-dimensional coboundaries of the complex N* [*resp.* Ω], *then for any* [*resp. any oriented*] *m-dimensional compact manifold M the number of points in* $\Sigma(f)$ *is even* [*resp. equal to zero, provided each of these points is counted with the coefficient* 1 *or* -1, *depending on whether the orientation of M coincides or not with the orientation that gives the coorientation of the class* Σ *as that point*].

A realization of assertions 4 and 5 of this theorem which results from computations in the complexes Ω and N is given in subsection 2.4 below.

2.3. Cohomology of the Complexes of R_0-Invariant Singularities, and Invariants of Foliations. The preceding theorem can be generalized to the case where instead of fibre bundles one takes arbitrary *foliations* (the definition of a foliation can be found in [130]), and the complexes of R_0-cooriented classes of functions are replaced by complexes of classes of maps into an arbitrary space $\mathbb{R}^p$.

Definition. Two cooriented foliations $\mathcal{F}$ and $\mathcal{F}'$ on manifolds E and E' are said to be *related* if there exist a manifold E_1 and a coorientable foliation $\mathcal{F}_1$ of the same dimension on E_1, such that: (i) $\partial E_1 = E' - E$, (ii) any leaf of $\mathcal{F}_1$ that intersects the submanifold E or E' is entirely contained in the latter and coincides with a leaf of $\mathcal{F}$ or $\mathcal{F}'$, respectively, and (iii) the coorientation of $\mathcal{F}$ at the points of E [resp. E'] is given by a normal frame which is obtained from an arbitrary coorienting frame of the foliation $\mathcal{F}$ [resp. $\mathcal{F}'$] by adjoining the inward [resp. outward] normal vector to E. Relatedness of noncooriented foliations is defined similarly. The relatedness relation for foliations generalizes the homotopy relation and has no connection with the notion of concordance (see [130]).

The group R_0 of germ of diffeomorphisms $(\mathbb{R}^n, 0) \to (\mathbb{R}^n, 0)$ acts on the space of jets of maps $(\mathbb{R}^n, 0) \to \mathbb{R}^p$ for any n and p. This permits us, by analogy with the foregoing, to define the complex $\Omega(n, p)$ of R_0-cooriented classes of such jets (in subsection 2.2 we considered the special case $p = 1$) and the complex $N(n, p)$. By considering restrictions of arbitrary maps $f: E \to \mathbb{R}^p$ to the leaves of the foliation $\mathcal{F}$ (cf. [327]) we obtain the following result.

Theorem. *For any value of p, any cooriented* [*resp. noncooriented*] *foliation of dimension n and codimension m on a compact manifold defines a homomorphism* $H^m(\Omega(n, p)) \to \mathbb{Z}$ [*resp.* $H^m(N(n, p)) \to \mathbb{Z}_2$]. *This homomorphism is an invariant of relatedness of cooriented* [*resp. noncooriented*] *foliations. When the given foliation is a fibre bundle, this homomorphism assigns to the element c the value of the class* $c(f)$ *on the fundamental cycle of the base (see the last theorem in subsection 2.2).*

The embeddings $\cdots \to \mathbb{R}^{p-1} \to \mathbb{R}^p \to \cdots$ yield a sequence of embeddings $\cdots \to \Omega(n, p-1) \to \Omega(n, p) \to \cdots$. Let $\Omega_n = \lim_{p\to\infty} \Omega(n, p)$. Then in the last theorem one can replace $\Omega(n, p)$ by Ω_n.

2.4. Computations in Complexes of Singularities of Functions. Geometric Consequences

Theorem. *All groups $H^i(\Omega(n))$ and $H^i(N(n))$ are finitely generated.*

This assertion follows from the following stabilization theorem. The limiting homomorphisms $\Omega_k(\Gamma) \to \Omega(n)$ induce homomorphisms

$$H^i(\Omega_k(\Gamma)) \to H^i(\Omega(n)). \tag{3}$$

Theorem ([361]). *For any r one can find a number k and a classification Γ such that the homomorphisms* (3) *are isomorphisms for $i \leqslant r$; the same holds for the complexes N.*

Example. Let $k \geqslant 10$ and let Γ be the classification obtained by decomposing the space $J_0^k(n)$ into the classes (2) and some other classes. Then the map (3) is an isomorphism for $i \leqslant 6$.

Theorem ([357]). *For a suitable choice of R_0-coorientation of the classes* (2) *the action of the coboundary operator on the groups $\Omega^l(\Gamma)$, $l = 0, 1, \ldots, 6$ is given by the following formulas:*

$$\begin{gathered}
\delta({}^rA_1) = -{}^{r-1}A_2 + {}^rA_2; \qquad \delta({}^rA_2) = 0,\\
\delta({}^rA_4) = -2{}^rA_5^+ + 2{}^rA_5^-,\\
\delta({}^rA_5^+) = -\delta({}^rA_5^-) = -{}^rA_6 + {}^{r-1}E_6^+ + {}^rE_6^-,\\
\delta({}^rA_6) = {}^{r-1}E_7 - {}^rE_7,\\
\delta({}^rE_6^+) = {}^rE_7 + 3{}^rP_8^1 + 3{}^rP_8^2,\\
\delta({}^rE_6^-) = -{}^rE_7 - 3{}^{r-1}P_8^1 - 3{}^{r-1}P_8^2,\\
\delta({}^rE_7) = 0,\\
\delta({}^rP_8^1) = -\delta({}^rP_8^2) = {}^rP_9^{\times+} + {}^rP_9^{\times-} + {}^rP_9^{\cdot+} + {}^rP_9^{\cdot-}.
\end{gathered} \tag{4}$$

Here ${}^r\Sigma = 0$ if $r < 0$ or if $r > n -$ corank Σ.

Corollary. *For $l \leqslant 6$ the groups $H^l(\Omega(n))$ are given by the following table*:

l	0	1	2	3	4	5	6
$n = 1$	$\mathbb{Z}$	0	0	0	$\mathbb{Z}_2$	0	0
$n = 2$	$\mathbb{Z}$	0	0	0	$\mathbb{Z}_2^2$	$\mathbb{Z}$	0
$n \geqslant 3$	$\mathbb{Z}$	0	0	0	$\mathbb{Z}_2^n$	$\mathbb{Z}$	$\mathbb{Z}_3^{n-2}$

In all cases indicated the generators of the group $H^4(\Omega(n))$ *are the elements* $({}^rA_5^+ + {}^rA_5^-)$, $r = 0, 1, \dots, n-1$. *The generator of* $H^5(\Omega(\geqslant 2))$ is the sum ${}^0A_6 + \dots + {}^{n-1}A_6$ *or the sum (cohomologous to it)* $({}^0E_6^+ + {}^0E_6^-) + \dots + ({}^{n-2}E_6^+ + {}^{n-2}E_6^-)$. *The generators of the group* $H^6(\Omega(\geqslant 3))$ *are the elements* $({}^rP_8^1 + {}^rP_8^2)$, $r = 0, 1, \dots, n-3$. *If* $n = 1$, *then* $H^m(\Omega(1))$ *is equal to* $\mathbb{Z}_2$ *for any* $m = 4l$ *and to* 0 *for* $m \neq 4l$.

Theorem ([357]). *The action of the differential in the complex* $N(n)$ *on the generators* (2) *is specified by the following conditions: on the cooriented generators it is obtained from formulas* (4) *by reduction* mod 2; *in addition,*

$$\begin{gathered}\delta({}^rA_3^+) = {}^rA_4 + {}^rD_4^+ + {}^rD_4^-; \qquad \delta({}^rA_3^-) = {}^rA_4 + {}^{r-1}D_4^+ + {}^{r-1}D_4^-;\\ \delta({}^rD_k^+) = \delta({}^rD_k^-) = {}^rD_{k+1}^+ + {}^rD_{k+1}^-, \qquad k = 4, 5, 6;\\ \delta({}^rD_7^+) = \delta({}^rD_7^-) = {}^rE_8 + {}^rD_8^+ + {}^rD_8^-;\\ \delta({}^rA_7^+) = {}^rA_8 + {}^rE_8; \qquad \delta({}^rA_7^-) = {}^rA_8 + {}^{r-1}E_8.\end{gathered} \tag{5}$$

Corollary. *For* $l = 0, 1, \dots, 6$ *the groups* $H^l(N(n))$ *are given by the table*

l	0	1	2	3	4	5	6
$n = 1$	$\mathbb{Z}_2$	0	$\mathbb{Z}_2$	0	$\mathbb{Z}_2$	0	$\mathbb{Z}_2$
$n \geqslant 2$	$\mathbb{Z}_2$	0	$\mathbb{Z}_2$	0	$\mathbb{Z}_2^n$	$\mathbb{Z}_2$	$\mathbb{Z}_2$

Almost all generators of these groups and obtained via reduction mod 2 *of the corresponding cohomology classes of the complex* Ω; *the exceptions are the generators* $({}^0A_3^+ + {}^0A_3^-) + \dots + ({}^{n-1}A_3^+ + {}^{n-1}A_3^-) \in H^2(N)$ *and* $({}^0A_7^+ + {}^0A_7^-) + \dots + ({}^{n-1}A_7^+ + {}^{n-1}A_7^-) \in H^6(N)$. *The group* $H^m(N(1))$ *is equal to* $\mathbb{Z}_2$ *for m even, and to* 0 *for m odd.*

According to subsections 2.2, 2.3, the cohomology groups given in the above table define invariants of bundles (and foliations). It follows from the results of section 3 below that all elements of the group $H^*(N(1))$ define trivial classes: they take only zero values on all bundles (but, maybe, not on foliations?)

Consequences of the coexistence of singularities. For any smooth bundle $E \to M$ with compact fibre and compact base and any generic function $f: E \to \mathbb{R}$ the following chains in M are (mod 2)-cycles homologous to zero (here instead of $\bar{\Sigma}(f)$ we simply write Σ): ${}^rA_2, {}^rA_4, {}^rD_k^+ \cup {}^rD_k^-$ $(k = 4, 5, 6, 7, 8)$, ${}^rA_6 \cup {}^{r-1}E_6^+ \cup {}^rE_6^-, {}^rE_7, {}^rP_8^1 \cup {}^rP_8^2, {}^rA_8, {}^rE_8$. In particular, if the dimension of M coincides with the codimension in $J_0^k(n)$ of one of the classes of singularities Σ listed above, then the number of points of the set $\Sigma(f)$ in M is even.

The following chains, endowed with the coorientation induced from the universal coorientation of the corresponding classes in $J_0^k(n)$, give cocycles homologous to zero: ${}^rA_2$, $2({}^rA_5^+ + {}^rA_5^-)$, $-{}^rA_6 + {}^{r-1}E_6^+ + {}^rE_6^-$, ${}^rE_7$, $3({}^rP_8^1 + {}^rP_8^2)$, ${}^rP_9$. For example, if M is one-dimensional, then the number of points ${}^rA_2$, counted with their sign, is equal to zero; if M is 7-dimensional and oriented, the same holds for the points ${}^rP_9$, for any r.

If the bundle $E \to M$ is a direct product (or is cobordant to one), then the following (cooriented) integer cycles are also homologous to zero: ${}^rA_5^+ + {}^rA_5^-$, ${}^0A_6 + \cdots + {}^{n-1}A_6$, $({}^0E_6^+ + {}^0E_6^-) + \cdots + ({}^{n-2}E_6^+ + {}^{n-2}E_6^-)$; in addition, the (mod 2)-cycles ${}^0A_3 \cup \cdots \cup {}^{n-1}A_3$ and ${}^0A_7 \cup \cdots \cup {}^{n-1}A_7$ are nonorientably homologous to zero.

2.5. Universal Complexes of Lagrangian and Legendrian Singularities. These complexes are constructed using classes of Lagrangian singularities and define Lagrangian characteristic classes, i.e., invariants of the relation of Lagrangian cobordism introduced in [26]. Specifically, for any m-dimensional cocycle $a_1\Sigma_1 + \cdots + a_t\Sigma_t$ of such a complex and any immersed Lagrangian submanifold $M \hookrightarrow T^*W$, in M there is defined the intersection index with the chain $a_1\Sigma_1(M) + \cdots + a_t\Sigma_t(M)$, where $\Sigma_i(M)$ denotes the set of singularities of class Σ_i of the Lagrangian projection $M \to W$. This intersection index defines a homomorphism of the m-dimensional oriented [resp. nonoriented] Lagrangian cobordism group into the m-dimensional cohomology group of the complex of cooriented [resp. noncooriented] Lagrangian singularities; see [357], [361].

The complex ω of cooriented classes of Lagrangian singularities is defined almost exactly in the same way as the complex Ω: the only difference is that R_0-equivalence must be replaced everywhere by stable R_0-equivalence. This is connected with the fact that to Lagrangian-equivalent germs of Lagrangian manifolds there correspond germs of generating functions that are merely stably equivalent.

Consider the disconnected union J_0^k of the spaces $J_0^k(1), J_0^k(2), \ldots$. The definition of the notion of *stable R_0-classification* on J_0^k is obtained from the definition of an R_0-classification given in subsection 2.2 upon replacing R_0-equivalence by stable R_0-equivalence in conditions 2) and 3). A stable R_0-coorientation of a stable class $\Sigma \subset J_0^k$ is defined to be an R_0-coorientation of all its components $\Sigma(n) = \Sigma \cap J_0^k(n)$ that is consistent with all possible embeddings $(J_0^k(n), \Sigma(n)) \to (J_0^k(n+1), \Sigma(n+1))$ that send a function $f = f(x_1, \ldots, x_n)$ into $f + x_{n+1}^2$ or $f - x_{n+1}^2$. The remaining part of the construction of the complexes ω and ν of cooriented and noncooriented classes copies the construction of the complexes Ω and N.

Example. Let us take all classes (2) differing only through the value of the left upper index r. The decomposition of the space J_0^k into the stable classes obtained in this manner and their complement Δ can be subdivided to yield a stable R_0-classification; to this end it suffices to subdivide only Δ. It turns out that any class $\Sigma = A_k, D_k^\pm, \ldots$ of the resulting classification is stably coorientable if and only if all classes ${}^r\Sigma(n)$ are coorientable; the coorientation of the classes ${}^r\Sigma(n)$ used in formulas (4) is stable, i.e., it is the restriction of some stable R_0-coorientation of the whole of Σ. Therefore, to compute the differentials in the complex ω [resp. v] it suffices to restrict formulas (4), (5) to the subcomplex of the complex $\bigoplus_{n=1}^{\infty} \Omega(n)$ [resp. $\bigoplus_{n=1}^{\infty} N(n)$] generated by the sums of the classes ${}^r\Sigma(n)$ over all r and n. This immediately yields the homology of this complex and its generators, as well as constraints on the coexistence of singularities on compact Lagrangian manifolds: as in subsection 2.4, these constraints are specified by coboundaries in our complexes. All this information is given in the following table.

Theory	i	1	2	3	4	5	6
Lagrangian, Legendrian in ST^*V or in $J^1(M, \mathbb{R})$	$H^i(\omega)$	$\mathbb{Z}$	0	0	$\mathbb{Z}_2$	$\mathbb{Z}$	$\mathbb{Z}$
	generators	A_2	—	—	A_5	A_6 or E_6	P_8
	zeros	—	—	—	$2A_5$	$A_6 - E_6$	$E_7 + 3P_8$
	$H^i(v)$	$\mathbb{Z}_2$	$\mathbb{Z}_2$	$\mathbb{Z}_2$	$\mathbb{Z}_2$	$\mathbb{Z}_2$	$\mathbb{Z}_2$
	generators	A_2	A_3	A_4 or D_4	A_5	A_6 or E_6	E_7 or P_8, A_7
	parity	—	—	$A_4 + D_4$	D_5	$A_6 + E_6$, D_6	$E_7 + P_8$, D_7
Legendrian in PT^*V	$H^i(\omega(K_0))$	0	0	0	$\mathbb{Z}_2$	0	$\mathbb{Z}_2 \oplus \mathbb{Z}$
	generators	—	—	—	A_5	—	$E_7 + 3P_8$, P_8
	zeros	—	—	—	$2A_5$	—	$2E_7 + 6P_8$
	$H^i(v(K_0))$	$\mathbb{Z}_2$	$\mathbb{Z}_2$	$\mathbb{Z}_2^2$	$\mathbb{Z}_2$	$\mathbb{Z}_2^3$	$\mathbb{Z}_2^3$
	generators	A_2	A_3	A_4, D_4	A_5	A_6, D_6, E_6	A_7, E_7, P_8
	parity	—	—	—	D_5	—	D_7

Theorem ([361]). *The characteristic classes given in the table satisfy the following multiplicative relations: in the complex* ω, $A_2 \cup A_6 = 3P_8$ mod Torsion; *in the complex* v, $A_2 \cup A_3 = A_4$, $A_2 \cup A_6 = E_7$, $A_2 \cup A_7 = A_8$, $A_3 \cup A_5 = P_8 = E_7$, $A_5 = A_2 \cup \mathrm{Sq}^1(A_3)$; *in both complexes the square of any element is equal to zero.*

Theorem (see [41]). *All three free generators of the complex* ω *indicated in the table are realized, i.e., they define nontrivial invariants of oriented Lagrangian cobordism. Specifically, the class* A_6 *[resp.* P_8*] can assume any value that is a multiple of* 3 *[resp. any even value] on suitable compact* 5*-dimensional [resp.* 6*-dimensional] Lagrangian manifolds.*

There are three types of universal Legendrian complexes, corresponding to the theories of Legendrian submanifolds (and Legendrian cobordisms) in the space of 1-jets of functions $J^1(M, \mathbb{R})$, in the projective bundles (PT^*V), and in the sphere bundles (ST^*V) associated with cotangent bundles T^*V of manifolds. The classification of Legendrian singularities in these spaces derives from the classification of the singularities of their generating functions up to stable D_0-, K_0-, and K_0^+-equivalence, respectively; here K_0 [resp. K_0^+] is the group of transformations of the space of germs of functions $(\mathbb{R}^n, 0) \to (\mathbb{R}, 0)$ which acts, besides via local diffeomorphisms $(\mathbb{R}^n, 0) \to (\mathbb{R}^n, 0)$, through multiplication by smooth functions which are different from zero [resp. positive] in a neighborhood of zero. It follows that in the case of Legendrian submanifolds in $J^1(M, \mathbb{R})$ one obtains exactly the same complexes $\omega(D_0)$ and $\nu(D_0)$ as in the Lagrangian theory. The complexes ω, $\nu(K_0^+)$ have the same cohomology as ω, $\nu(D_0)$. The complexes ω, $\nu(K_0)$ are simplifications of ω, $\nu(K_0^+)$: the action of the additional elements of the group K_0 glues together certain D_0-classes (A_{2l+1}^+ and A_{2l+1}^-, D_{2l+1}^+ and D_{2l+1}^-, E_6^+ and $-E_6^-$, $P_9^{\times +}$ and $P_9^{\times -}$, X_9^+ and X_9^-) and destroys the coorientation on A_{4l+2} and E_8. The action of the coboundary operator in the resulting complexes is again obtained from formulas (4), (5) upon taking into account the indicated simplifications. The results on cohomology and relations obtained in this manner are shown in the lower part of the table given above.

2.6. On Universal Complexes of General Maps of Manifolds. All constructions of 2.2 and 2.5 carry over to the case of general maps $M \to N$, and each classification of such maps gives rise to its own universal complex. Thus, one can consider maps up to $\mathscr{A}$- or $\mathscr{K}$-equivalence (see §3.1), the admissible changes of coordinates may be required to preserve the orientation of the manifold M, or N, or both, one can consider only maps that satisfy some differential relations (see [164]), and so on. All these complexes define invariants of the corresponding (co)bordism theories; concrete computations in these complexes are an object of future research.

§3. Multiple Points and Multisingularities

3.1. A Formula for Multiple Points of Immersions, and Embedding Obstructions for Manifolds. Let $m < n$ and let f be a smooth map of an m-dimensional compact manifold M into an n-dimensional manifold N. Let N_k denote the set of the k-fold points of the set $f(M)$, i.e., the points $y \in N$ such that $\# f^{-1}(y) = k$, and let $M_k = f^{-1}(N_k)$. If f is a generic immersion, the closures of M_k and N_k are $(n - (n - m)k)$-dimensional cycles in M and N, respectively. The elements of $H^*(M; \mathbb{Z}_2)$ and $H^*(N; \mathbb{Z}_2)$ dual to these cycles will be denoted by m_k and n_k. Note that they not change under deformations of the immersion f. Also, let $e \in H^{n-m}(M; \mathbb{Z}_2)$ be the $(n - m)$th Stiefel-Whitney class of the normal bundle $\nu = f^*TN/TM$ over M.

Theorem (see [285]). *Let f be a generic immersion. Then*

$$m_k = f^*(n_{k-1}) - e \cup m_{k-1}. \tag{6}$$

If, in addition, the number $n - m$ is even and the bundle ν is orientable (and hence its Euler class $e \in H^{n-m}(M; \mathbb{Z})$ is well defined, see [249]), then all the m_k and n_k can also be defined as integer cohomology classes and formula (6) remains valid in $H^*(M; \mathbb{Z})$.

Example. For any immersion of M in a Euclidean space, $w(TM) \cup w(\nu) = 1$, i.e., $w(\nu) = (w(TM))^{-1} \in A(M)$. From this it follows that if the $(n - m)$-dimensional homogeneous component of the class $(w(TM))^{-1}$ is different from zero, then M cannot be embedded in $\mathbb{R}^n$. In fact, for any immersion $M \to \mathbb{R}^n$, in formula (6) with $k = 2$ the class $n_1 = 0$, and consequently the set M_2 of double points of f not only is nonempty, but even yields a nontrivial homology class of M (dual to $w_{n-m}(\nu)$).

Example (see [45]). Let f be a generic immersion of a compact surface M in $\mathbb{R}^3$. Then the number of triple points of f is congruent mod 2 to the Euler characteristic of M. In fact, by (6) this number is congruent mod 2 to the value of the square of the first Stiefel-Whitney class of the normal bundle to M on the fundamental cycle of M, and that value always coincides with the $\mathbb{Z}_2$-Euler characteristic.

3.2. Triple Points of Singular Surfaces. Suppose now that $f: M^2 \to \mathbb{R}^3$ is not necessarily an immersion, i.e., f has some singular points. If f is generic then these singular points are Whitney umbrellas (see subsection 3.1.2). Let us displace slightly the tip of each umbrella from the surface $f(M)$ in the direction indicated by the self-intersection line (Fig. 49). We obtain in this manner a set of points W in $\mathbb{R}^3 \setminus f(M)$. The number of points in W is even: a line of self-intersection of $f(M)$, emanating from an umbrella, can terminate only at another umbrella. Consequently, $W = 0$ in $H_0(\mathbb{R}^3; \mathbb{Z}_2)$.

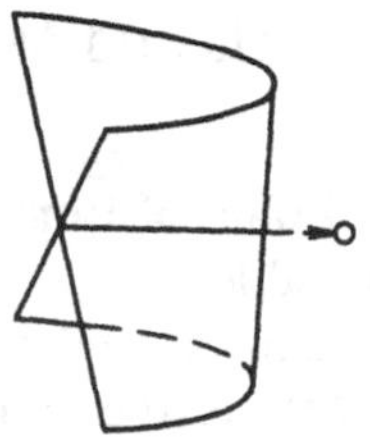

Fig. 49

Theorem ([321]). *The number of triple points of a generic map of a compact surface M into $\mathbb{R}^3$ is congruent* mod 2 *to the sum of the Euler characteristic of M and the linking index of the cycles $f(M)$ and W in $\mathbb{R}^3$.*

3.3. Multiple Points of Complex Maps. Let $f: M \to N$ be an algebraic map (not necessarily an immersion) of complex algebraic manifolds. In the present case the classes m_k are defined from the very beginning in the integer cohomology of M. Let $c(\nu) = 1 + c_1(\nu) + c_2(\nu) + \cdots$ denote the total Chern class of the normal bundle $\nu = f^*TN/TM$; as in subsection 1.4, we let $f_*: H^i(M) \to H^{i+(n-m)}(N)$ denote the Gysin homomorphism.

Theorem (see [197]). *If the proper algebraic map $f: M \to N$ is sufficiently general*[4], *then*

$$m_2 = f^*f_*(1) - c_{n-m}(v), \qquad (7)$$
$$m_3 = f^*f_*(m_2) - 2c_{n-m}(v)m_2 + \sum_{i=0}^{n-m+1} 2^{n-m-i}c_i(v)c_{2(n-m)-i}(v);$$

if $n = m + 1$, then

$$m_4 = f^*f_*m_3 - 3c_1(v)m_3 + 6c_2(v)m_2 - 6c_1(v)c_2(v) - 12c_3(v).$$

Note that the first of these formulas repeats formula (6) for $k = 2$, but already a blind transfer of the second formula to the real case would lead to a contradiction with the second example given in § 3.1.

3.4. Self-Intersections of Lagrangian Manifolds

Theorem (see [26]). *Let $M^{2l} \to (W^{4l}, \omega)$ be a generic Lagrangian immersion of a compact manifold in a symplectic manifold. Then*

$$M \times M = \chi(M) + 2\#, \qquad (8)$$

where $M \times M$ is the intersection index in the homology of W, χ is the Euler characteristic, and $\#$ is the number of self-intersection points. (In the case where M is not orientable, formula (8) *should be interpreted as a congruence* mod 2).

In particular, an orientable manifold with nonzero Euler characteristic does not admit Lagrangian immersions in $\mathbb{C}^{2l}$.

This theorem is very close to formulas (6) (the case $k = 2$) and (7): $M \times M$ plays the role of the term $f^*f_*(1)$, $2\#$ that of m_2, and $\chi(M)$ that of $e(v)$ (here it is important that for Lagrangian immersions the normal bundle is isomorphic to the cotangent one).

Givental' [149] generalized formula (8) to the case of an isotropic map $M^2 \to (W^4, \omega)$ possessing certain singular points – *"unfolded Whitney umbrellas"*. (Apparently, almost all isotropic maps $M^2 \to W^4$ have only such singularities.) Namely, in this case in the right-hand side of formula (8) one has to add the number of such singular points.

3.5. Complexes of Multisingularities. We have already seen that the number of Whitney umbrellas of a map of a compact surface into a manifold N is even, being equal to the number of endpoints of the lines of self-intersection of the

[4] Here "sufficiently general" means that the jet extensions of the map are assumed to be non-degenerate, and hence, in contrast to the real case, is not the same as "almost any": for example, in general a compact manifold does not admit "sufficiently general" maps into $\mathbb{C}^n$. An analogy with the real case arises here only in the case of Stein manifolds.

surface. Similarly, on a compact two-dimensional wavefront the number of swallowtails is even. These remarks can be generalized to higher dimensions by means of the complex of multisingularities of maps. This complex provides new bordism invariants and new constraints on the coexistence of singularities.

Suppose one is given a classification Γ of singularities of maps $\mathbb{R}^m \to \mathbb{R}^n$. Then it is natural to classify the k-fold points $y \in N_k$ of an arbitrary map $f: M \to N$ also according to the Γ-classes of singularities of f at the points of the set $f^{-1}(y)$; the classes of such a classification Γ^k on the set N_k are in correspondence with the unordered collections of k classes of the initial classification Γ. (In particular, Γ^0 reduces to a single class, which corresponds to the set $N \setminus f(M)$.) Such collections are called *classes* of *multisingularities.* If Γ satisfies certain regularity conditions (more restrictive than those imposed in subsection 2.2), then for almost any map $f: M \to N$ the partition $N = N_0 \cup N_1 \cup \cdots$ of the manifold N into the points of all possible classes of the classification Γ^k, $k = 0, 1, \ldots$, is a regular decomposition, and from its elements one can construct chains and cycles in N. The classes that they define in the homology of N correspond to the cohomology of the formal complex of multisingularities $\mathfrak{N}(\Gamma)$.

This complex is defined as the free $\mathbb{Z}_2$-module generated by all possible multiclasses $\{\sigma_1, \ldots, \sigma_k\}$, $k = 0, 1, \ldots$ of the classification Γ. The dimension of such a generator is equal to the sum of the codimensions in N of the subsets $f(\sigma_i)$, $i = 1, \ldots, k$, for a typical function f possessing such singularities. The coboundary operators are defined based on the geometry of the mutual adjacencies of multiclasses: the incidence coefficient of two multiclasses σ^1, σ^2 of codimensions l, $l + 1$ is equal to the parity of the number of components of the set $\{\sigma^1\} \subset N$ near any point of the set $\{\sigma^2\}$ of a typical map $f: M \to N$.

Example. In the case of maps $M^2 \to N^3$ the single generator of the group $\mathfrak{N}^2$ corresponds to self-intersection lines; the generators in $\mathfrak{N}^3$ are Whitney's umbrella and the triple point. Then δ (self-intersection line) = (Whitney's umbrella). In fact, near a triple point the set of double points is represented by an even number (six) of branches, and so the incidence coefficient [self-intersection line, triple point] is equal to zero. For Legendrian maps $M \to N$, the generators of the group $\mathfrak{N}^2$ are the self-intersection line and the cuspidal edge, while those of the group $\mathfrak{N}^3$ are the triple point, the transverse intersection of the cuspidal edge with the nonsingular part of the front, and the swallowtail. Here δ (cuspidal edge) = 0 and δ (self-intersection line) = (swallowtail).

Basic Theorem ([361]). 1. *Suppose the sum of l multisingularities $\sigma^1 + \cdots + \sigma^l$, $\sigma^i = \{\sigma^i_1, \ldots, \sigma^i_{k(i)}\}$, is a q-dimensional cocycle of the complex of multisingularities. Then for any nondegenerate map of a compact manifold M into N^n the union of the corresponding multisingularities in N^n forms an $(n\text{-}q)$-dimensional cycle* mod 2. *The class of this cycle in $H_{n-q}(N; \mathbb{Z}_2)$ does not change under homotopies of the map.*

2. *If $n = q$, i.e., the cycle in question is a collection of points, then its total parity is a bordism invariant of maps into N^n.*

3. *If, in addition,* $\sigma^1 + \cdots + \sigma^l$ *is a coboundary in the complex* $\mathfrak{N}(\Gamma)$, *then the number of these points is necessarily even.*

This theorem is valid for any theory of singularities and smooth bordisms (Lagrangian, Legendrian, ordinary smooth maps, etc.); the only restriction is that $\dim M \leqslant \dim N$. Assertions 4 and 5 of the main theorem of subsection 2.2, concerning bundle cobordisms, can also be generalized to the case of multi-singularities (and then in them one has to use all possible intersections of sets $\Sigma(f)$, defined for functions $E \to \mathbb{R}^n$); however, for foliations this method fails to produce invariants.

The cohomology and generators of the complexes $\mathfrak{N}(R_0) \cong \mathfrak{N}(K_0^+)$ and $\mathfrak{N}(K_0)$ of multisingularities of Legendrian manifolds in $J^1(W, \mathbb{R})$, or in ST^*N and in PT^*N are given in the following table.

i	1	2	3	4	5	6	7
$\dim H^i(\mathfrak{N}(K_0^+))$ generators	1 A_1	1 A_2	1 A_1A_2	1 $A_4 - D_4$	1 $A_1A_4 = A_1D_4$	2 $A_6 = E_6$, $A_2A_4 = A_2D_4$	3 $E_7 = P_8$, $A_1A_6 = A_1A_6$, $A_1A_2A_4 = A_1A_2D_4$
$\dim H^i(\mathfrak{N}(K_0))$ generators	1 A_1	1 A_2	1 A_1A_2	2 A_4, D_4	2 A_1A_4, A_1D_4	5 A_6, D_6, E_6, A_2A_4, A_2D_4	7 E_7, P_8, A_1A_6, A_1D_6, A_1E_6, $A_1A_2A_4, A_1A_2D_4$

Computations in the above complexes yield the following conclusions.

Theorem ([358], [361]). 1. *On a general wavefront in* N^3 [*resp.,* N^5, N^7] *the number of points of type* A_3 [*resp.* A_5, A_7] *is even.*

2. *On a wavefront in* N^6 *the number of points* E_6 *is congruent* mod 2 *to the number of points at which the one-dimensional stratum* A_5 *pierces the front.*

3. *The parity of the number of multisingularities* A_1A_2, A_1A_4, A_1D_4, A_2A_4, A_2D_4, A_1A_6, A_1D_6, A_1E_6, $A_1A_2A_4$, $A_1A_2D_4$ *is a cobordism invariant for wavefronts in manifolds of respective dimensions* 3, 5, 5, 6, 6, 7, 7, 7, 7, 7. *In the case of fronts in* $J^1(W^{n-1}, \mathbb{R})$ *(or, generally, equipped* $(n-1)$*-dimensional fronts, see* [26]), *the parities of the numbers of multisingularities* A_1A_2, $A_1A_4 \simeq A_1D_4$, $A_2A_4 \simeq A_2D_4$, $A_1A_6 \simeq A_1E_6$, $A_1A_2A_4 \simeq A_1A_2D_4$ *for the respective values* 3, 5, 6, 7, 7 *of* n *are cobordism invariants.*

The characteristic number A_1A_2 is realized, for example, if $N^3 = \mathbb{R}P^2 \times S^1$ and the front consists of two components: (a semicubic in $\mathbb{R}P^2$) $\times S^1$ and $\mathbb{R}P^2 \times$ (a point in S^1).

Assertions 1 and 2 correspond to the following formulas for the differentials of the complex: $\delta(A_1A_1) = A_3$, $\delta(A_2A_2) = A_5$, $\delta(A_3A_3) = A_7$, $\delta(A_1A_2A_2) = E_6 + A_1A_5$.

Assertion 1 concerning wavefronts of Legendrian maps carries over word-for-word to the case of caustics of Lagrangian maps via the following theorem, where p and p' denote the canonical projections $T^*N \to N$ and $J^1(N, \mathbb{R}) \to N$.

Theorem ([361]). *For any Lagrangian immersion* $i: M \to T^*N$ *there exists a Legendrian immersion* $i': M \to J^1(N, \mathbb{R})$ *with the property that the maps* $p \circ i$ *and* $p' \circ i'$ *coincide, and consequently all Lagrangian singularities of the manifold* $i(M)$ *coincide with the Legendrian singularities of the same type of the manifold* $i'(M)$.

3.6. Multisingularities and Multiplication in the Cohomology of the Target Space of a Map. The characteristic number $A_1 A_2$ is no longer realized on fronts in $J^0(M, \mathbb{R})$ or in $\mathbb{R}^3$: it is even for any such compact front (see [6]). In fact, the closure of the cuspidal edge A_2 can be displaced from the front so that it will intersect the front only near points at which the front intersected the original edge transversely (see Fig. 50). Consequently, the number of piercing points is congruent mod 2 to the cup product of the one- and two-dimensional cohomology classes of the space $J^0(M, \mathbb{R})$ (or $\mathbb{R}^3$) that are dual to the front and the cuspidal edge, respectively. But the cup product in the cohomology of these two spaces is trivial, which proves our assertion.

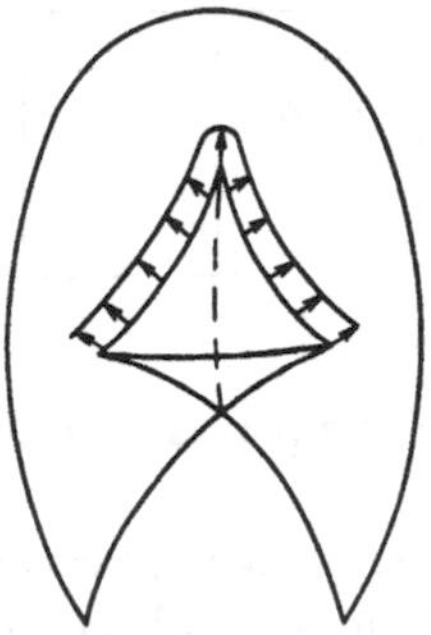

Fig. 50

In the case of more complex singularities Σ, the cocycle $A_1\Sigma$ dual to the set of transverse intersection of Σ with the front, does not necessarily coincide with the product of the cocycles dual to the front A_1 and the cycle Σ. More precisely, let N be the base of the Legendrian bundle, i.e., the manifold containing the front in question. Let $[\Sigma]$ denote the difference, taken in $H^*(N; \mathbb{Z}_2)$, of the cocycles $A_1\Sigma$ and $A_1 \cup \Sigma$.

Theorem ([361]). $[A_2] = 0$, $[A_3] = 0$, $[A_4] = 0$, $[A_5] = E_6$, $[A_6] = E_7$, $[A_7] = 0$, $[D_4] = 0$, $[D_5] = 0$, $[D_6] = E_7$, $[D_7] = 0$, $[E_6] = E_7$, $[P_8] = 0$.

Example. Suppose $\dim N = 7$, and consequently the closure of any of the sets A_6, D_6, E_6 is a curve in N. If, in addition, the multiplication $H^1(N; \mathbb{Z}_2) \otimes H^6(N; \mathbb{Z}_2) \to H^7(N; \mathbb{Z}_2)$ is trivial, then the number of points at which any of these lines pierces the front is congruent mod 2 to the number of singularities of type E_7. This follows from the fact that any of these lines can be displaced from the front so that it will intersect the front only near previous points of transverse intersection and in addition will have an odd [resp. even] number of new intersections near each point of type E_7 [resp. near the other points of codimension 7, i.e., A_7, D_7 and P_8].

§4. Spaces of Functions with Critical Points of Mild Complexity

The natural stratification of the space of germs of smooth functions generates a large number of invariants of smooth manifolds. In fact, let us fix some stratum (or set of strata) S and consider the space $\mathscr{F}_{<S}(M)$ of all smooth functions on the manifold M which have only critical points "simpler than S" (i.e., such that their germs at any point do not belong to S or its closure). The homotopy invariants of the spaces $\mathscr{F}_{<S}(M)$ (and of the natural mappings between such spaces with different choices of S) are smooth invariants of the manifold M. Analogous constructions can be carried out for multisingularities.

Despite the obvious importance of these invariants, little has been done in this direction because their computation involves considerable difficulties (even the problems concerning the space of Morse functions on M turned out to be extremely difficult, see [300]). We formulate below some of the simplest results on spaces of functions on one-dimensional manifolds ($M = \mathbb{R}^1$ or S^1) and on their local analogues (the spaces $\mathbb{R}^m \setminus \Sigma_{\geqslant S}$ of polynomials $x^{m+2} + \lambda_1 x^m + \cdots + \lambda_m x$ with critical points simpler than S in the real and complex domain); see also [32].

4.1. Functions with Singularities Simpler than A_3. Let us consider the spaces of C^∞-functions on the line that coincide for $|x| \geqslant 1$ either with x (denoted by $\mathscr{F}^{\mathrm{ev}}$) or with x^2 (denoted by $\mathscr{F}^{\mathrm{odd}}$).

Theorem. *$\pi_1(\mathscr{F}_{<A_3}(\mathbb{R})) \simeq \mathbb{Z}$ in both cases* ev *and* odd; *$\pi_1(\mathscr{F}_{<A_3}(S^1)) \simeq \mathbb{Z}$.*

To formulate the local analogue of this theorem, we consider the space $\mathbb{R}^m - A_3$ of polynomials with critical points of multiplicity $\leqslant 2$ (i.e., of complexity at most that of x^3).

Theorem. *$\pi_1(\mathbb{R}^m \setminus A_3) \simeq \mathbb{Z}$ for $m \geqslant 2$.*

Example. For $m = 3$ the set A_3 is the cuspidal edge of a swallowtail in $\mathbb{R}^3$, and consequently its complement is homotopy equivalent to a circle.

The set A_3 is transversely oriented (in the local case as well as in the global case of a one-dimensional manifold M); consequently, the "number of rotations" – the linking coefficient with A_3 of a closed path that goes around this cycle, $\mathrm{ind} \in H^1(F_{<A_3}; \mathbb{Z})$ [resp. $H^1(\mathbb{R}^m - A_3; \mathbb{Z})$] is defined.

Theorem. *The Maxwell stratum in the space of functions on the line [resp. in the space of polynomials, $\mathbb{R}^m$] is coorientable. Its boundary (as a chain with closed supports), considered with its coorientation, is the naturally cooriented cycle A_3 of the functions [resp. polynomials] with a critical point of multiplicity higher than 2 [of type $x^{\geq 4}$].*

The index of a path that goes around A_3 is equal to the intersection index of the curve with the naturally cooriented Maxwell stratum.

The natural coorientation of the Maxwell stratum is defined as follows. If $x < y$ are two Morse critical points of f with common critical value, then on one side of the stratum the value of f at a critical point close to x is larger than at a critical point close to y. That side is declared to be positive if the points x and y are of the same type (i.e., both are maxima or minima), and negative if they are of distinct type.

Example. In the case $m = 3$ the Maxwell stratum intersects the plane $\lambda_1 = -1$ along an arc of curve with two cuspidal points and one self-intersection point, and with the endpoints in A_3. The recipe described above yields the same coorientation as $\partial/\partial\lambda_3$ on the arc of curve between the cuspidal points, and the opposite coorientation from the cuspidal points to the endpoints of the arc (Fig. 51). The compatibility of the coorientations in neighborhoods of the cuspidal points in this example proves the coorientability of the Maxwell stratum of functions on the line.

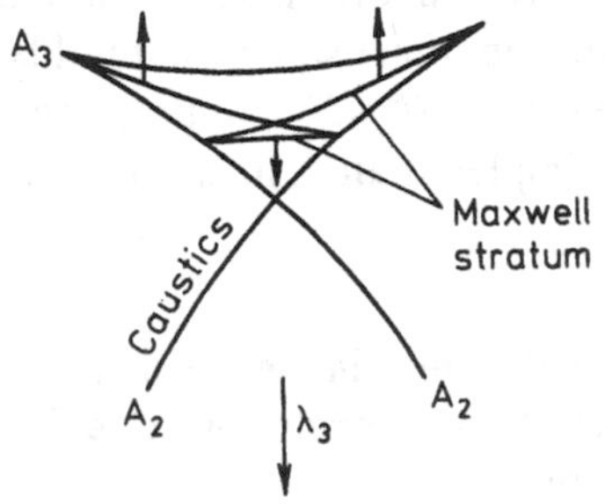

Fig. 51

4.2. The Group of Curves Without Horizontal Inflexional Tangents. The groups π_1 and H^1 of the complement of A_3 in spaces of functions on the line may be interpreted as groups of plane curves without horizontal inflexional tangents. In fact, any path $t \mapsto f_t$ in the complement of A_3 in a space of functions on the line defines a function of two variables, $F(t, x) = f_t(x)$. The equation $\partial F/\partial x = 0$ defines a curve in the plane fibred over the t-axis. One can choose the path so that it starts and terminates at the point x (for the case $\mathscr{F}^{\text{ev}}$) or x^2 (for the case $\mathscr{F}^{\text{odd}}$). Then in the first case the corresponding curve is closed, and in the second case one can assume that for $x^2 + t^2 > R$ it coincides with the axis $x = 0$.

To paths that avoid the points of A_3 there correspond curves without points with horizontal inflexional tangents. We say that two such curves are *equivalent* if they are carried into one another by isotopies that are the identity map outside a sufficiently large disk (in the class of curves without horizontal inflexional tangents), and Morse perestroikas (without horizontal separatrices of saddles). The resulting equivalence classes naturally form a group (for $\mathscr{F}^{\mathrm{ev}}$ it is a group of classes of closed curves), in which the group operation is that of taking the disjoint union of curves lying in the half-planes $t \geqslant 0$ and $t \leqslant 0$.

Theorem. *The groups of classes of plane curves defined above are isomorphic to $\mathbb{Z}$. For $\mathscr{F}^{\mathrm{ev}}$ a generator is the class of the kidney-shaped curve $t = x^2 \pm \sqrt{1 - x^2}$, and for $\mathscr{F}^{\mathrm{odd}}$ a generator is the disjoint union of the kidney-shaped curve and the axis $x = 0$* (Fig. 52).

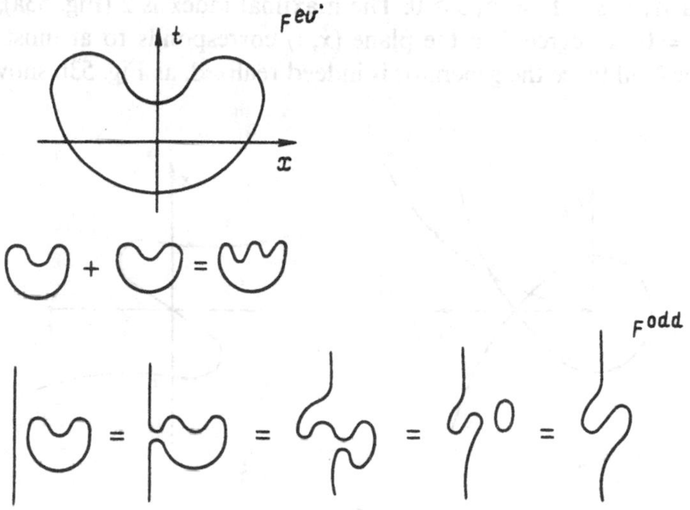

Fig. 52

To compute which multiple of the generator a given curve is (i.e., the index of the corresponding path in the function space) one can proceed as follows. Consider, for the sake of simplicity, the case of closed curves, $\mathscr{F}^{\mathrm{ev}}$. Pick a function $\Phi(t, x)$ that is negative at infinity and has a simple zero along the given curve (which is assumed to be smooth and to have no inflexional tangents parallel to the x-axis). Now form the "*double*" – the closed compact surface $z = \Phi(t, x)$ in $\mathbb{R}^3$.

Theorem. *The curve $\Phi = 0$ is equivalent to m times the kidney-shaped generator, where m is the difference of the numbers of local maxima and local minima of the function t on the double.*

(An analogous description is available for the case $\mathscr{F}^{\mathrm{odd}}$).

Example. For the generator the number in question is $2 - 1 = 1$.

Remark 1. The foregoing theory can be used not only to compute the number of rotations of a curve around the stratum A_3 or its intersection index with the Maxwell stratum, but also, taking the opposite direction, to obtain estimates in real algebraic geometry. Specifically, via a suitable interpretation of the difference of the number of maxima and the number of minima of the function t on the double, constructed for a plane algebraic curve of given degree, it becomes possible to estimate this number from above as the number *ind* of rotations of the corresponding algebraic curve in the space of polynomials in one variable of the fixed degree.

Example. When $\Phi = 0$ is a curve of degree 3, the problem reduces to the determination of the maximal intersection index of a curve $\Lambda_1 = p_2(t)$, $\Lambda_3 = p_3(t)$ (where p_r is a polynomial of degree r) with half of the semicubic parabola $\Lambda_1 = s^2$, $\Lambda_2 = s^3$, $s \geqslant 0$. The maximal index is 2 (Fig. 53a), and so a curve $\Phi = 0$ of degree 3 in the plane (x, t) corresponds to at most twice the generator (and twice the generator is indeed realized, as Fig. 53b shows).

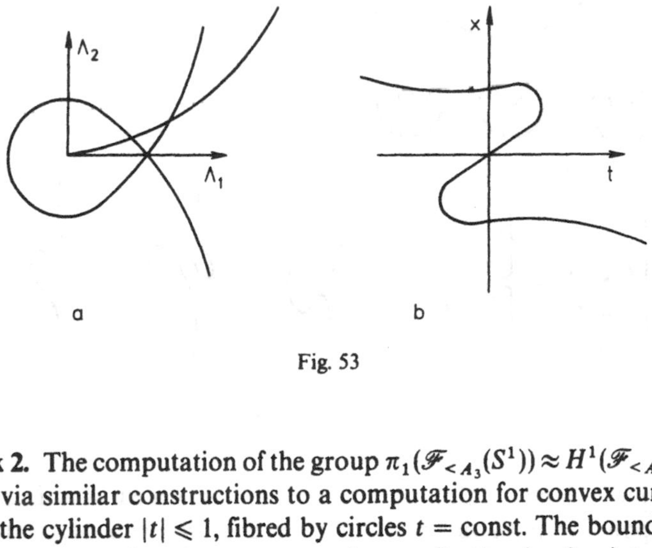

Fig. 53

Remark 2. The computation of the group $\pi_1(\mathcal{F}_{<A_3}(S^1)) \approx H^1(\mathcal{F}_{<A_3}(S^1); \mathbb{Z}) \approx \mathbb{Z}$ reduces via similar constructions to a computation for convex curves on the surface of the cylinder $|t| \leqslant 1$, fibred by circles $t = \text{const}$. The boundary conditions for $t = \pm 1$ are that the curve must have a fixed pair of points on each of these two circles.

This computation shows, in particular, that the four-vertex theorem (which asserts that the caustic of a plane convex curve has at least four cuspidal points, cf. [31]) does not carry over to the case of Lagrangian cylinders in $T^*\mathbb{R}^2$ with boundary conditions of the cylinder being unit vectors normal to the curve: the Lagrangian projection map of a cylinder in general position with the same boundary conditions can have less than four cusps.

Incidentally, we should mention that there remains the possibility of estimating from below the number of cuspidal points of caustics (envelopes of the family of geodesics normal to a curve) of Riemannian metrics (or, more generally, the number of cusps on an optical Lagrangian manifold; see [30], [76]). To obtain such an estimate it suffices to establish some analogue of Bennequin's inequality for contact structures on the solid torus $S^1 \times D^2$.

4.3. Homotopy Properties of the Complements of Unfurled Swallowtails

Definition. The *unfurled swallowtail* $\Sigma^n(\mathbb{R}^N)$ is the n-dimensional manifold in the N-dimensional space of the polynomials

$$x^{N+1} + \lambda_1 x^{N-1} + \cdots + \lambda_N, \quad \lambda \in \mathbb{R}^N, \quad N \geqslant 2n,$$

consisting of the polynomials that have a root of comultiplicity $\leqslant n$ (i.e., admit a representation $(x - A_0)^{N+1-n}(x^n + A_1 x^{n-1} + \cdots + A_n)$, with $N \geqslant 2n$ and $A_0 \in \mathbb{R}$).

Theorem. *The complement of the unfurled swallowtail Σ^n in $\mathbb{R}^N$ is homotopy equivalent to a sphere S^{N-n-1} linked with it and is diffeomorphic to $\mathbb{R}^N \setminus \mathbb{R}^n$.*

Example. The curve cut on the three-dimensional sphere $|\lambda|^2 = 1$ by the two-dimensional unfurled swallowtail $\Sigma^2(\mathbb{R}^4) = \{\lambda \in \mathbb{R}^4 \colon x^5 + \lambda_1 x^3 + \lambda_2 x^2 + \lambda_3 x + \lambda_4 = (x - a)^3(x^2 + 3ax + b)\}$ is not knotted.

Remark 1. All unfurled swallowtails of a fixed codimension are diffeomorphic (this is a theorem of Givental' [146]); in particular, they are diffeomorphic to $\Sigma^n(\mathbb{R}^{2n})$ and homeomorphic to $\mathbb{R}^n$. However, the fact that the embedding of Σ^n in $\mathbb{R}^{2n}$ (more precisely, of the link $L^{n-1} = \Sigma^n \cap S^{2n-1}$ in S^{2n-1}) is not "knotted" is not obvious; its proof is based on the construction of a polynomial diffeomorphism of $\mathbb{R}^N$ which takes Σ^n into the graph of a continuous map $\mathbb{R}^n \to \mathbb{R}^{N-n}$, and on the connection between the polynomials of degree $2n + 1$ with a root of multiplicity $n + 1$ and the polynomials of degree $2n$ with n pairs of multiple roots (cf. [354, Theorem 10]).

Remark 2. It follows from the preceding theorem that the homotopy groups of the complement of the manifold of polynomials in one variable with a critical point of multiplicity ν equal to or larger than half the degree d are precisely the groups $\pi_i(S^{d-\nu-2})$. The rich geometric structure of the stratified manifold Σ^n can be exploited to introduce additional structures in $\{\pi_i(S^j)\}$. The homotopy classes of maps in $\mathbb{R}^m - A_\mu$ can be interpreted as cobordisms, so that these spaces are analogous to Thom spectra.

Remark 3. Apparently, one has the homotopy equivalence $\mathbb{C}^N \setminus \Sigma^n(\mathbb{C}^N) \sim S^{2(N-n)-1}$ for $N \geqslant 2n$, $N \geqslant 3$.

Remark 4. Apparently, the natural embeddings of versal deformations generate a stabilization of the homotopy groups and cohomology rings $\pi_i(\mathbb{R}^N \setminus A_\mu) \to \pi(i, \mu)$, $H^*(\mathbb{R}^N \setminus A_\mu) \to H^*(\mu)$ as $N \to \infty$ (and an analogous stabilization in the complex case);

$\pi_i(\mathbb{R}^N \setminus A_\mu) = 0$ for $i < \mu - 2$, and $\mathbb{Z}$ for $i = \mu - 2$;

$H^i(\mathbb{R}^N - A_\mu) = 0$ for $i \neq r(\mu - 2)$ (see also [32] and subsection 5.7 below).

§ 5. Elimination of Singularities and Solution of Differential Conditions

This section is devoted to problems of the following kind. Suppose one is given a smooth map $f: M \to N$. Does there exist a map homotopic to f which has no singularities of some prescribed type Σ? A negative answer to this question is usually connected with the nontriviality of the class $[\Sigma]$ dual to Σ in the cohomology of M, see §§ 1, 2. One can ask whether this obstruction is the only one: if $[\Sigma] = 0$, can one eliminate the singularity Σ by means of a smooth homotopy? Does a map $M \to N$ without singularities of type Σ exist at all? For example, does an immersion $M \to \mathbb{R}^n$ exist if all obstructions described in subsection 1.4 are equal to 0? What is the structure of the set of all maps that have only singularities of prescribed types?

5.1. Cancellation of Whitney Umbrellas and Cusps. The Immersion Problem

Theorem (Whitney's strong immersion theorem [377]). *Any m-dimensional compact manifold M, $m > 1$, can be immersed in $\mathbb{R}^{2m-1}$.*

In fact, one can readily construct an embedding of M in a Euclidean space of sufficiently high dimension; then, projecting the resulting submanifold parallel to an $(N - 2m + 1)$-dimensional plane in general position one obtains a map $f: M \to \mathbb{R}^{2m-1}$ that fails to be an immersion at only a finite number of points. Near the corresponding points in $\mathbb{R}^{2m-1}$ the surface $f(M)$ is a generalized Whitney umbrella; see § 3.1. From each such point there emanates a self-intersection line, which can terminate only at another umbrella; see Fig. 49. The map f can then be deformed near the preimage of this self-intersection line so that the umbrellas will "attract" each other closer along that line, after which they can be eliminated through a standard perestroika.

The next example is the elimination of Whitney cusps. A generic map $f: M^m \to V^2$ has only folds Σ^1 and cusps $\Sigma^{1,1}$; the codimensions of these sets are $m - 1$ and m, respectively. Suppose that M and V are orientable and that M is closed (i.e., compact and $\partial M = \varnothing$). Then, according to [324], the dual mod 2 cohomology classes coincide with the $(m - 1)$st and the mth Stiefel-Whitney classes of M, respectively; in particular, the number of cusps is mod 2 congruent to the Euler characteristic $\chi(M)$.

Theorem ([215]). A. *If $m > 2$, then f is homotopic to a map that has exactly one cusp for $\chi(M)$ odd and has no cusps for $\chi(M)$ even (in particular, for odd m).*

B. *If $m = 2$, then f is homotopic to a map with the property that on the closure of each fold line there are at most two cusps.*

This theorem, too, is proved by means of mutual "cancellation" of unnecessary cusps. The modern algebraic-topological technique allows one to generalize considerably the results given above; in particular, in [122] it is proved that assertion A of the last theorem is also valid for $m = 2$.

Theorem ([86]). *Any compact m-dimensional manifold can be immersed in* $\mathbb{R}^n$ *provided* $n \geqslant 2m - d(m)$, *where* $d(m)$ *is the number of units in the dyadic expansion of* m (cf. §2.6.2).

This last estimate is sharp: if $m = 2^a + 2^b + \cdots$, then the obstruction described in subsection 1.4 prohibits an immersion of the manifold $\mathbb{R}P^{2^a} \times \mathbb{R}P^{2^b} \times \cdots$ in $\mathbb{R}^{2m-d(m)-1}$. We shall return to the problem of elimination of singularities in subsection 5.5.

5.2. The Smale-Hirsch Theorem

Definition. A *homomorphism of tangent bundles* $TM \to TN$ consists of a continuous map $h: M \to N$ and a homomorphism of linear spaces $T_xM \to T_{h(x)}N$ given for each point $x \in M$ and depending continuously on x, such that the corresponding map $TM \to TN$ of the total spaces is continuous and linear on fibres. A homomorphism of tangent bundles is called a *monomorphism* if the homomorphism between tangent spaces that it defines is a monomorphism at any point x.

For example, the 1-jet extension of any smooth map $f: M \to N$ is a homomorphism of tangent bundles. The homomorphisms obtained in this manner are said to be integrable. According to the definition, an integrable homomorphism is a monomorphism if and only if the map f is an immersion.

A homotopy of maps $M \to N$ [resp. of immersions, or of homomorphisms or monomorphisms $TM \to TN$] is, by definition, a continuous map of the segment [0, 1] into the space of maps [resp. of immersions, or of homomorphisms or monomorphisms], equipped with a suitable topology.

Theorem. *If* $m < n$ *and the manifold* N *has no boundary, then the canonical embedding* $f \mapsto j^1 f$ *of the space of immersions* $M \to N$ *in the space of monomorphisms* $TM \to TN$ *is a weak homotopy equivalence* (*i.e., it induces an isomorphism of the i-dimensional homotopy groups of these spaces for all i*).

In the case where M is a sphere and $N = \mathbb{R}^n$, this is proved in [311]. The generalized statement given above is due to Hirsch [180]. This theorem settles the differential-topological aspect of the immersion problem and reduces to a purely algebraic problem, formulated as follows.

If the manifold M admits an immersion in the Euclidean space $\mathbb{R}^n$, then the tangent bundle TM has n continuous sections which span the tangent space T_xM at each point x: these sections are given by the orthogonal projections of the unit basis vectors in $\mathbb{R}^n$ on the tangent plane to M. From the Smale-Hirsch theorem it follows that, conversely, the existence of such a set of sections guarantees the existence of an immersion. For example, in the last theorem of subsection 5.1, it would have been enough to verify the existence of such sections for any manifold.

5.3. The w.h.e.- and h-Principles. Suppose that a subset Ω is singled out in the space $J^k(M, N)$ of k-jets of maps $M \to N$. Let $B(\Omega)$ be the space of the sections of the projection $J^k(M, N) \to M$ that map M into Ω, and let $A(\Omega)$ be the space of the smooth maps $M \to N$ whose k-jet extensions belong to $B(\Omega)$. The image of $A(\Omega)$ under the k-jet extension mapping is called the *set of integrable sections* in $B(\Omega)$.

Definition ([164]). A *differential condition* (or *relation*) of order k is an arbitrary subset $\Omega \subset J^k(M, N)$. The condition Ω is said to satisfy the *w.h.e.-principle* [resp. *h-principle*] if the embedding $A(\Omega) \to B(\Omega)$ is a weak homotopy equivalence [resp. induces a group epimorphism $\pi_0(A(\Omega)) \to \pi_0(B(\Omega))$, i.e., in any homotopy class of sections from $B(\Omega)$ there is an integrable section].

For example, the Smale-Hirsch theorem for immersions in $\mathbb{R}^n$ can be reformulated as follows: the condition $\Omega \subset J^1(M, \mathbb{R}^n)$ consisting of the monomorphisms $T_xM \to T_y\mathbb{R}^n$ for all $x \in M$, $y \in \mathbb{R}^n$ satisfies the w.h.e.-principle if $\dim M < n$.

The validity of the w.h.e.- or h-principles for a multitude of differential conditions was established by Feit [126], Phillips [265], [266], Poénaru [274], Gromov [163]–[166], Gromov and Eliashberg [167], Eliashberg [122]–[124], and Lees [211]; a list of the results in this direction obtained by 1970 is given in [164]; a detailed exposition of these results and their development is contained in the monograph [166]. We give some of them below.

Theorem. *The w.h.e.-principle holds for maps $M \to N$ that have rank at least k at any point $x \in M$ if $k <$ dim N, or if $k =$ dim N and the manifold M is open (has no closed connected components).*

In the case where M is open this was established in [181] for $k < \dim N$ and in [265] for $k = \dim N$; the proof for the case $k < \dim N$ and M arbitrary was given in [126]. For $k = \dim N$ the assumption that M is open is essential: for example, there exist "nonintegrable maps" $S^1 \to \mathbb{R}^1$ with no critical points.

Theorem ([167]). *Suppose given a vector subbundle $S \subset TM$ and assume that either $k <$ dim N or $k =$ dim N and M is open. Then the h-principle holds for maps $f: M \to N$ such that the restriction of the differential of f to S has everywhere* rank $\geqslant k$.

Theorem ([122]). *Suppose M is connected, C is a nonempty codimension-one submanifold of M, and $m = n$. Then the h-principle holds for maps $M \to N$ that have a fold $\Sigma^{1,0}$ on C and no other singularities on M.*

In the last case the epimorphism $\pi_0(A(\Omega)) \to \pi_0(B(\Omega))$ is not an isomorphism. To show this we use an example, due to Milnor, of two nonhomotopic immersions of the disk D^2 in $\mathbb{R}^2$ which coincide on ∂D^2, see Fig. 54. Assume the sphere S^2 is standardly embedded in $\mathbb{R}^3$. Let φ be the orthogonal projection $S^2 \to \mathbb{R}^2$ and let ψ be the map whose restriction to the upper [resp. lower] hemisphere is the composition of φ and the immersion shown in Fig. 54a [resp. 54b].

Proposition ([122]). *The maps φ and ψ are not homotopic in the class of all maps that have a fold on the equator of the sphere and no other singularities, but their 2-jet extensions are homotopic in the class of all sections of the*

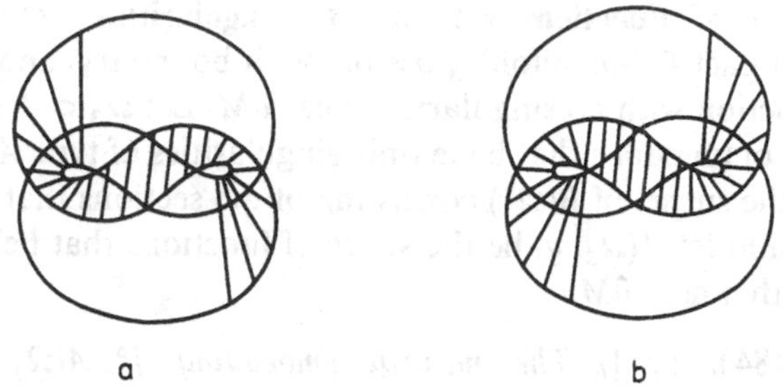

Fig. 54

bundle $J^2(S^2, \mathbb{R}^2) \to S^2$ *that map the equator into the set* $\Sigma^{1,0}$ *and are nonsingular elsewhere.*

Theorem ([266], [163]). *Suppose that* M *is open and that a smooth foliation* $\mathscr{F}$ *is given on the manifold* N. *Let* $\Omega \subset J^1(M, N)$ *denote the set of* 1*-jets of the maps that are transverse to* $\mathscr{F}$. *Then the w.h.e.-principle holds for* Ω.

5.4. The Gromov-Lees Theorem on Lagrangian Immersions. Assume that $\dim M = \dim X$, and let T^*X be the total space of the cotangent bundle of X, equipped with the canonical 1-form ω^1 ("*pdq*", see [15]) and the symplectic form $\omega^2 = d\omega^1$. A homomorphism $h\colon TM \to T(T^*X)$ is called a *Lagrangian homomorphism* if for any point $x \in M$ the corresponding linear map $T_xM \to T_{h(x)}(T^*X)$ is injective and its image is a Lagrangian plane (i.e., the form ω^2 vanishes on the image). An immersion $\varphi\colon M \to T^*X$ is called a *Lagrangian immersion* if its 1-jet extension is a Lagrangian homomorphism, and is called an *exact Lagrangian immersion* if, in addition, the integral of the form $\varphi^*(\omega^1)$ along any closed path in M is equal to zero.

Theorem ([164], [211]). *The obvious embedding of the space of exact Lagrangian immersions* $M \to T^*X$ *in the space of Lagrangian homomorphisms* $TM \to T(T^*X)$ *establishes a one-to-one correspondence between the connected components of these spaces.*

5.5. Elimination of Thom-Boardman Singularities. Let us order the Thom-Boardman symbols $I = \{i_1, \ldots, i_k\}$ lexicographically. Denote by $\Omega^I \subset J^k(M, N)$ the set of jets of maps that have no singularities of type Σ^L for $L \geqslant I$.

Definition (see [270]). The differential condition Ω^I is said to be *prolongable* if there exists an open, $\mathscr{A}$-invariant set $\Omega \subset J^k(M \times \mathbb{R}, N)$ that is taken into Ω^I by the canonical restriction mapping $J^k(M \times \mathbb{R}, N) \to J^k(M, N)$.

Theorem ([270]). *If the condition* Ω^I *is prolongable, then it satisfies the w.h.e.-principle in the case of closed (as well as open) manifolds* M.

Theorem ([269]). *The condition* Ω^I, $I = \{i_1, \ldots, i_k\}$, *is prolongable if* $i_k > \dim M - \dim N - d^I$, *where* $d^I = \sum_{s=1}^{k-1} \alpha_s$ *and* α_s *is equal to* 1 *if* $i_s - i_{s+1} > 1$ *and to* 0 *otherwise.*

5.6. The Space of Functions with no A_3 Singularities. Let M be an m-dimensional compact C^∞-manifold (possibly with boundary) and let $g: M \to \mathbb{R}$ be a smooth function with no singularities near ∂M. Let $\Omega_3 \subset J^3(M, \mathbb{R})$ denote the space of jets of functions that have only singularities of type A_1 or A_2. Also, let $B(\Omega_3, g)$ be the subset of $B(\Omega_3)$ consisting of the sections that coincide with $j^3(g)$ near ∂M, and let $A(\Omega_3, g)$ be the space of functions that belong to $A(\Omega_3)$ and coincide with g near ∂M.

Theorem ([184], [77]). *The natural embedding $j^3: A(\Omega_3, g) \to B(\Omega_3, g)$ induces an isomorphism of the i-dimensional homotopy groups of these spaces for $i < m$, and an epimorphism for $i = m$.*

Example. Let $M = S^1$. Then the fibration $\Omega_3 \to S^1$ is homotopy equivalent to the trivial fibration with fibre S^2, and $\pi_1(B(\Omega_3)) \cong \mathbb{Z}$. From the theorem of subsection 4.1 it follows that in this case the epimorphism $\pi_1(A(\Omega_3)) \to \pi_1(B(\Omega_3))$ is an isomorphism.

5.7. Let us state some more recent results on the theme of §4.4, 4.5, see [363], [364].

Theorem. *For any $k \geqslant 3$ the space $\mathscr{F}_{<A_k}(\mathbb{R}^1)$ is homology equivalent to the loop space ΩS^{k-1}, and the space $\mathscr{F}_{<A_k}(S^1)$ is homology equivalent to the space of continuous maps $S^1 \to S^{k-1}$. Moreover, for k24 these homology equivalences are the corollaries of the weak homotopy equivalences of these spaces. For any $m \geqslant k \geqslant 3$ there exists an embedding of the space $\mathbb{R}^m - A_k$ in $\mathscr{F}_{<A_k}(\mathbb{R}^1)$ which induces an isomorphism of the homology groups of dimension less than or equal to $(k-2)[(m+1)/k]$.*

Theorem. *Under the assumptions of the theorem of the preceding subsection* 5.6, *the embedding j^3 is a homology equivalence.*

Remark. In [363], [364] a stronger statement was announced, namely that j^3 in the last theorem is a weak homotopy equivalence. This assertion seems to be plausible – and is generally believed to be true – but in fact it is not proved: in [364] it is based on a too optimistic use of Whitehead's theorem.

For other results of this type see also [364 c].

§ 6. Tangential Singularities and Vanishing Inflexions

6.1. The Calculus of Tangential Singularities. Let V be a smooth k-dimensional surface in $\mathbb{C}P^n$ or $\mathbb{R}P^n$. By assigning to each point of V the tangent space to V at that point we obtain the Gauss map of the manifold V in the Grassmann manifold G_{n+1}^{k+1} of all k-dimensional subspaces of P^n. This map may have singularities: these correspond to the points at which the curvature form of V degenerates. In the simplest case of curves in P^2, the singularities are the points of inflexion of the given curve, and their images in the Grassmannian $G_3^2 \cong P^2$ are the cusp points of the dual curve. The singular points of Gauss maps form cycles in V, and one is led to the problem of computing the cohomology classes of V dual to these cycles.

In the case of plane curves the answer is given by the following formula (see, e.g., [197]): the number of points of inflexion of a nonsingular curve of degree d in $\mathbb{C}P^2$, counting multiplicities, is equal to $3d(d-2)$.

In the next case according to complexity, that of surfaces in P^3, the classification of tangential singularities was obtained by O.A. Platonova and O.P. Shcherbak (see [27]). As it turned out, on a generic surface there are only the following singularities: a curve P_1 of parabolic points; a finite set P_2 of points at which the curve P_1 is tangent to an asymptotic direction; a curve H_2 of points of inflexion of asymptotic lines; a finite set H_3 of self-intersection points of H_2; and a finite set H_4 of points at which the curve H_2 is tangent to an asymptotic direction.

Since the space of all nondegenerate algebraic surfaces of a fixed degree in $\mathbb{C}P^3$ is connected, the degrees of the curves P_1, H_2 and the number of points in the sets P_2, H_3, and H_4 depend only on the degree of the surface.

Theorem (see [201]). *For a general surface of degree d in $\mathbb{C}P^3$, $\deg P_1 = 4d(d-2)$, $\deg H_2 = d(11d-24)$, $\# P_2 = 2d(d-2)(11d-24)$, $\# H_3 = 5d(7d^2-28d+30)$, and $\# H_4 = 5d(d-4)(7d-12)$.*

Modulo 2 these results hold also for real surfaces in $\mathbb{R}P^3$. For example, the curve P_1 is always homologous to zero, the number of points in P_2 is even, and, in the case of surfaces of odd degree, the numbers of points in H_3 and H_4 are odd, and, in particular, different from zero. For a more detailed discussion of this subject the reader is referred to [39, § 3]. Analogous results for hypersurfaces in P^4 can be found in [244].

6.2. Vanishing Inflexions: The Case of Plane Curves. When a curve in $\mathbb{C}P^2$ degenerates the number of its points of inflexion decreases. For example, if the curve has only the simplest type of degenerations – δ self-intersections and k semicubic cusps – then the number of its points of inflexion is equal to $3d(d-2) - 6\delta - 8k$ (see, e.g., [197]). The reason for this is that, as one moves in the space of curves of degree d, at the moment a degeneracy occurs, 6 or 8 points of inflexion coalesce at the corresponding singular point. Similarly, at each singular point of a curve that contains no straight lines and multiple components, several points of inflexion coalesce; the number of coalescing points is determined as follows (see Fig. 55).

Fig. 55

For each locally irreducible component of the curve, in a neighborhood of the singular point investigated, we introduce affine coordinates with the center at that point, such that the component is tangent to one of the coordinate axes. Then the component admits the parametric representation $x = t^a$, $y = \lambda_1 t^{b_1} + \lambda_2 t^{b_2} + \cdots, = 0, a < b_1 < b_2 \ldots$.

Theorem ([303]). *At each singular point exactly*

$$3\mu - 3 + \sum (a + b_1) \tag{9}$$

points of inflexion vanish, where the sum of the numbers a, b_1 runs over all possible local components of the curve at the singular point and μ is its Milnor number.

For cusps and self-intersections the number (9) is 8 and 6, respectively.

Thus, the number of points of inflexion of a projective curve of degree d which contains no straight lines and multiple components is equal to $3d(d - 2)$ minus the sum of the numbers (9), taken for all the singular points of the curve.

6.3. Inflexions that Vanish at a Morse Singular Point. Let $f: (\mathbb{C}^n, 0) \to (\mathbb{C}, 0)$ be a function-germ with a Morse singularity at 0, and let $\tilde{f}$ be a small non-singular perturbation of f (for example, of the form $f + \varepsilon$). Consider the Gauss (tangent) map of the manifold $V = \{\tilde{f} = 0\}$, which acts from V into the space of all affine hyperplanes in $\mathbb{C}^n$. The singular set Σ^2 of this map is empty in a neighborhood of 0. In fact, Σ^2 corresponds to the points as which the Hessian of the restriction of the function $\tilde{f}$ to the tangent hyperplane to V has defect $\geqslant 2$. But the rank of the Hessian is semicontinuous, and at the original singular point of the original function f its value for any tangent hyperplane is $\geqslant n - 2$. Thus, all singularities of the Gauss map of the surface V near 0 are Morin singularities S_k. Moreover, in the generic case the codimension of the singular set S_k in V equals k; in particular, S_n is empty, and S_{n-1} is represented by a finite number of points. To each of these points there corresponds a germ of the dual hypersurface which is diffeomorphic to the swallowtail $\{\lambda$: the polynomial $z^{n+1} + \lambda_1 z^{n-1} + \cdots + \lambda_n$ has a multiple root$\}$. The points of the set $S_{n-1} \subset V$ lying close to 0 will be referred to as the *vanishing inflexions* of the surface V.

Theorem ([28]). *The number of inflexions that vanish at a generic Morse singular point is equal to* $(n + 1)!$

Let us give the real-case variant of this theorem, in which we consider only perturbations of the function f of the form $f + \varepsilon$. Let N_+ [resp. N_-] denote the number of vanishing inflexions on the real surface $\{f + \varepsilon = 0\}$ for $\varepsilon > 0$ [resp. $\varepsilon < 0$].

Theorem (Yu.M. Baryshnikov, 1987). *The following relations hold for almost any original Morse singularity f: $N_+ + N_- \leqslant 2(n!)$; if n is even, then $N_+ = N_-$; for any n, N_+ and N_- are even and $N_+ + N_-$ is a multiple of 4.*

6.4. Integration with Respect to the Euler Characteristic, and its Applications. In the theory of algebraic curves one often employs the following elementary assertion.

Theorem (the Riemann-Hurwitz formula). *Let X and Y be nonsingular complex algebraic curves, and let $f: X \to Y$ be a d-sheeted ramified covering. Then $\chi(X) = d \cdot \chi(Y) - \sum (b_i - 1)$, where the sum is taken over all ramification points, b_i is the ramification order at the i-th point, and $\chi(\cdot)$ is the Euler characteristic.*

Higher-dimensional generalizations of this theorem are formulated in terms of the concept of integration with respect to the Euler characteristic. This integration theory was developed by Viro (see [366]), also cf. [190], and turned out to be very useful in problems connected with projective duality. Let us describe this construction in the particular case of projective semialgebraic sets.

Recall that a set is said to be *semialgebraic* if in a neighborhood of any of its points it can be represented as a finite union of sets which are given by a finite number of algebraic equations and strict algebraic inequalities. Let X and Y be algebraic sets; $\mathbb{Z}$ denotes the ring of integers. A function $\varphi: X \to \mathbb{Z}$ is said to be *constructive* if there exists a finite semialgebraic partition $X = \bigcup_i X$ (i.e., the sets X_i are semialgebraic and $X_i \cap X_j = \varnothing$ for $i \neq j$) such that the restriction of φ to each X_i is constant. If $f_*: X \to Y$ is an algebraic map and $\psi: Y \to \mathbb{Z}$ is a constructive function, then $f^*\psi$ is also constructive. Any semialgebraic set has a well-defined Euler characteristic. The *integral* of a constructive function φ *with respect to the Euler characteristic* is defined to be the sum $\sum_i \varphi(X_i)\chi(X_i)$, and is denoted by $\int_X \varphi(x)\, d\chi(x)$. It is readily seen that its value does not depend on the choice of a partition of X into sets on which φ is constant.

Theorem (see [366]). *Let $f: X \to Y$ be an algebraic map of algebraic sets, and let $b: X \to \mathbb{Z}$ be a constructive function with the property that the value of integral $\int_{f^{-1}(y)} b(x)\, d\chi(y)$ does not depend on the point $y \in Y$; let D denote this common value. Then $\chi(X) = D\chi(Y) - \int_X (b(x) - 1)\, d\chi(x)$.*

(For the role of such a function $b(x)$ one can always take the Euler characteristic of the set $U \cap f^{-1}(y)$, where U is a regular neighborhood of x in X and y is a generic point in Y that lies sufficiently close to $f(x)$.)

For any constructive function $\varphi: \mathbb{C}P^n \to \mathbb{Z}$ there is defined its *Radon transform* (with respect to the Euler characteristic) $\varphi^\vee: \mathbb{C}P^{n^\vee} \to \mathbb{Z}$, where $\mathbb{C}P^{n^\vee}$ denotes the space of hyperplanes in $\mathbb{C}P^n$. Specifically, for any hyperplane $H \in \mathbb{C}P^{n^\vee}$, $\varphi^\vee(H) = \int_H \varphi(x)\, d\chi(x)$.

Proposition ([366]). $\varphi^{\vee\vee}(x) = \varphi(x) + (n-1)\int_{\mathbb{C}P^n} \varphi(x)\, d\chi(x)$.

Thus, the Radon transform yields a duality pairing of the spaces of constructive functions, factored by the subspaces of constants.

Now let $X \subset \mathbb{C}P^n$ be an algebraic hypersurface (possibly with singularities). The dual variety $X^\vee \subset \mathbb{C}P^{\vee n}$ is defined to be the closure of the set of hyperplanes tangent to X at its nonsingular points. For each point $x \in \mathbb{C}P^n$ let $m_X(x)$ denote the intersection index at x of the divisor X and a line in general position passing through x (for example, $m_X(x) = 0 \Leftrightarrow x \notin X$).

Theorem (the generalized Klein formula, see [366]). *For any complex projective curve X,*

$$\deg X - \int_{X \cap \mathbf{R}P^2} m_X(x)\, d\chi(x) = \deg X^\vee - \int_{X^\vee \cap \mathbf{R}P^{2\vee}} m_{X^\vee}(\xi)\, d\chi(\xi).$$

Analogous formulas hold in higher dimensions; see [366].

References*

The list given below includes, besides references quoted in the present volume, a number of textbooks. For a first acquaintance with the subject of the theory of singularities we recommend [8], [247], [65], and [278]. A systematic exposition of various aspects of the theory of singularities can be found in the books [37], [38], [150], [71], and [153]. To form a picture about the circle of problems that are currently under intensive investigation, the reader may consult the collections of papers [309], [87], [88], the survey [369] and the recent book "Theory of Singularities and its Applications" (ed. V.I. Arnold); Advances in Soviet Math., Vol. 1, Amer. Math. Soc., 1990. A number of important connections of the theory of singularities with other areas of mathematics are examined in volumes 1, 4, and 5 of the series "Dynamical systems". As sources of further references we indicate [37], [38], [278], and [369].

[1] A'Campo, N.: Le nombre de Lefschetz d'une monodromie. Indagationes Math. *35*, 113–118 (1973). Zbl.276.14004

[2] A'Campo, N.: Le groupe de monodromie du déploiement des singularités isolées de courbes planes. I. Math. Ann. *213*, 1–32 (1975). Zbl.316.14011

[3] A'Campo, N.: Le groupe de monodromie du déploiement des singularités isolées de courbes planes. II. *Proc. Int. Congr. Math.*, Vancouver 1974, Vol. I, 395–404 (1975). Zbl.352.14011

[4] A'Campo, N.: La fonction zeta d'une monodromie. Comment. Math. Helv. *50*, 233–248 (1975). Zbl.333.14008

[5] A'Campo, N.: Tresses, monodromie et le groupe symplectique. Comment. Math. Helv. *54*, 318–327 (1979). Zbl.441.32004

[6] Alekseev, A.V.: Parity of the number of points of intersection of cuspidal edges of a front and its surface. Usp. Mat. Nauk *39*, No. 3, 227–228 (1984). English transl.: Russ. Math. Surv. *39*, No. 3, 195–196 (1984). Zbl.606.58007

[7] Arnol'd, V.I.: On a characteristic class entering into conditions of quantization. Funkts. Anal. Prilozh. *1*, No. 1, 1–14 (1967). English transl.: Funct. Anal. Appl. *1*, No. 1, 1–13 (1967). Zbl.175,203

[8] Arnol'd, V.I.: Singularities of smooth mappings. Usp. Mat. Nauk *23*, No. 1, 3–44 (1968). English transl.: Russ. Math. Surv. *23*, No. 1, 1–43 (1968). Zbl. 159,536

[9] Arnol'd, V.I.: Some topological invariants of algebraic functions. Tr. Mosk. Mat. O.-va *21*, 27–46 (1970). English transl.: Trans. Mosc. Math. Soc. *21*, 30–52 (1970). Zbl.208,240

[10] Arnol'd, V.I.: Topological invariants of algebraic functions. II. Funkts. Anal. Prilozh. *4*, No. 2, 1–9 (1970). English transl.: Funct. Anal. Appl. *4*, 91–98 (1970). Zbl.239.14012

[11] Arnol'd, V.I.: Lectures on bifurcations and versal families. Usp. Mat. Nauk *27*, No. 5, 119–184 (1972). English transl.: Russ. Math. Surv. *27*, No. 5, 54–123 (1972). Zbl.248.58001

[12] Arnol'd, V.I.: Normal forms of functions near degenerate critical points, the Weyl groups A_k, D_k, E_k, and Lagrangian singularities. Funkts. Anal. Prilozh. *6*, No. 4, 3–25 (1972). English transl.: Funct. Anal. Appl. *6*, 254–272 (1972). Zbl.278.57011

[13] Arnol'd, V.I.: A classification of the unimodal critical points of functions. Funkts. Anal. Prilozh. 7, No. 3, 75–76 (1973). English transl.: Funct. Anal. Appl. 7, 230–231 (1973). Zbl.294.57018

[14] Arnol'd, V.I.: Remarks on the method of stationary phase and on the Coxeter numbers. Usp. Mat. Nauk *28*, No. 5, 17–44 (1973). English transl.: Russ. Math. Surv. *28*, No. 5, 19–48 (1973). Zbl.285.40002

* For the convenience of the reader, references to reviews in Zentralblatt für Mathematik (Zbl.), compiled using the MATH database, and Jahrbuch über die Fortschritte der Mathematik (Jbuch) have, as far as possible, been included in this bibliography.

[15] Arnol'd, V.I.: Mathematical Methods of Classical Mechanics. Nauka: Moscow 1974. English transl.: Graduate Texts Math. 60, Springer: New York 1978. Zbl.386.70001

[16] Arnol'd, V.I.: Normal forms of functions in neighborhoods of degenerate critical points. Usp. Mat. Nauk *29*, No. 2, 11–49 (1974). English transl: Russ. Math. Surv. *29*, No. 2, 10–50 (1974). Zbl.298.57022

[17] Arnol'd, V.I.: Critical points of smooth functions, and their normal forms. Usp. Mat. Nauk *30*, No. 5, 3–65 (1975). English transl: Russ. Math. Surv. *30*, No. 5, 1–75 (1975). Zbl.338.58004

[18] Arnol'd, V.I.: Spectral sequences for the reduction of functions to normal forms. Zadachy Mekh. Mat. Fiz., Moscow, 7–20 (1976). English transl.: Sel. Math. Sov. *1*, No. 1, 3–18 (1981). Zbl.472.58006

[19] Arnol'd, V.I.: Wave front evolution and equivariant Morse lemma. Commun. Pure Appl. Math. *29*, No. 6, 557–582 (1976). Zbl.343.58003

[20] Arnol'd, V.I.: Critical points of smooth functions. *Proc. Int. Congr. Math.*, Vancouver 1974, Vol. I, 19–39. (1975). Zbl.343.58002

[21] Arnol'd, V.I.: Critical points of functions on a manifold with boundary, the simple Lie groups B_k, C_k, F_4, and singularities of evolutes. Usp. Mat. Nauk *33*, No. 5, 91–105 (1978). English transl.: Russ. Math. Surv. *33*, 99–116 (1978). Zbl.408.58009

[22] Arnol'd, V.I.: The index of a singular point of a vector field, the Petrovskiĭ-Oleĭnik inequalities, and mixed Hodge structures. Funkts. Anal. Prilozh. *12*, No. 1, 1–14 (1978). English transl.: Funct. Anal. Appl. *12*, 1–12 (1978). Zbl.398.57031

[23] Arnol'd, V.I.: Indices of singular points of 1-forms on manifolds with boundary, convolutions of invariants of groups generated by reflections, and singular projections of smooth surfaces. Usp. Mat. Nauk *34*, No. 2, 3–38 (1978). English trans.: Russ. Math. Surv. *34*, No. 2, 1–42 (1979). Zbl.405.58019

[24] Arnol'd, V.I.: *Catastrophe Theory*. Priroda, No. 10, 54–63 (1979) (Russian)

[25] Arnol'd, V.I.: On some problems in singularity theory. In *Geometry and Analysis*, Pap. dedic. V. K. Patodi, 1–9 (1981). Zbl.492.58006

[26] Arnol'd, V.I.: Lagrange and Legendre cobordisms. I. Funkts. Anal. Prilozh. *14*, No. 3, 1–13 (1980). English transl.: Funct. Anal. Appl. *14*, 167–177 (1981) Zbl.448.57017; II. Funkts. Anal. Prilozh. *14*, No. 4, 8–17 (1980). English transl.: Funct. Anal. Appl. *14*, 252–260 (1981). Zbl.472.55002

[27] Arnol'd, V.I.: Singularities of ray systems. Usp. Mat. Nauk *38*, No. 2, 77–147 (1983). English transl.: Russ. Math. Surv. *38*, No. 2, 87–176 (1983). Zbl.522.58007

[28] Arnol'd, V.I.: Vanishing inflections. Funkts. Anal. Prilozh. *18*, No. 2, 51–52 (1984). English transl.: Funct. Anal. Appl. *18*, 128–130 (1984). Zbl.565.32009

[29] Arnol'd, V.I.: Catastrophe theory. Itogi Nauki Tekh., Ser. Sovrem. Probl. Mat., Fundam. Napravleniya *5*, 219–277. (1986). English transl.: Encycl. Math. Sci., *5*, *Dynamical Systems* V, Springer: Berlin, to appear 1993. Zbl.623.00023

[30] Arnol'd, V.I.: First steps in symplectic topology. Usp. Mat. Nauk *41*, No. 6, 3–18 (1986). English transl.: Russ. Math. Surv. *41*, No. 6, 1–21 (1986). Zbl. 618.58021

[31] Arnol'd, V.I.: The branched covering $\mathbb{C}P^2 \to S^4$, hyperbolicity and projective topology. Sib. Mat. Zh. *29*, No. 5, 36–47 (1988). English transl.: Sib. Math. J. *29*, No. 5, 717–726 (1988). Zbl.668.57003

[32] Arnol'd, V.I.: Spaces of functions with mild singularities. Funkts. Anal. Prilozh. *23*, No. 3, 1–10 (1989). English transl.: Funct. Anal. Appl. *23*, 169–177 (1989). Zbl.717.58008

[33] Arnol'd, V.I., Afraĭmovich, V.S., Il'yashenko, Yu. S., Shil'nikov, L.P.: Bifurcation theory. Itogi Nauki Tekh., Ser. Sovrem. Probl. Mat., Fundam. Napravleniya *5*, 5–218. (1986). English transl.: Encycl. Math. Sci. *5*, *Dynamical Systems V*, Springer: Berlin, to appear 1992. Zbl.639.00033

[34] Arnol'd, V.I., Givental', A.B.: Symplectic geometry. Itogi Nauki Tekh., Ser. Sovrem. Probl. Mat., Fundam. Napravleniya *4*, 7–139. (1985). English transl.: Encycl. Math. Sci. *4*, *Dynamical Systems* IV, Springer: Berlin 1990. Zbl.592.58030

[35] Arnol'd, V.I., Il'yashenko, Yu. S.: Ordinary differential equations. Itogi Nauki Tekh., Ser.

Sovrem. Probl. Mat., Fundam. Napravleniya I, 7–149. (1985). English transl.: Encycl. Math. Sci. *1*, *Dynamical Systems I*. Springer: Berlin 1988. Zbl.602.58020

[36] Arnol'd, V.I., Oleĭnik, O. A.: Topology of real algebraic manifolds. Vestn. Mosk. Univ., Ser. I, No. 6, 7–17 (1979). English transl.: Mosc. Univ. Math. Bull. *34*, No. 6, 5–17 (1979). Zbl.423.14013

[37] Arnol'd, V.I., Varchenko, A.N., Guseĭn-Zade, S.M.: *Singularities of Differentiable Mappings*, Vol. I. Nauka: Moscow 1982. English transl.: Birkhäuser: Basel 1986. Zbl.513.58001

[38] Arnol'd, V.I.: Varchenko, A.N., Guseĭn-Zade, S.M.: *Singularities of Differentiable Mappings*, Vol. 2. Nauka: Moscow 1984. English transl.: Birkhäuser: Basel 1988. Zbl.545.58001

[39] Arnol'd, V.I., Vasil'ev, V.A., Goryunov, V.V., Lyashko, O.V.: Singularities. II. Itogi Nauki Tekh., Ser. Sovrem. Probl. Mat., Fundam. Napravleniya *39*, 5–254. (1989). English transl.: Encycl. Math. Sci. *39*, *Dynamical Systems* VIII, Springer: Berlin, 1993. Zbl.671.00010

[40] Atiyah, M.F., Singer, I.M.: The index of elliptic operators. III. Ann. Math., II. Ser. *87*, 546–604 (1968). Zbl. 164, 243

[41] Audin, M.: *Cobordismes d'immersions lagrangiennes et legendriennes*. Trav. en Cours, 20. Hermann: Paris 1987. Zbl.615.57001

[42] Bakhtin, V.I.: Topological equivalence of germs of functions. Mat. Zametki *35*, No. 4, 579–588 (1984). English transl.: Math. Notes *35*, 306–310 (1984). Zbl.543.58011

[43] Bakhtin, V.I.: Topological normal forms of transformations of caustics of the series D_μ. Vestn. Mosk. Univ., Ser. I, No. 4, 58–61 (1987). English transl.: Mosc. Univ. Math. Bull. *42*, No. 4, 63–66 (1987). Zbl.633.58005

[44] Bakhtin, V.I.: Transversality of mappings, preimages of mappings, and the representation of sets of degenerate matrices as images. Usp. Mat. Nauk *42*, No. 2, 217–218 (1987). English transl.: Russ. Math. Surv. *42*, No. 2, 271–272 (1987). Zbl.694.58004

[45] Banchoff, T.: Triple points and singularities of projections of smoothly immersed surfaces. Proc. Am. Math. Soc. *46*, 402–406 (1974). Zbl.309.57016

[46] Basset, A.B.: The maximum number of double points on a surface. Nature *73*, 1–246 (1906). Jbuch 37, 646

[47] Beauville, A.: Sur le nombre maximum de points doubles d'une surface dans P^3 ($\mu(5) = 31$). *Journées de géometrie algébrique d'Angers*, pp. 207–215. Sijthoff and Noordhoff: Amsterdam (1980). Zbl.445.14016

[48] Belitskiĭ, G.R.: Equivalence and normal forms of germs of smooth mappings. Usp. Mat. Nauk *33*, No. 1, 95–155 (1978). English transl.: Russ. Math. Surv. *33*, No. 1, 107–177 (1978). Zbl.385.58007

[49] Benson, M.: Analytic equivalence of isolated hypersurface singularities defined by homogeneous polynomials. *Singularities*, pp. 111–118, Proc. Symp. Pure Math. *40*, Part I. Amer. Math. Soc.: Providence (1983). Zbl.521.32006

[50] Bernshtein, D.N.: The number of roots of a system of equations. Funkts. Anal. Prilozh. *9*, No. 3, 1–4 (1975). English transl.: Funct. Anal. Appl. *9*, 183–185 (1975). Zbl.328.32001

[51] Bernshtein, D.N., Kushnirenko, A.G., Khovanskiĭ, A.G.: Newton polyhedra. Usp. Mat. Nauk *31*, No. 3, 201–202 (1976), (Russian). Zbl.354.14001

[52] Bierstone, E.: Local properties of smooth maps equivariant with respect to finite group actions. J. Differ. Geom. *10*, 523–540 (1975). Zbl.317.58004

[53] Bierstone, E.: *The Structure of Orbit Spaces and Singularities of Equivariant Mappings*. Monograf. Mat. 35. Inst. Mat. Pura Apl.: Rio de Janeiro 1980. Zbl.501.57001

[54] Boardman, J.M.: Singularities of differentiable maps. Publ. Math., Inst. Hautes Etud. Sci. *33*, 21–57 (1967). Zbl.165,568

[55] Borel, A., Haefliger, A.: La classe d'homologie fondamentale d'une espace analytique. Bull. Soc. Math. Fr. *89*, 461–513 (1961). Zbl.102,385

[56] Bourbaki, N.: *Éléménts de mathématiques. Fasc. XXXIV. Groupes et algèbres de Lie. Chapitres IV, V, VI*. Actual. Sci. Ind., 1337. Hermann: Paris 1968. Zbl.186,330

[57] Brieskorn, E.: Über die Auflösung gewisser Singularitäten von holomorphen Abbildungen. Math. Ann. *166*, 76–102 (1966). Zbl.145,94

[58] Brieskorn, E.: Die Monodromie der isolierten Singularitäten von Hyperflächen. Manuscr. Math. *2*, 103–161 (1970). Zbl.186,261

[59] Brieskorn, E.: Singular elements of semi-simple algebraic groups. *Actes Congr. Internat. Math.*, Vol. 2, pp. 279–284. Gauthier-Villars: Paris 1971. Zbl.223.22012

[60] Brieskorn, E.: Sur les groupes de tresses (d'après V.I. Arnold). *Séminaire Bourbaki* (1971/1972), Exp. No. 401, pp. 21–44, Lect. Notes Math. 317. Springer: Berlin 1973. Zbl.277.55003

[61] Brieskorn, E.: *Special Singularities—Resolution, Deformation and Monodromy.* Lecture notes prepared in connection with Am. Math. Soc. Summer Inst. on algebraic geometry (Humboldt State Univ., Arcata, CA) 1974.

[62] Brieskorn, E.: Die Hierarchie der 1-modularen Singularitäten. Manuscr. Math. *27*, 183–219 (1979). Zbl.406.58009

[63] Brieskorn, E.: Die Milnorgitter der exzeptionellen unimodularen Singularitäten. Bonn. Math. Schr. *150*. Inst. Math. Univ. Bonn: Bonn (1983). Zbl.525.14002

[64] Brieskorn, E.: Milnor lattices and Dynkin diagrams. In: Singularities, pp. 153–165, Proc. Symp. Pure Math. *40*, Part I. Amer. Math. Soc.: Providence (1983). Zbl.527.14006

[65] Bröcker, T., Lander, L.: *Differentiable Germs and Catastrophes.* English transl.: Lond. Math. Soc. Lect. Note Ser. *17*. Cambridge Univ. Press: Cambridge 1975. Zbl.302.58006

[66] Bruce, J.W.: A stratification of the space of cubic surfaces. Math. Proc. Camb. Philos. Soc. *87*, 427–441 (1980). Zbl.434.58009

[67] Bruce, J.W.: An upper bound for the number of singularities on a projective hypersurface. Bull. Lond. Math. Soc. *13*, 47–50 (1981). Zbl.454.14018

[68] Bruce, J.W.: Counting singularities. Proc. Roy. Soc. Edinb., Sect. A *93*, 137–159 (1982). Zbl.509.58010

[69] Bruce, J.W., Gaffney, T., du Plessis, A.A.: On left equivalence of map germs. Bull. Lond. Math. Soc. *16*, 303–306 (1984). Zbl.513.58010

[70] Bruce, J.W., Giblin, P.J.: A stratification of the space of plane quartic curves. Proc. Lond. Math. Soc., III. Ser. *42*, 270–298 (1981). Zbl.403.14004

[71] Bruce, J.W., Giblin, P.J.: *Curves and Singularities.* Cambridge Univ. Press: Cambridge 1984. Zbl.534.58008

[72] Bruce, J.W., du Plessis, A.A., Wall, C.T.C.: Determinacy and unipotency. Invent. Math. *88*, 521–554 (1987). Zbl.596.58005

[73] Bruce, J.W., Wall, C.T.C.: On the classification of cubic surfaces. J. Lond. Math. Soc., II. Ser. *19*, 245–256 (1979). Zbl.393.14007

[74] Bryzgalova, L.N.: Singularities of the maximum of a function that depends on parameters. Funkts. Anal. Prilozh. *11*, No. 1, 59–60 (1977). English transl.: Funct. Anal. Appl. *11*, 49–51 (1977). Zbl.348.58005

[75] Bryzgalova, L.N.: The maximum functions of a family of functions that depend on parameters. Funkts. Anal. Prilozh. *12*, No. 1, 66–67 (1978). English transl.: Funct. Anal. Appl. *12*, 50–51 (1978). Zbl.377.58007

[76] Byalyĭ, M.L., Polterovich, L.V.: Geodesic flows on the two-dimensional torus and "commensurability-incommensurability" phase transitions. Funkts. Anal. Prilozh. *20*, No. 4, 9–16 (1986). English transl.: Funct. Anal. Appl. *20*, 260–266 (1986). Zbl.641.58032

[77] Cerf, J.: Suppression des singularités de codimension plus grande que 1 dans les familles de fonctions différentiables réeles (d'après K. Igusa). *Séminaire Bourbaki*, 36e année, Vol. 1983/84, Exp. No. 627. Zbl.555.57013

[78] Chern, S.S.: *Complex Manifolds.* Mimeographed notes. Univ. Chicago: Chicago 1956. Zbl.74,303

[79] Chess, D.S.: A note on the classes $[S_1^k(f)]$. *Singularities*, pp. 221–224, Proc. Symp. Pure Math. 40, Part I. Amer. Math. Soc.: Providence (1983). Zbl.523.58011

[80] Chislenko, Yu. S.: Decompositions of simple singularities of real functions. Funkts. Anal. Prilozh. *22*, No. 4, 52–67 (1988). English transl.: Funct. Anal. Appl. *22*, No. 4, 297–310 (1988). Zbl.667.58004

[81] Chmutov, S.V.: Monodromy groups of singularities of functions of two variables. Funkts.

Anal. Prilozh. *15*, No. 1, 61–66 (1981). English transl.: Funct. Anal. Appl. *15*, 48–52 (1981). Zbl.466.32002

[82] Chmutov, S.V.: On the monodromy of isolated singularities. *XVI All-Union Algebra Conference*, Part II, pp. 172–173 (1981). Akad. Nauk SSSR, Mat. Inst. Leningrad. Otdel.: Leningrad (Russian). Zbl.525.16001

[83] Chmutov, S.V.: Monodromy groups of critical points of functions. Invent. Math. *67*, 123–131 (1982). Zbl.468.57024

[84] Chmutov, S.V.: The monodromy groups of critical points of functions. II. Invent. Math. *73*, 491–510 (1983). Zbl.536.32006

[85] Cohen, F.R., Lada, T.J., May, J.P.: *The Homology of Iterated Loop Spaces*. Lect. Notes Math. 533. Springer: Berlin 1976. Zbl.334.55009

[86] Cohen, R.L.: The homotopy theory of immersions. *Proc. Int. Congr. Math.* 1983, pp. 627–639 (1984). PWN: Warsaw. Zbl.574.57015

[87] Current problems in mathematics (Itogi Nauki Tekh., Ser. Sovrem. Probl. Mat.) *22*. VINITI: Moscow (1983). English transl.: J. Sov. Math. *27*, 2679–2830 (1984). Zbl.534.58014; Zbl.537.58012; Zbl.543.58008; Zbl.544.14006; Zbl.548.58024; Zbl.555.58009

[88] Current problems in mathematics (Itogi Nauki Tekh., Ser. Sovrem. Probl. Mat., Noveish. Dostizheniya) *33*. VINITI: Moscow (1988). English transl.: Journal of Soviet Math. *52*, No. 4, 1990

[89] Damon, J.: A partial topological classification for stable map germs. Bull. Amer. Math. Soc. *82*, 105–107 (1976). Zbl.318.57036

[90] Damon, J.: Topological properties of discrete algebra types. I. The Hilbert-Samuel function. Adv. Math., Suppl. Stud. *5*, 83–118 (1979). Zbl.471.58005; II. Real and complex algebras. Amer. J. Math. *101*, 1219–1248 (1979). Zbl.498.58005

[91] Damon, J.: Topological properties of real simple germs, curves, and nice dimensions $n > p$. Math. Proc. Camb. Philos. Soc. *89*, 457–472 (1981). Zbl.516.58013

[92] Damon, J.: Finite determinacy and topological triviality. I. Invent. Math. *62*, 299–324 (1980). Zbl.489.58003; II. Compos. Math. *47*, 101–132 (1982). Zbl.523.58005

[93] Damon, J.: Deformations of sections of singularities and Gorenstein surface singularities. Amer. J. Math. *109*, 695–722 (1987). Zbl.628.14003

[94] Damon, J.: The unfolding and determinacy theorems for subgroups of $\mathscr{A}$ and $\mathscr{K}$. *Singularities*, pp. 233–254, Proc. Symp. Pure Math. *40*, Part I. Amer. Math. Soc.: Providence (1983). Zbl.519.58014

[95] Damon, J.: Topological triviality in versal unfoldings. *Singularities*, pp. 255–266, Pro. Symp. Pure Math. 40, Part I. Amer. Math. Soc.: Providence (1983). Zbl.523.58006

[96] Damon, J.: The unfolding and determinacy theorems for subgroups of $\mathscr{A}$ and $\mathscr{K}$, Mem. Amer. Math. Soc. No. 306 (1984). Zbl.545.58010

[97] Damon, J.: Topological equivalence for nonisolated singularities and global affine hypersurfaces. Contemp. Math. *90*, 21–53 (1989). Zbl.677.58011

[98] Damon, J.: Topological triviality of versal unfoldings of complete intersections. Ann. Inst. Fourier *34*, 225–251 (1984). Zbl.534.58010

[99] Damon, J.: Topological triviality and versality for subgroups of $\mathscr{A}$ and $\mathscr{K}$. Mem. Amer. Math. Soc. No. 389 (1988). Zbl.665,58005

[100] Damon, J.: Topological equivalence of bifurcation problems. Nonlinearity *1*, No. 2, 311–331 (1988). Zbl.651.58024

[101] Damon, J., Gaffney, T.: Topological triviality of deformations of functions and Newton filtrations. Invent. Math. *72*, 335–358 (1983). Zbl.519.58021

[102] Danilov, V.I.: Newton polyhedra and vanishing cohomology. Funkts. Anal. Prilozh. *13*, No. 2, 32–47 (1979). English transl.: Funct. Anal. Appl. *13*, 103–115 (1979). Zbl.427.14006

[103] Danilov, V.I., Khovanskiĭ, A.G.: Newton polyhedra and an algorithm for calculating Hodge-Deligne numbers. Izv. Akad. Nauk SSSR, Ser. Mat. *50*, No. 5, 925–945 (1986). English transl.; Math. USSR, Izv. *29*, 279–298 (1987). Zbl.669.14012

[104] Deligne, P.: *Équations différentielles à points singuliers réguliers*. Lect. Notes Math. 163. Springer: Berlin 1970. Zbl.244.14004

[105] Deligne, P.: Théorie de Hodge. I. *Actes Congr. Int. Math. 1970*, Vol. I, pp. 425–430. Gauthier-Villars: Paris 1971. Zbl.219.14006; II. Publ. Math., Inst. Hautes Étud. Sci. *40*, 5–57 (1971). Zbl.219.14007; III. Publ. Math., Inst. Hautes Étud. Sci. *44*, 5–77 (1974). Zbl.237.14003

[106] Deligne. P.: Les immeubles des groupes des tresses généralisés. Invent. Math. *17*, 273–302 (1972). Zbl.238.20034

[107] Demazure, M.: Classification des germs à point critique isolé et à nombres de modules 0 ou 1 (d'après V.I. Arnol'd). *Séminaire Bourbaki*, 26e année, Vol. 1973/74, Exp. No. 443, pp. 124–142, Lect. Notes Math. 431. Springer: Berlin 1975. Zbl.359.57012

[108] Dimca, A., Gibson, C.G.: Classification of equidimensional contact unimodular map germs. Math. Scand. *56*, 15–28 (1985). Zbl.579.32014

[109] Dolgachev, I.V.: Conic quotient singularities of complex hypersurfaces. Funkts. Anal. Prilozh. *8*, No. 2, 75–76 (1974). English transl.: Funct. Anal. Appl. *8*, 160–161 (1974). Zbl.295.14017

[110] Dolgachev, I.V.: Automorphic forms and quasihomogeneous singularities. Funkts. Anal. Prilozh. *9*, No. 2, 67–68 (1975). English transl.: Funct. Anal. Appl. *9*, 149–151 (1975). Zbl.321.14003

[111] Dolgachev, I.V., Nikulin, V.V.: V.I. Arnold's exceptional singularities and K-3 surfaces. *VII All-Union Topology Conference*. Abstracts. Akad. Nauk BSSR, Inst. Mat.: Minsk (1977), (Russian)

[112] Dufour, J.-P.: Déploiements de cascades d'applications différentiables. C.R. Acad. Sci., Paris, Sér. A *281*, 31–34 (1975). Zbl.317.58005

[113] Dufour, J.-P.: Sur la stabilité des diagrammes d'applications différentiables. Ann. Sci. Éc. Norm. Supér., IV. Sér. *10*, 153–174 (1977). Zbl.354.58011

[114] Durfee, A.H.: A naive guide to mixed Hodge theory. *Singularities*, pp. 313–320. Proc. Symp. Pure Math. *40*, Part I. Amer. Math. Soc.: Providence 1983. Zbl.521.14003

[115] Dyer, E., Lashof, R.K.: Homology of iterated loop spaces. Amer. J. Math. *84*, 35–88 (1962). Zbl.119,182

[116] Ebeling, W.: Quadratische Formen und Monodromiegruppen von Singularitäten. Math. Ann. *255*, 463–498 (1981). Zbl.438.32004

[117] Ebeling, W.: Arithmetic monodromy groups. Math. Ann. *264*, 241–255 (1983). Zbl.501.14018

[118] Ebeling, W.: Milnor lattices and geometric bases of some special singularities. Enseign. Math., II. Sér. *29*, 263–280 (1983). Zbl.536.14002

[119] Ebeling, W.: On the monodromy groups of singularities. *Singularities*, pp. 327–336, Proc. Symp. Pure Math. *40*, Part I. Amer. Math. Soc.: Providence 1983. Zbl.518.32007

[120] Ebeling, W.: An arithmetic characterization of the symmetric monodromy groups of singularities. Invent. Math. *77*, 85–99 (1984). Zbl.527.14031

[121] Eisenbud, D., Levine, H.: An algebraic formula for the degree of a C^∞ map germ. Ann. Math., II. Ser. *106*, 19–44 (1977). Zbl.398.57020

[122] Eliashberg, Ya.M.: Singularities of fold type. Izv. Akad. Nauk SSSR, Ser. Mat. *34*, No. 5, 1110–1126 (1970). English transl.: Math. USSR. Izv. *4*, 1119–1134 (1970). Zbl.202,548

[123] Eliashberg, Ya.M.: Surgery of singularities of smooth mappings. Izv. Akad. Nauk SSSR, Ser. Mat. *36*, No. 2, 1321–1347 (1972). English transl.: Math. USSR, Izv. *6*, 1302–1326 (1972). Zbl.254.57019

[124] Eliashberg, Ya.M.: Cobordisme des solutions de relations différentiels. *Séminaire sud-rhodanien de géométrie*, I, pp. 17–31, Trav. en Cours. Hermann: Paris 1984. Zbl.542.57024

[125] Elkik, R.: Singularités rationelles et déformations. Invent. Math. *47*, 139–147 (1978). Zbl.363.14002

[126] Feit, S.: k-mersions of manifolds. Acta Math. *122*, 173–195 (1969). Zbl.179,521

[127] Fuks, B.A.: *Special Chapters on the Theory of Analytic Functions of Several Complex Variables*. Fizmatgiz: Moscow 1963. English transl.: Transl. Math. Monogr. *14*, Amer. Math. Soc.: Providence 1965. Zbl.146,308

[128] Fuks, D.B.: Cohomology of the braid group mod 2. Funkts. Anal. Prilozh. *4*, No. 2, 62–73 (1970). English transl.: Funct. Anal. Appl. *4*, 143–151 (1970). Zbl.222.57031

[129] Fuks, D.B.: Quillenization and bordisms. Funkts. Anal. Prilozh. *8*, No. 1, 36–42 (1974). English transl.: Funct. Anal. Appl. *8*, 31–36 (1974). Zbl.324.57024

[130] Fuks, D.B.: Cohomology of infinite-dimensional Lie algebras and characteristic classes of foliations. Itogi Nauki Tekh., Ser. Sovrem. Probl. Mat. *10*, pp. 179–285. VINITI: Moscow (1978). English transl.: J. Sov. Math. *11*, 922–980 (1979). Zbl.499.57001

[131] Fukuda, M., Fukuda, T.: Algebras $Q(f)$ determine the topological types of generic map germs. Invent. Math. *51*, 231–237 (1979). Zbl.408.58012

[132] Gabrielov, A.M.: Intersection matrices for certain singularities. Funkts. Anal. Prilozh. 7, No. 3, 18–32 (1973). English transl.: Funct. Anal. Appl. 7, 182–193 (1973). Zbl.288.32011

[133] Gabrielov, A.M.: Bifurcations, Dynkin diagrams, and the modality of isolated singularities. Funkts. Anal. Prilozh. *8*, No. 2, 7–12 (1974). English transl.: Funct. Anal. Appl. *8*, 94–98 (1974). Zbl.344.32007

[134] Gabrielov, A.M.: Dynkin diagrams of unimodal singularities. Funkts. Anal. Prilozh. *8*, No. 3, 1–6 (1974). English transl.: Funct. Anal. Appl. *8*, 192–196 (1974). Zbl.304.14010

[135] Gabrielov, A.M.: Polar curves and intersection matrices of singularities. Invent. Math. *54*, 15–22 (1979). Zbl.421.32011

[136] Gabrielov, A.M., Kushnirenko, A.G.: Description of deformations with constant Milnor number for homogeneous functions. Funkts. Anal. Prilozh. *9*, No. 4, 67–68 (1975). English transl.: Funct. Anal. Appl. *9*, 329–331 (1975). Zbl.319.32023

[137] Gaffney, T.: Properties of finitely determined germs. Ph.D. Thesis. Brandeis Univ.: Waltham (1975).

[138] Gaffney, T.: On the order of determination of a finitely determined germ. Invent. Math. *37*, 83–92 (1976). Zbl.354.58012

[139] Gaffney, T.: A note on the order of determination of a finitely determined germ. Invent. Math. *52*, 127–130 (1979). Zbl.419.58004

[140] Gaffney, T.: The Thom polynomial of $\overline{\Sigma^{1111}}$. *Singularities*, pp. 399–408, Proc. Symp. Pure Math. *40*, Part I. Amer. Math. Soc.: Providence 1983. Zbl.531.58012

[141] Gaffney, T.: The structure of $TA(f)$, classification and an application to differential geometry. *Singularities*, pp. 409–427, Proc. Symp. Pure Math. *40*, Part I. Amer. Math. Soc.: Providence 1983. Zbl. 523.58008

[142] Gaffney, T., du Plessis, A.: More on the determinacy of smooth map germs. Invent. Math. *66*, 137–163 (1982). Zbl.489.58004

[143] Gibson, C.G.: *Singular Points of Smooth Mappings*. Res. Notes Math. *25*. Pitman: London 1979. Zbl.426.58001

[144] Gibson, C.G., Wirthmüller, K., du Plessis, A.A., Looijenga, E.J.N.: *Topological Stability of Smooth Mappings*. Lect. Notes Math. 552. Springer: Berlin 1976. Zbl.377.58006

[145] Givental', A.B.: Convolution of invariants of groups generated by reflections that are associated to simple singularities of functions. Funkts. Anal. Prilozh. *14*, No. 2, 4–14 (1980). English transl.: Funct. Anal. Appl. *14*, 81–89 (1980). Zbl.463.58010

[146] Givental', A.B.: Manifolds of polynomials that have a root of fixed comultiplicity, and the generalized Newton equation. Funkts. Anal. Prilozh. *16*, No. 1, 13–18 (1982). English Transl.: Funct. Anal. Appl. *16*, 10–14 (1982). Zbl.508.58020

[147] Givental', A.B.: Asymptotics of the intersection form of a quasihomogeneous singularity of functions. Funkts. Anal. Prilozh. *16*, No. 4, 63–65 (1982). English transl.: Funct. Anal. Appl. *16*, 294–297 (1982). Zbl.509.32002

[148] Givental', A.B.: The maximum number of singular points on a projective hypersurface. Funkts. Anal. Prilozh. *17*, No. 3, 73–74 (1983). English transl.: Funct. Anal. Appl. *17*, 223–225 (1983). Zbl.529.14022

[149] Givental', A.B.: Lagrangian imbeddings of surfaces and the unfolded Whitney umbrella. Funkts. Anal. Prilozh. *20*, No. 3, 35–41 (1986). English transl.: Funct. Anal. Appl. *20*, 197–203 (1986). Zbl.621.58025

[150] Golubitsky, M., Guillemin, V.: *Stable Mappings and Their Singularities*. Grad. Texts Math. 14. Springer: New York 1973. Zbl.294.58004

[151] Golubitsky, M., Schaeffer, D.: A theory for imperfect bifurcation via singularity theory. Commun. Pure Appl. Math. *32*, 21–98. Zbl.409.58007

[152] Golubitsky, M., Schaeffer, D.: Imperfect bifurcation in the presence of symmetry. Commun. Math. Phys. *67*, 205–232 (1979). Zbl.467.58019

[153] Golubitsky, M., Schaeffer, D.: *Singularities and Groups in Bifurcation Theory*, Vol. I. Appl. Math. Sci. 55. Springer: New York 1985. Zbl.607.35004

[154] Goryunov, V.V.: Cohomology of braid groups of series C and D and certain stratifications. Funkts. Anal. Prilozh. *12*, No. 2, 76–77 (1978). English transl.: Funct. Anal. Appl. *12*, 139–140 (1978). Zbl.401.20034

[155] Goryunov, V.V.: Adjacencies of the spectra of certain singularities. Vestn. Mosk. Univ., Ser. I, No. 4, 19–22 (1981). English transl.: Mosc. Univ. Math. Bull. *36*, No. 4, 22–25 (1981). Zbl.514.32004

[156] Goryunov, V.V.: Cohomology of braid groups of series C and D. Tr. Mosk. Mat. O.-va *42*, 234–242 (1981). English transl.: Trans. Mosc. Math. Soc., No. 2, 233–241 (1982). Zbl. 547.55016

[157] Goryunov, V.V.: Singularities of projections of complete intersections. Itogi Nauki Tekh., Ser. Sovrem. Probl. Mat. *22*, 167–206. VINITI: Moscow (1983). English transl.: J. Sov. Math. *27*, 2785–2811 (1984). Zbl.555.58009

[158] Greuel, G.M.: Der Gauss-Manin-Zusammenhang isolierter Singularitäten von vollständigen Durchschnitten. Thesis. Göttingen (1973). Appeared as: Math. Ann. *214*, 235–266 (1975). Zbl.285.14002

[159] Griffiths, P.: *Monodromy of Homology and Periods of Integrals on Algebraic Manifolds*. Mimeographed notes. Princeton Univ.: Princeton 1978.

[160] Griffiths, P.: Periods of integrals on algebraic manifolds: summary of main results and discussion of open problems. Bull. Amer. Math. Soc. *76*, 228–296 (1970). Zbl.214,198

[161] Griffiths, P., Harris, J.: *Principles of Algebraic Geometry*. Wiley: New York 1978. Zbl.408.14001

[162] Griffiths, P., Schmid, W.: Recent developments in Hodge theory: a discussion of techniques and results. *Discrete Subgroups of Lie Groups and Applications to Moduli*, Bombay 1973, pp. 31–127. Oxford Univ. Press: London 1975. Zbl.355.14003

[163] Gromov, M.L.: Stable mappings of foliations into manifolds. Izv. Akad. Nauk SSSR, Ser. Mat. *33*, No. 4, 707–734 (1969). English transl.: Math. USSR, Izv. *3*, 671–694 (1969). Zbl.197,204

[164] Gromov, M.L.: A topological technique for the construction of solutions of differential equations and inequalities. *Actes Congr. Int. Math. 1970*, Vol. 2, pp. 221–225. Gauthier-Villars: Paris 1971. Zbl.237.57019

[165] Gromov, M.L.: Convex integration of differential relations. I. Izv. Akad. Nauk SSSR, Ser. Mat. *37*, No. 2, 329–343 (1973). English transl.: Math. USSR, Izv. *7*, 329–343 (1973). Zbl.254.58001

[166] Gromov, M.L.: *Partial Differential Relations*. Springer: Berlin 1986. Zbl.651,53001

[167] Gromov, M.L., Eliashberg, Ya. M.: Elimination of singularities of smooth mappings. Izv. Akad. Nauk SSSR, Ser. Mat. *35*, No. 4, 600–626 (1971). English transl.: Math. USSR, Izv. *5*, 615–639 (1971). Zbl.221.58009

[168] Gudkov, D.A.: The topology of real projective algebraic varieties. Usp. Mat. Nauk. *29*, No. 4, 3–79 (1974). English transl.: Russ. Math. Surv. *29*, No. 4, 1–79 (1974). Zbl.316.14018

[169] Gunning, R.C., Rossi, H.: *Analytic Functions of Several Complex Variables*. Prentice-Hall: Englewood Cliffs 1965. Zbl.141,86

[170] Guseĭn-Zade, S. M.: Intersection matrices for certain singularities of functions of two variables. Funkts. Anal. Prilozh. *8*, No. 1, 11–15 (1974). English transl.: Funct. Anal. Appl. *8*, 10–13 (1974). Zbl.304.14009

[171] Guseĭn-Zade, S.M.: Dynkin diagrams of the singularities of functions of two variables. Funkts. Anal. Prilozh. *8*, No. 4, 23–30 (1974). English transl.: Funct. Anal. Appl. *8*, 295–300 (1974). Zbl.309.14006

[172] Guseĭn-Zade, S.M.: Monodromy groups of isolated singularities of hypersurfaces. Usp. Mat. Nauk *32*, No. 2, 23–65 (1977). English transl.: Russ. Math. Surv. *32*, No. 2, 23–69 (1977). Zbl.363.32010

[173] Guseĭn-Zade, S.M.: The index of a singular point of a gradient vector field. Funkts. Anal. Prilozh. *18*, No. 1, 7–12 (1984). English transl.: Funct. Anal. Appl. *18*, 6–10 (1984). Zbl.555.32006

[174] Guseĭn-Zade, S.M., Nekhoroshev, N.N.: Adjacencies of singularities A_k to points of the μ = const stratum of a singularity. Funkts. Anal. Prilozh. *17*, No. 4, 82–83 (1983). English transl.: Funct. Anal. Appl. *17*, 312–313 (1983). Zbl.555.32008

[175] Haefliger, A., Kosiński, A.: Un théorème de Thom sur les singularités des applications différentiables. *Séminaire Henri Cartan, 9e* année: 1956/57, Exp. No. 8. Secr. Math.: Paris (1958). Zbl.178.266

[176] Hartman, P.: *Ordinary Differential Equations.* Wiley: New York 1964. Zbl.125,321

[177] Hartshorne, R.: *Algebraic Geometry.* Grad. Texts Math. 52. Springer: New York 1977. Zbl.367.14001

[178] Hervé, M.: *Several Complex Variables, Local Theory.* Oxford Univ. Press: London 1963. Zbl.113,290

[179] Hironaka, H.: Resolution of singularities of an algebraic variety over a field of characteristic zero. I, II. Ann. Math., II. Ser. *79*, No. 1, 109–203; No. 2, 205–326 (1964). Zbl.122,386

[180] Hirsch, M.: Immersions of manifolds. Trans. Amer. Math. Soc. *93*, 242–276 (1959). Zbl.113,172

[181] Hirsch, M.: On embedding differentiable manifolds in Euclidean space. Ann. Math., II. Ser. 73, 566–571 (1961). Zbl.123,167

[182] Hung, N.H.V.: The mod 2 cohomology algebras of symmetric groups. Acta Math. Vietnam. *6*, 41–48 (1981). Zbl.518.20044

[183] Husemoller, D.: *Fibre Bundles.* McGraw-Hill: New York 1966. Zbl.144,448

[184] Igusa, K.: Higher singularities of smooth functions are unnecessary. Ann. Math., II. Ser. *119*, 1–58 (1984). Zbl. 548.58005

[185] Janssen, W.: Skew-symmetric vanishing lattices and their monodromy groups. Math. Ann. *266*, 115–133 (1983). Zbl. 537.14005

[186] Janssen, W.: Skew symmetric vanishing lattices. Math. Ann. *272*, 17–22 (1985). Zbl.592.10015

[187] Jaworski, P.: Distribution of critical values of miniversal deformations of parabolio singularities. Invent. Math. *86*, 19–33 (1986). Zbl.578.32037

[188] Jaworski, P.: Decompositions of parabolic singularitis on one level. Vestn. Mosk. Univ., Ser. I, No. 2, 49–53 (1987). English transl.: Mosc. Univ. Math. Bull. *42*, No. 2, 30–35 (1987). Zbl.648.58006

[189] Jaworski, P.: Decompositions of parabolic singularities. Bull. Sci. Math., II. Sér. *112*, 143–176 (1988). Zbl. 674.14002

[190] Kapranov, M.M.: Integration with respect to the Euler characteristic. Mosk. Gos. Univ.: Moscow (1981) (Russian)

[191] Katz, N.: Nilpotent connections and the monodromy theorem. Applications of a result of Turrittin. Publ. Math., Inst. Hautes Étud. Sci. *39*, 175–232 (1970). Zbl.221.14007

[192] Kempf, G., Knudsen, F., Mumford, D., Saint-Donat B.: *Toroidal Embeddings.* I. Lect. Notes Math. 339. Springer: Berlin 1973. Zbl.271.14017

[193] Khimshiashvili, G.N.: The local degree of a smooth mapping. Soobshch. Akad. Nauk Gruz. SSR *85*, No. 2, 309–312 (1977) (Russian). Zbl.346.55008

[194] Khovanskiĭ, A.G.: Newton polyhedra and toroidal varieties. Funkts. Anal. Prilozh. *11*, No. 4, 56–64 (1977). English transl.: Funct. Anal. Appl. *11*, 289–296 (1978). Zbl.445.14019

[195] Khovanskiĭ, A.G.: Newton polyhedra and the genus of complete intersections. Funkts. Anal. Prilozh. *12*, No. 1, 51–61 (1978). English transl.: Funct. Anal. Appl. *12*, 38–46 (1978). Zbl.387.14012

[196] Khovanskiĭ, A.G.: The index of a polynomial vector field. Funkts. Anal. Prilozh. *13*, No. 1, 49–58 (1979). English transl: Funct. Anal. Appl. *13*, 38–45 (1979). Zbl.422.57009

[197] Kleiman, S.L.: The enumerative theory of singularities. *Real and Complex Singularities*, pp. 297–396. Sijthoff and Noordhoff: Alphen aan den Rijn 1977. Zbl. 385.14018.

[198] Klein, F.: *Vorlesungen über das Ikosaeder und die Auflösung der Gleichungen vom Fünften Grade.* Teubner: Leipzig 1884. Jbuch.16,61

[199] Kneser, M.: Erzeugung ganzzahliger orthogonaler Gruppen durch Spiegelungen. Math. Ann. *255*, 453–462 (1981). Zbl.439.10016

[200] Kulikov, V.S.: Degeneration of elliptic curves, and resolution of uni- and bi-modal singularities. Funkts. Anal. Prilozh. *9*, No. 1, 72–73 (1975). English transl.: Funct. Anal. Appl. *9*, 69–70 (1975). Zbl.317.14013

[201] Kulikov, V.S.: Calculation of singularities of an imbedding of a generic algebraic surface in the projective space P^3. Funkts. Anal. Prilozh. *17*, No. 3, 15–27 (1983). English transl.: Funct. Anal. Appl. *17*, 176–186 (1983). Zbl.589.14036

[202] Kuo, T.C.: On C^0-sufficiency of jets of potential functions. Topology *8*, 167–171 (1969). Zbl.183,46

[203] Kushnirenko, A.G.: Polyèdres de Newton et nombres de Milnor. Invent. Math. *32*, 1–31 (1976). Zbl.328.32007

[204] Kushnirenko, A.G.: Multiplicity of the solution of a system of holomorphic equations. *Optimal Control (Mathematical Models of Production Control)*, No. 70, pp. 62–65. Mosk. Gos. Univ.: Moscow 1977 (Russian)

[205] Lamotke, K.: Die Homologie isolierter Singularitäten. Math. Z. *143*, 27–44 (1975). Zbl.302.32013

[206] Landis, E.E.: Tangential singularities. Funkts. Anal. Prilozh. *15*, No. 2, 36–49 (1981). English transl.: Funct. Anal. Appl. *15*, 103–114 (1981). Zbl.493.58005

[207] Lazzeri, F.: A theorem on the monodromy of isolated singularities. Astérisque, No. 7–8, 269–275 (1973). Zbl.301.32011

[208] Lazzeri, F.: Some remarks on the Picard-Lefschetz monodromy. *Quelques journées singulières*, 10 pp. (1974) Centre Math. Éc. Polytech.: Paris. Zbl.277.32015

[209] Le Dung Tráng: Une application d'un théorème d'A'Campo à l'équisingularité. Indagationes Math. *35*, 403–409 (1973). Zbl.271.14001

[210] Le Dung Tráng, Ramanujam, C.P.: The invariance of Milnor numbers implies the invariance of the topological type. Amer. J. Math. *98*, 67–78 (1976). Zbl.351.32009

[211] Lees, J.A.: On the classification of Lagrange immersions. Duke Math. J. *43*, 217–224 (1976). Zbl.329.58006

[212] Lefschetz, S.: *L'analyse situs et la géométrie algébrique*. Gauthier-Villars: Paris 1950. Zbl.35,102

[213] Leray, J.: Le calcul différential et intégral sur une variété analytique complexe (Problème de Cauchy III). Bull. Soc. Math. Fr. *87*, 81–180 (1959). Zbl.199,412

[214] Levine, H.: The singularities S_1^q. Ill. J. Math. *8*, 152–168 (1964). Zbl.124,388

[215] Levine, H.: Elimination of cusps. Topology *3*, Suppl. 2, 263–296 (1965). Zbl.146,200

[216] Lin, V. Ya.: Superpositions of algebraic functions. Funkts. Anal. Prilozh. *6*, No. 3, 77–78 (1972). English transl.: Funct. Anal. Appl. *6*, 240–241 (1972). Zbl.272.12103

[217] Lin, V. Ya.: Superpositions of algebraic functions. Funkts. Anal. Prilozh. *10*, No. 1, 37–45 (1976). English transl.: Funct. Anal. Appl. *10*, 32–38 (1976). Zbl.346.32009

[218] Lipman, J.: Rational singularities with applications to algebraic surfaces and unique factorization. Publ. Math., Inst. Hautes Étud. Sci. *36*, 195–279 (1969). Zbl.181,489

[219] Looijenga, E.J.N.: The complement of the bifurcation variety of a simple singularity. Invent. Math. *23*, 105–116 (1974). Zbl.278.32008

[220] Looijenga, E.J.N.: A period mapping for certain semi-universal deformations. Compos. Math. *30*, 299–316 (1975). Zbl.312.14006

[221] Looijenga, E.J.N.: On the semi-universal deformation of a simple elliptic hypersurface singularity. I. Unimodality. Topology *16*, 257–262 (1977); Zbl.373.32004; II. The discriminant. Topology *17*, 23–40 (1978). Zbl.392.57013

[222] Luengo, I.: The μ-constant stratum is not smooth. Invent. Math. *90*, 139–152 (1987). Zbl.627.32018

[223] Luna, D.: Fonctions différentiables invariantes sous l'opération d'un groupe réductif. Ann. Inst. Fourier *26*, 33–49 (1976). Zbl.315.20039

[224] Lyashko, O.V.: Decompositions of simple singularities of functions. Funkts. Anal. Prilozh. *10*, No. 2, 49–56 (1976). English transl.: Funct. Anal. Appl. *10*, 122–128 (1976). Zbl.351.32013

[225] Lyashko, O.V.: The geometry of bifurcation diagrams. Usp. Mat. Nauk *34*, No. 3, 205–206 (1979). English transl.: Russ. Math. Surv. *32*, No. 3, 209–210 (1979). Zbl.434.32020

[226] Lyashko, O.V.: Geometry of bifurcation diagrams. Itogi Nauki Tekh., Ser. Sovrem. Probl. Mat. *22*, 94–129 (1983). VINITI: Moscow. English transl.: J. Sov. Math. *27*, 2736–2759 (1984). Zbl.548.58024

[227] Mac Lane, S.: *Homology*. Springer: Berlin 1963. Zbl.133,265

[228] Malgrange, B.: *Ideals of Differentiable Functions*. Oxford Univ. Press: London 1966. Zbl.177,179

[229] Malgrange, B.: Intégrales asymptotiques et monodromie. Ann. Sci. Éc. Norm. Super., *IV*. Sér. 7, 405–430 (1974). Zbl.305.32008

[230] Manin, Yu. I.: Algebraic curves over fields with differentiation. Izv. Akad. Nauk SSSR, Ser. Mat. *22*, No. 6, 737–756 (1958). English transl.: Transl., II. Ser., Amer. Math. Soc. *37*, 59–78 (1964). Zbl.151,275

[231] Manin, Yu. I.: Rational points of algebraic curves over function fields. Izv. Akad. Nauk SSSR, Ser. Mat. *27*, No. 6, 1395–1440 (1963). English transl.: Transl., II. Ser., Amer. Math. Soc. *50*, 189–234 (1966). Zbl.166,169

[232] Markushevich, A.I.: *Introduction to the Classical Theory of Abelian Functions*. Nauka: Moscow 1979 (Russian). Zbl.493.14023

[233] Martinet, J.: Déploiements versels des applications différentiables et classification des applications stables. *Singularités d'applications différentiables*, pp. 1–44, Lect. Notes Math. 535. Springer: Berlin 1976. Zbl.362.58004

[234] Martinet, J.: *Singularities of Smooth Functions and Maps*. Lond. Math. Soc. Lect. Note Ser., *58*. Cambridge Univ. Press: Cambridge 1982. Zbl.522.58006

[235] Maslov, V.P.: *Perturbation Theory and Asymptotic Methods*. Mosk. Gos. Univ.: Moscow 1965. French transl.: Ganthier-Villars: Paris 1972. Zbl.247.47010

[236] Mather, J.: Stability of C^∞-mappings. I. Ann. Math., II. Ser. *87*, 89–104 (1968); Zbl.159,249; II. Ann. Math., II. Ser. *89*, 254–291 (1969); Zbl.177,260; III. Publ. Math., Inst. Hautes Étud. Sci. *35*, 127–156 (1969); Zbl.159,250; IV. Publ. Math., Inst. Hautes Étud. Sci. *37*, 223–248 (1970); Zbl.202,551; V. Adv. Math. *4*, 301–336 (1970); Zbl.207,543; VI. *Proceedings of Liverpool Singularities Symposium*, I, pp. 207–253, Lect. Notes Math. 192. Springer: Berlin 1971. Zbl.211,561

[237] Mather, J.: Generic projections. Ann. Math., II. Ser. *98*, 226–245 (1973). Zbl.267.58005

[238] Mather, J.: Stratifications and mappings. *Dynamical Systems*, pp. 195–232. Academic Press: New York 1973. Zbl.286.58003

[239] Mather, J.: On Thom-Boardman singularities. *Dynamical Systems*, pp. 232–248. Academic Press: New York 1973. Zbl.292.58004

[240] Mather, J.: How to stratify mappings and jet spaces. *Singularités d'applications différentiables*, pp. 128–176, Lect. Notes Math. 535. Springer: Berlin 1976. Zbl.398.58008

[241] Mather, J.: Differentiable invariants. Topology *16*, 145–155 (1977). Zbl.376.58002

[242] Matov, V.I.: Topological classification of germs of the maximum and minimax functions of families of functions in general position. Usp. Mat. Nauk *37*, No. 4, 167–168 (1982). English transl.: Russ. Math. Surv. *37*, No. 4, 127–128 (1982). Zbl.506.58004

[243] Matov, V.I.: Extremum functions of finite families of convex homogeneous functions. Funkts. Anal. Prilozh. *21*, No. 1, 51–62 (1987). English transl.: Funct. Anal. Appl. *21*, 42–52 (1987). Zbl.629.26008

[244] McCrory, C., Shifrin, T., Varley, R.: The Gauss map of a generic hypersurface in P^4. J. Differ. Geom. *30*, 689–759 (1989). Zbl.646.14032

[245] McKay, J.: Cartan Matrices, finite groups of quaternions, and Kleinian singularities. Proc. Amer. Math. Soc. *81*, 153–154 (1981). Zbl.477.20006

[246] Milnor J.: *Morse Theory*. Ann. Math. Stud. *51*. Princeton Univ. Press: Princeton 1963. Zbl.108,104

[247] Milnor J.: *Singular Points of Complex Hypersurfaces*. Ann. Math. Stud. *61*. Princeton Univ. Press: Princeton 1968. Zbl.184,484

[248] Milnor J., Orlik, P.: Isolated singularities defined by weighted homogeneous polynomials. Topology *9*, 385–393 (1970). Zbl.204,565

[249] Milnor J., Stasheff, J.D.: *Characteristic Classes*. Ann. Math. Stud. *76*. Princeton Univ. Press: Princeton 1974. Zbl.298.57008

[250] Morin, B.: Formes canoniques des singularités d'une application différentiable. C.R. Acad. Sci., Paris *260*, 5662–5665; 6503–6506 (1965). Zbl.178,268

[251] Morin, B.: Calcul jacobien. Ann. Sci. Éc. Norm. Supér., IV. Sér. *8*, 1–98 (1975). Zbl.359.57013

[252] Nilsson, N.: Some growth and ramification properties of certain integrals on algebraic manifolds. Ark. Mat. *5*, 463–476 (1965). Zbl.168,420

[253] Nilsson, N.: Monodromy and asymptotic properties of certain multiple integrals. Ark. Mat. *18*, 181–198 (1980). Zbl.483.32008

[254] Orlik, P., Wagreich, P.: Isolated singularities of algebraic surfaces with C^* action. Ann. Math., II. Ser. *93*, 205–228 (1971). Zbl.212,537

[255] Palamodov, V.P.: On the multiplicity of a holomorphic mapping. Funkts. Anal. Prilozh. *1*, No. 3, 54–65 (1967). English transl.: Funct. Anal. Appl. *1*, 218–226 (1967). Zbl.164,92

[256] Palamodov, V.P.: Remarks on finite-multiplicity differentiable mappings. Funkts. Anal. Prilozh. *6*, No. 2, 52–61 (1972). English transl.: Funct. Anal. Appl. *6*, 128–135 (1972). Zbl.264.58004

[257] Pellikaan, R.: Hypersurface singularities and resolutions of Jacobi modules. Thesis. Rijksuniv. Utrecht: Utrecht (1985). Zbl.589.32017

[258] Pellikaan, R.: On hypersurface singularities which are stems. Rep. No. 326. Vrije Univ.: Amsterdam (1987)

[259] Perron, B.: "μ const" implique "type topologique constant" en dimension complexe trois. C.R. Acad. Sci., Paris, Sér. I *295*, 735–738 (1982). Zbl.526.32009

[260] Petrovskiĭ, I.G.: On the topology of real plane algebraic curves. Ann. Math., II. Ser. *39*, 189–209 (1938). Zbl.18,270

[261] Petrovskiĭ, I.G., Oleĭnik, O.A.: On the topology of real algebraic surfaces. Izv. Akad. Nauk SSSR, Ser. Mat. *13*, No. 5, 389–402 (1949). English transl.: Transl., Amer. Math. Soc. *70* (1952). Zbl.35,102

[262] Pham, F.: Formules de Picard-Lefschetz généralisées et ramification des intégrales. Bull. Soc. Math. Fr. *93*, 333–367 (1965). Zbl.192,297

[263] Pham, F.: Remarque sur l'équisingularité universelle. Fac. Sci.: Nice (1970)

[264] Pham, F.: *Singularités des systèmes différentiels de Gauss-Manin*. Birkhäuser: Boston 1979. Zbl.524.32015

[265] Phillips, A.: Submersions of open manifolds. Topology *6*, 170–206 (1967). Zbl.204,237

[266] Phillips, A.: Foliations on open manifolds. II. Comment. Math. Helv. *44*, 367–370 (1969). Zbl.179,520

[267] Picard, E., Simart, G.: *Théorie des fonctions algébriques de deux variables indépendantes*, Vol. 1. Gauthier-Villars: Paris 1897. Jbuch. 28,327

[268] Pinkham, H.: Groupe de monodromie des singularités unimodulaires exceptionnelles. C.R. Acad. Sci., Paris, Sér. A *284*, 1515–1518 (1977). Zbl.391.14005

[269] du Plessis, A.: Maps without certain singularities. Comment. Math. Helv. *50*, 363–382 (1975). Zbl.313.58010

[270] du Plessis, A.: Homotopy classification of regular sections. Compos. Math. *32*, 301–333 (1976). Zbl.333.58004

[271] du Plessis, A.: On the determinacy of smooth map germs. Invent. Math. *58*, 107–160 (1980). Zbl.446.58004

[272] du Plessis, A.: On the genericity of topologically finitely-determined map germs. Topology *21*, 131–156 (1982). Zbl.499.58007

[273] du Plessis, A.: Genericity and smooth finite determinacy. *Singularities*, pp. 295–312, Proc. Symp. Pure Math. *40*, Part I. Amer. Math. Soc.: Providence 1983. Zbl.523.58009

[274] Poénaru, V.: On regular homotopy in codimension 1. Ann. Math., II. Ser. *83*, 257–265 (1966). Zbl.142,411

[275] Poénaru, V.: *Singularités C^∞ en présence de symétrie.* Lect. Notes Math. 510. Springer: Berlin 1976. Zbl.325.57008

[276] Porteous, I.R.: Simple singularities of maps. *Proceedings of Liverpool Singularities-Symposium*, pp. 286–307, Lect. Notes Math. 192. Springer: Berlin 1971. Zbl.221.57016

[277] Porteous, I.R.: The second-order decomposition of Σ^2. Topology *11*, 325–334 (1972). Zbl.263.57014

[278] Poston, T., Stewart, I.: *Catastrophe Theory and its Applications.* Pitman: London 1978. Zbl.382.58006

[279] *Real and Complex Singularities. Proc. Nordic Summer School*, Oslo, 1976. Sijthoff and Noordhoff: Alphen aan den Rijn 1977. Zbl.357.00011

[280] Roberts, M.: Finite determinacy of equivariant map germs. Thesis. Univ. Liverpool: Liverpool (1982).

[281] Rokhlin, V.A., Fuks, D.B.: *Beginner's Course in Topology.* Nauka: Moscow 1977. English transl.: Graduate Texts Math., Springer: Berlin 1984. Zbl.417.55002

[282] Ronga, F.: Le calcul des classes duales aux singularités de Boardman d'ordre deux. Comment. Math. Helv. *47*, 15–35 (1972). Zbl.236.58003

[283] Ronga, F.: Stabilité locale des applications équivariantes. *Differential Topology and Geometry*, pp. 23–35, Lect. Notes Math. 484. Springer: Berlin 1975. Zbl.355.58005

[284] Ronga, F.: Une application topologiquement stable qui ne peut pas être approchée par une application différentiablement stable. C.R. Acad. Sci., Paris, Sér. A *287*, 779–782 (1975). Zbl.397.58009

[285] Ronga, F.: On multiple points of smooth immersions. Comment. Math. Helv. *55*, 521–527 (1980). Zbl.457.57013

[286] Saito, K.: Einfach-elliptische Singularitäten. Invent. Math. *23*, 289–325 (1974). Zbl.296.14019

[287] Saito, K.: Primitive forms for universal unfolding of a function with an isolated critical point. J. Fac. Sci., Univ. Tokyo, Sect. I A *28*, 775–792 (1981). Zbl.523.32015

[288] Saito, K.: A characterization of the intersection form of a Milnor's fiber for a function with an isolated critical point. Proc. Japan Acad., Ser. A *58*, 79–81 (1982). Zbl.539.57011

[289] Saito, M.: Exponents of a reduced and irreducible plane curve, Preprint RIMS, Kyoto Univ.: Kyoto (1982).

[290] Saito, M.: On the exponents and the geometric genus of an isolated hypersurface singularity. *Singularities*, pp. 465–471, Proc. Symp. Pure Math. *40*, Part II, Amer. Math. Soc. Providence 1983. Zbl.545.14031

[291] Saito, M.: Exponents and Newton polyhedra of isolated hypersurface singularities. Math. Ann. *281*, 411–417 (1988). Zbl.628.32038

[292] Samoĭlenko, A.M.: The equivalence of a smooth function to a Taylor polynomial in a neighborhood of a critical point of finite type. Funkts. Anal. Prilozh. *2*, No. 4, 63–69 (1968). English transl.: Funct. Anal. Appl. *2*, 318–323 (1968). Zbl.183,149

[293] Schmid, W.: Variation of Hodge structure: the singularities of the period mapping. Invent. Math. *22*, 211–319 (1973). Zbl.278.14003

[294] Schwarz, G.: Smooth functions invariant under the action of a compact Lie group. Topology *14*, 63–68 (1975). Zbl.297.57015

[295] Sebastiani, M.: Preuve d'une conjecture de Brieskorn. Manuscr. Math. *2*, 301–308 (1970). Zbl.194,114

[296] Sebastiani, M., Thom, R.: Un résultat sur la monodromie. Invent. Math. *13*, 90–96 (1971). Zbl.233.32025

[297] Segal, G.B.: Configuration-spaces and iterated loop-spaces. Invent. Math. *21*, 213–221 (1973). Zbl.267.55020

[298] Serre, J-P.: Quelques problèmes globaux relatifs aux variétés de Stein. *Colloque sur les fonctions de plusiers variables, tenu a Bruxelles*, pp. 57–68. Georges Thone: Liege (1953). Zbl.53,53

[299] Serre, J-P.: *Algèbres de Lie semi-simples complexes.* Benjamin: New York 1966. Zbl.144,21

[300] Sharko, V.V.: Minimal Morse functions. *Geometry and Topology in Global Nonlinear Prob-*

lems, pp. 123–141. Voronezh. Gos. Univ.: Voronezh 1984. Zbl.535.58008. English transl.: Lect. Notes Math. 1108, Springer: Berlin, 218–234 (1984)

[301] Shvarts, A.S.: The genus of a fibered space. Tr. Mosk. Mat. O.-va 10, 217–272 (1961). English transl.: Transl., II. Ser., Amer. Math. Soc. *55*, 49–140 (1966). Zbl.178,262

[302] Shoshitaĭshvili, A.N.: Functions with isomorphic Jacobian ideals. Funkts. Anal. Prilozh. *10*, No. 2, 57–62 (1976). English transl.: Funct. Anal. Appl. *10*, 128–133 (1976). Zbl.346.32023

[303] Shustin, E.I.: Invariants of singular points of algebraic curves. Mat. Zametki *34*, No. 6, 929–931 (1983). English transl.: Math. Notes *34*, 962–963 (1983). Zbl.549.14011

[304] Siersma, D.: Classification and deformation of singularities. Doctoral dissertation. Univ. of Amsterdam, Amsterdam (1974). Zbl.283.57012

[305] Siersma, D.: Isolated line singularities. *Singularities*, pp. 485–496, Proc. Symp. Pure Math. *40*, Part II. Amer. Math. Soc.: Providence 1983. Zbl.514.32007

[306] Siersma, D.: Hypersurfaces with singular locus a plane curve and transversal type A_1. Banach Cent. Publ. *20*, 397–410 (1988). Zbl.662.32011

[307] Siersma, D.: Singularities with critical locus a one-dimensional complete intersection and transversal type A_1. Topology Appl. *27*, 51–73 (1987). Zbl.635.32006

[308] Siersma, D.: Quasihomogeneous singularities with transversal type A_1. *Singularities*, Contemp. Math. 90, 261–294, Amer. Math. Soc.: Providence 1989. Zbl.682.32012

[309] *Singularities*. Proc. Symp. Pure Math. *40*, Part I, II. Amer. Math. Soc.: Providence 1983. Zbl.509.00008

[310] Slodowy, P.: Platonic solids, Kleinian singularities and Lie groups. *Algebraic Geometry*, pp. 102–138, Lect. Notes Math. 1008, Springer: Berlin 1983. Zbl.516.14002

[311] Smale, S.: The classification of immersions of spheres in Euclidean space. Ann. Math., II. Ser. *69*, 327–344 (1959). Zbl.89,182

[312] Smale, S.: On the topology of algorithms. I. J. Complexity *3*, 81–89 (1987). Zbl.639.68042

[313] Solomon, L.: Invariants of finite reflection groups. Nagoya Math. J. *22*, 57–64 (1963). Zbl.117,271

[314] Springer, T.A.: *Invariant Theory*. Lect. Notes Math. 585. Springer: Berlin 1977. Zbl.346.20020

[315] Stagnaro, E.: Sul massimo numero di punti doppi isolati di una supeficie algebrica di P^3. Rend. Semin. Mat. Univ. Padova *59*, 179–208 (1978). Zbl.427.14012

[316] Steenbrink, J.H.M.: Limits of Hodge structures. Invent. Math. *31*, 229–257 (1975). Zbl.302.14002

[317] Steenbrink, J.H.M.: Mixed Hodge structure on the vanishing cohomology. *Real and Complex Singularities*, pp. 525–563. Sijthoff and Nordhoof: Alphen aan den Rijn (1977). Zbl.373.14007

[318] Steenbrink, J.H.M.: Intersection form for quasihomogeneous singularities. Compos. Math. *34*, 211–223 (1977). Zbl.347.14001

[319] Steenbrink, J.H.M.: Mixed Hodge structures associated with isolated singularities. *Singularities*, pp. 513–536, Proc. Symp. Pure Math. *40*, Part II. Amer. Math. Soc.: Providence 1983. Zbl.515.14003

[320] Steenbrink, J.H.M.: Semicontinuity of the singularity spectrum. Invent. Math. *79*, 557–566 (1985). Zbl.568.14021

[321] Szücs, A.: Surfaces in R^3. Bull. Lond. Math. Soc. *18*, 60–66 (1986). Zbl.563.57015

[322] Teissier, B.: Cycles évanescents, sections planes et conditions de Whitney. Astérisque 7–8, 285–362 (1973). Zbl.295.14003

[323] Teissier, B.: Variétés polaires. I. Invariants polaires des singularités d'hypersurfaces. Invent. Math. *40*, 267–292 (1977). Zbl.446.32002

[324] Thom, R.: Les singularités des applications différentiables. Ann. Inst. Fourier *6*, 43–87 (1956). Zbl.75,321

[325] Thom, R.: Ensembles et morphismes stratifiés. Bull. Amer. Math. Soc. *75*, 240–284 (1969). Zbl.197,205

[326] Thom, R.: La stabilité topologique des applications polynomiales. Enseign. Math., II. Sér. *8*, 24–33 (1962). Zbl.109,400

[327] Thom, R.: Généralisation de la théorie de Morse aux variétés feuilletées. Ann. Inst. Fourier *14*, 173–190 (1964). Zbl.178,266

[328] Thom, R.: *Stabilité structurelle et morphogénèse.* Benjamin: New York 1973. Zbl.294.92001

[329] Thom, R.: *Modèles mathématiques de la morphogénèse.* Acad. Naz. Lincei: Pisa 1971. Zbl.347.58003

[330] Timourian, J.G.: The invariance of Milnor's number implies topological triviality. Amer. J. Math. *99*, 437–446 (1977). Zbl.373.32003

[331] Togliatti, E.: Sulle forme cubiche dello spazio a cinque dimensioni aventi il massimo numero finito di punti doppi. *Scritti offerti a L. Berzolari*, pp. 577–593. Ist. Mat. R. Univ. Pavia: Pavia (1936). Zbl.16,221

[332] Tougeron, J.-C.: Idéaux des fonctions différentiables. Ann. Inst. Fourier *18*, 177–240 (1968). Zbl.188,451. *20*, 179–233 (1970). Zbl.188,452

[333] Trotman, D.: Counterexamples in stratification theory: two discordant horns. *Real and Complex Singularities*, pp. 679–686. Sijthoff and Noordhoff: Alphen aan den Rijn 1977. Zbl.378.57012

[334] Trotman, D.: Geometric versions of Whitney regularity for smooth stratifications. Ann. Sci. Éc. Norm. Supér., IV. Sér. *12*, 453–463 (1979). Zbl.456.58002

[335] Trotman, D.: Comparing regularity conditions on stratifications. *Singularities*, pp. 575–586, Proc. Symp. Pure Math. *40*, Part II. Amer. Math. Soc.: Providence 1983. Zbl.519.58009

[336] Tyurina, G.N.: Topological properties of isolated singularities of complex spaces of codimension one. Izv. Akad. Nauk SSSR, Ser. Mat. *32*, No. 3, 605–620 (1968). English transl.: Math. USSR, Izv. *2*, 557–571 (1968). Zbl.176,509

[337] Tyurina, G.N.: Locally semi-universal plane deformations of isolated singularities of complex spaces. Izv. Akad. Nauk SSSR, Ser. Mat. *33*, No. 5, 1026–1058 (1969). English transl.: Math. USSR, Izv. *3*, 967–999 (1969). Zbl.196,97

[338] Vaĭnshteĭn, F.V.: The cohomology of braid groups. Funkts. Anal. Prilozh. *12*, No. 2, 72–73 (1978). English transl.: Funct. Anal. Appl. *12*, 135–137 (1978). Zbl.413.55013

[339] du Val, P.: On isolated singularities of surfaces that do not affect the conditions of adjunction. I, II, III, Proc. Camb. Philos. Soc. *30*, 453–459, 460–465, 483–491 (1934). Zbl.10,176–177

[340] Varchenko, A.N.: Local topological properties of analytic mappings. Izv. Akad. Nauk SSSR, Ser. Mat. *37*, No. 4, 883–916 (1973). English transl.: Math. USSR, Izv. 7, 883–917 (1973). Zbl.285.32006

[341] Varchenko, A.N.: Local topological properties of smooth mappings. Izv. Akad. Nauk SSSR, Ser. Mat. *38*, No. 5, 1037–1090 (1974). English transl.: Math. USSR, Izv. *8*, 1033–1082 (1974). Zbl.313.58009

[342] Varchenko, A.N.: A theorem on versal topological deformations. Izv. Akad. Nauk SSSR, Ser. Mat. *39*, No. 2, 294–314 (1975). English transl.: Math. USSR, Izv. *9*, 277–296 (1975). Zbl.333.32005

[343] Varchenko, A.N.: Zeta-function of monodromy and Newton's diagram. Invent. Math. *37*, 253–262 (1976). Zbl.333.14007

[344] Varchenko, A.N.: Asymptotic behavior of holomorphic forms determines a mixed Hodge structure. Dokl. Akad. Nauk SSSR *255*, No. 5, 1035–1038 (1980). English transl.: Sov. Math., Dokl. *22*, 772–775 (1980). Zbl.516.14007

[345] Varchenko, A.N.: Hodge properties of the Gauss-Manin connection. Funkts. Anal. Prilozh. *14*, No. 1, 46–47 (1980). English transl.: Funct. Anal. Appl. *14*, 36–37 (1980). Zbl.457.32008

[346] Varchenko, A.N.: Asymptotic Hodge structure on vanishing cohomology. Izv. Akad. Nauk SSSR, Ser. Mat. *45*, No. 3, 540–591. (1981). English transl.: Math. USSR, Izv. *18*, 469–512 (1982). Zbl.476.14002

[347] Varchenko, A.N.: A lower bound for the codimension of the μ = const stratum in terms of the mixed Hodge structure. Vestn. Mosk. Univ., Ser. I No. 6, 28–31 (1982). English transl.: Mosc. Univ. Math. Bull. *37*, No. 6, 30–33 (1982). Zbl.511.32004

[348] Varchenko, A.N.: The complex singularity index does not change along the stratum μ = const. Funkts. Anal. Prilozh. *6*, No. 1, 1–12 (1982). English transl.: Funct. Anal. Appl. *16*, 1–9 (1982). Zbl.498.32010

[349] Varchenko, A.N.: Changes in the discrete characteristics of critical points of functions under deformations. Usp. Mat. Nauk *38*, No. 5, 126–127 (1983) (Russian)

[350] Varchenko, A.N.: Semicontinuity of the spectrum and an upper bound for the number of singular points of a projective hypersurface. Dokl. Akad. Nauk SSSR *270*, No. 6, 1294–1297 (1983). English transl.: Sov. Math., Dokl. *27*, 735–739 (1983). Zbl.537.14003

[351] Varchenko, A.N.: Local residue and the intersection form in vanishing cohomology. Izv. Akad. Nauk SSSR, Ser. Mat. *49*, No. 1, 32–54 (1985). English transl.: Math. USSR, Izv. *26*, 31–52 (1986). Zbl.571.32004

[352] Varchenko, A.N., Chmutov, S.V.: Finite irreducible groups generated by reflections are the monodromy groups of suitable singularities. Funkts. Anal. Prilozh. *18*, No. 3, 1–13 (1984). English transl.: Funct. Anal. Appl. *18*, 171–183 (1984). Zbl.573.32012

[353] Varchenko, A.N., Chmutov, S.V.: On the tangent cone to the μ = const. stratum. Vestn. Mosk. Univ., Ser. Mat. No. 1, 6–9 (1985). English transl.: Mosc. Univ. Math. Bull. *40*, No. 1, 7–12 (1985). Zbl.593.58002

[354] Varchenko, A.N., Givental', A.B.: The period mapping and the intersection form. Funkts. Anal. Prilozh. *16*, No. 2, 7–20 (1982). English transl.: Funct. Anal. Appl. *16*, 83–93 (1982). Zbl.497.32008

[355] Varchenko, A.N., Khovanskiĭ, A.G.: Asymptotics of integrals along the vanishing cycles and the Newton polyhedron. Dokl. Akad. Nauk SSSR *283*, No. 3, 521–525 (1985). English transl.: Sov. Math., Dokl. *32*, 122–127 (1985). Zbl.595.32012

[356] Vasil'ev, V.A.: Affineness of normal forms of the μ = const strata of germs of smooth functions. Funkts. Anal. Prilozh. *12*, No. 3, 72–73 (1978). English transl.: Funct. Anal. Appl. *12*, 218–220 (1978). Zbl.391.58007

[357] Vasil'ev, V.A.: Characteristic classes of Lagrangian and Legendrian manifolds that are dual to singularities of caustics and wave fronts. Funkts. Anal. Prilozh. *15*, No. 3, 10–22 (1981). English transl: Funct. Anal. Appl. *15*, 164–173 (1981). Zbl.493.57008

[358] Vasil'ev, V.A.: Self-intersections of wave fronts and Legendrian (Lagrangian) characteristic numbers. Funkts. Anal. Prilozh. *16*, No. 2, 68–69 (1982). English transl.: Funct. Anal. Appl. *16*, 131–133 (1982). Zbl.556.58007

[359] Vasil'ev, V.A.: Stable cohomology of the complements to the discriminant varieties of singularities of holomorphic functions. Usp. Mat. Nauk *42*, No. 2, 219–220 (1987). English transl.: Russ. Math. Surv. *42*, No. 2, 307–308 (1987). Zbl.625.32019

[360] Vasil'ev, V.A.: Cohomology of braid groups and the complexity of algorithms. Funkts. Anal. Prilozh. *22*, No. 3, 15–24 (1988). English transl.: Funct. Anal. Appl. *22*, No. 3, 182–190 (1988). Zbl.659.68071

[361] Vasil'ev, V.A.: *Lagrange and Legendre Characteristic Classes.* Gordon & Breach: New York 1988. Zbl.715.53001. Second edition to appear

[362] Vasil'ev, V.A.: Stable cohomology of complements to the discriminants of deformations of singularities of smooth functions. Itogi Nauki Tekh., Ser. Sovrem. Probl. Mat., Noveĭshie Dostizheniya *33*, pp. 3–29. VINITI: Moscow (1988). English transl.: J. Sov. Math. *52*, 3217–3230 (1990). Zbl.643.00012

[363] Vasil'ev, V.A.: On the topology of spaces of functions without complex singularities. Usp. Mat. Nauk *44*, No. 3, 149–150 (1989). English transl.: Russ. Math. Surv. *44*, No. 3, 218–219 (1989). Zbl.691.57010

[364] Vasil'ev, V.A.: Topology of spaces of functions without complex singularities. Funkts. Anal. Prilozh. *23*, No. 4, 14–26 (1989). English transl.: Funct. Anal. Appl. *23*, 277–286 (1989). Zbl.702.57013

[364a] Vasil'ev, V.A.: Topology of complements to discriminants and loop spaces, Theory of Singularities and its Applications (ed. V.I. Arnold). Advances in Soviet Math., Vol. 1, Amer. Math. Soc., 1990, 9–21

[364b] Vasil'ev, V.A.: Cohomology of knot spaces. The same issue as for [364a], 23–69

[364c] Vasil'ev, V.A.: Complements of Discriminants of Smooth Maps: Topology and Applications. Amer. Math. Soc. 1992

[365] Viro, O.Ya.: Curves of degree 7, curves of degree 8 and the Ragsdale conjecture. Dokl. Akad. Nauk SSSR *254*, No. 6, 1306–1310 (1980). English transl.: Sov. Math., Dokl. *22*, 566–570 (1980). Zbl.422.14032

[366] Viro, O.Ya.: *Some Integral Calculus Based on Euler Characteristic*. Lect. Notes Math. 1346, 127–138. Springer: Berlin 1988. Zbl.686.14019

[367] Wajnryb, B.: On the monodromy group of plane curve singularities. Math. Ann. *246*, 141–154 (1980). Zbl.399.14020

[368] Wall, C.T.C.: Affine cubic functions. IV. Functions on C^3, nonsingular at infinity. Philos. Trans. Roy. Soc. Lond., A *302*, 415–455 (1981). Zbl.451.14009

[369] Wall, C.T.C.: Finite determinacy of smooth map-germs. Bull. Lond. Math. Soc. *13*, 481–539 (1981). Zbl.451.58009

[370] Wall, C.T.C.: On finite C^k left determinacy. Invent. Math. *70*, 399–405 (1983). Zbl.515.58005

[371] Wall, C.T.C.: Determination of the semi-nice dimensions. Math. Proc. Camb. Philos. Soc. *97*, 79–88 (1985). Zbl.568.58008

[372] Wassermann, G.: Stability of unfoldings in space and time. Acta Math. *135*, 57–128 (1975). Zbl.315.58010

[373] Wassermann, G.: (r, s)-stability of unfoldings. Preprint. Regensburg (1976).

[374] Weinstein, A.: Singularities of families of functions. *Differentialgeometrie im Grossen*, pp. 323–330. Bibliographisches Inst.: Mannheim 1971. Zbl.221.58008

[375] Wells, R.O.: *Differential Analysis on Complex Manifolds*. Prentice-Hall: Englewood Cliffs 1973. Zbl.262.32005

[376] Whitney, H.: On the abstract properties of linear dependence. Amer. J. Math. *57*, 509–533 (1935). Zbl.12,4

[377] Whitney, H.: The singularities of smooth n-manifolds in $2n$-space and $(2n$-$1)$-space. Ann. Math., II. Ser. *45*, 220–293 (1944).

[378] Whitney, H.: On singularities of mappings of Euclidean spaces. I. Ann. Math., II. Ser. *62*, 374–410 (1955). Zbl.68,371

[379] Whitney, H.: Local properties of analytic varieties. *Differential and Combinatorial Topology*, pp. 205–244. Princeton Univ. Press: Princeton 1965. Zbl.129,394

[380] Whitney, H.: Tangents to an analytic variety. Ann Math., II. Ser. *81*, 496–549 (1965). Zbl.152,277

[381] Wilson, L.C.: Infinitely determined map germs. Can. J. Math. *33*, 671–684 (1981). Zbl.476.58005

[382] Wilson, L.C.: Map germs infinitely determined with respect to right-left equivalence. Pacific J. Math. *102*, 235–245, (1982). Zbl.454.58005

[383] Wirthmüller, K.: Universale topologische triviale Deformationen. Thesis. Univ. Regensburg: Regensburg (1978).

[384] Zakalyukin, V.M.: Perestroikas of wave fronts that depend on a parameter. Funkts. Anal. Prilozh. *10*, No. 2, 69–70 (1976). English transl.: Funct. Anal. Appl. *10*, 139–140 (1976). Zbl.329.58008

[385] Zakalyukin. V.M.: Legendrian mappings in Hamiltonian systems. *Some Problems in Mechanics*, pp. 11–16. Mosk. Aviats. Inst.: Moscow 1977 (Russian)

[386] Zakalyukin, V.M.: Perestroikas of fronts and caustics depending on a parameter, and versality of mappings. Itogi Nauki Tekh., Ser. Sovrem. Probl. Mat. *22*, 56–93. VINITI: Moscow (1983). English transl.: J. Sov. Math. *27*, 2713–2735 (1984). Zbl.534.58014

Author Index

Subject Index

Druck- und Bindearbeiten: Legoprint, Italien